T0139851

Big Data Management

The big data paradigm presents a number of challenges for university curricula on big data or data science related topics. On the one hand, new research, tools and technologies are currently being developed to harness the increasingly large quantities of data being generated within our society. On the other, big data curricula at universities are still based on the computer science knowledge systems established in the 1960s and 70s. The gap between the theories and applications is becoming larger, as a result of which current education programs cannot meet the industry's demands for big data talents.

This series aims to refresh and complement the theory and knowledge framework for data management and analytics, reflect the latest research and applications in big data, and highlight key computational tools and techniques currently in development. Its goal is to publish a broad range of textbooks, research monographs, and edited volumes that will:

– Present a systematic and comprehensive knowledge structure for big data and data science research and education
– Supply lectures on big data and data science education with timely and practical reference materials to be used in courses
– Provide introductory and advanced instructional and reference material for students and professionals in computational science and big data
– Familiarize researchers with the latest discoveries and resources they need to advance the field
– Offer assistance to interdisciplinary researchers and practitioners seeking to learn more about big data

The scope of the series includes, but is not limited to, titles in the areas of database management, data mining, data analytics, search engines, data integration, NLP, knowledge graphs, information retrieval, social networks, etc. Other relevant topics will also be considered.

Alejandro Peña-Ayala

Editor

Educational Data Science: Essentials, Approaches, and Tendencies

Proactive Education based on Empirical Big Data Evidence

 Springer

Editor
Alejandro Peña-Ayala (iD)
Artificial Intelligence in Education Lab, WOLNM &
Sección de Estudios de Posgrado e
Investigación, ESIME-Z, Instituto
Politécnico Nacional
CDMX, Mexico

ISSN 2522-0179   ISSN 2522-0187   (electronic)
Big Data Management
ISBN 978-981-99-0028-2   ISBN 978-981-99-0026-8   (eBook)
https://doi.org/10.1007/978-981-99-0026-8

This Springer imprint is published by the registered company Springer Nature Singapore Pte Ltd.
The registered company address is: 152 Beach Road, #21-01/04 Gateway East, Singapore 189721,
Singapore

Preface

As a result of the broad use of computer systems, the spread of Internet applications, and the deployment of artificial intelligence approaches in a wide variety of human daily life matters, among other technological trends, an accelerated datafication of digital data spreads worldwide. Such a revolutionary dynamic wave has demanded the emergence of novel paradigms to gather, storage, manage, and examine huge volume of distributed data that hides rich information and knowledge that is valuable and useful to lead the development of data-driven solutions.

In this context, *Data Science* is one of those disruptive domains that leverages the best of mature disciplines (e.g., statistics, data analysis, and machine learning) to build a baseline, which grounds specialized approaches that focus on data, as the main target of study to characterize, analyze, and interpret the phenology of symbolic representations of abstract and concrete entities and events that live and occur in the world and outer space. Hence, Data Science appears as part of the arrowhead of novel technologies, whose strength relies on the power to give meaning to the big data collected from diverse human activity fields as the *education*.

Educational Data Science is a newborn domain, whose nature reveals a transdisciplinary flavor that is benefited of Data Science and its underlying disciplines, in addition to the relatively new domains of data mining, knowledge discovery in databases, and analytics. Inclusive, some approaches also take into account some constructs from biology, psychology, neurology, and learning science. Fostered by a robust baseline, Educational Data Science pursues to gain a spot in the arena of specialized domains that strive for seeking valuable patterns and produce useful insights form educational big data to find out covert features and tendencies that characterize teaching and learning endeavors with the aim of interpreting them and lead adaptive and personalized processes to enhance educational practices.

With this in mind, this edited book is devoted to those that are in charge of educational management, educators, pedagogues, academics, computer technologists, researchers, and postgraduate students, who pursue to acquire a conceptual, formal, and practical perspective of how to use, develop, and deploy Educational

Data Science practices to build proactive, prescriptive, and reactive applications that personalize education, enhance teaching, and improve learning!

In order to reach such an aim, this publication compiles a sample of the labor recently achieved by several groups of researchers in three continents, who are willing in contributing to extend the domain, proposing new constructs to ground Educational Data Science labor, drawing specialized landscapes of the domain, and briefing research experiences that are useful to lead future projects and practices.

As a result of the edition process, which embraces the submission of proposals and their respective evaluation, as well as the edition of the complete manuscript with the corresponding revision, tuning, and decision according to Springer quality criteria, seven works were accepted, edited as chapters, and organized to compose a new volume of the Springer Book Series in Big Data Management, whose content is described by means of the following profile:

Chapter 1: Describes and applies Learning Engineering to engage students in effectively studying and learning in digital settings with the support of educational data science that fosters student-centered approaches that examine large-scale datasets to inspire data-driven teaching and learning practices.

Chapter 2: Sketches a review of clustering models to support educational data science with the aim of characterizing learners and generating cohorts of students that reveal similar features under the principle of avoiding discrimination of those that show singular characteristics of behavior.

Chapter 3: Outlines a broad view of educational data science to reveal its nature and state-of-the-art according to a proposed taxonomy that classifies related works into three categories: baseline, introduction, and approaches, which ground the domain, define its essence, and report practical experiences.

Chapter 4: Applies educational data science for designing a data-driven end-to-end quality assurance process for offering credit distance education delivered by means of massive open online courses, whose data about courses, teaching content, and learner feedback is examined to offer valuable insights.

Chapter 5: Pursues to understand the effect of text cohesion, a critical feature for supporting comprehension, in the abstracts of academic papers using educational data science methods of text mining and hybrid classifiers that apply fuzzy reasoning to detect how referential cohesion traits bias writing clarity.

Chapter 6: Develops temporal analysis to study learning behaviors by means of an educational data science approach that generates sequential pattern mining, which facilitates the evaluation of instructional interventions and the generation of features for prediction models that back a recommender system.

Chapter 7: Pursues to improve students' engagement and support teacher lessons by means of generating a cluster heat map of students' commitment, facilitating content sync ratio, and examining results of outlier detection from tracking logs, clickstreams, quiz scores, and Mahalanobis' generalized distance.

I express my gratitude to authors, reviewers, and the Springer Editorial board, where participate: *Editor-in-Chief*: Xiaofeng Meng; *Editorial Board Members*: Daniel Dajun Zeng, Hai Jin, Haixun Wang, Huan Liu, X. Sean Wang, and Weiyi

Meng; *Advisory Editors*: Jiawei Han, Masaru Kitsuregawa, Philip S. Yu, Tieniu Tan, and Wen Gao; and particularly to *Publishing Editor*: Nick Zhu. All of them developed a valuable collaboration to accomplish the edition of this book.

I also acknowledge the support given by the Consejo Nacional de Ciencia y Tecnología (CONACYT) and the Instituto Politécnico Nacional (IPN), both institutions pertain to the Mexican Federal Government, which authorized the following grants: CONACYT–SNI-36453, IPN-SIBE-2023-2024, IPN-SIP-EDI, IPN-SIP 2022-0803, and IPN-SIP 2023. Moreover, a special mention and gratitude is given to master student José Morales–Ramirez for his valuable contribution to develop this edited book.

Last but not least, I acknowledge and express testimony of the lead, unction and strength given by my Father, Brother Jesus, and Helper, as part of the research projects performed by World Outreach Light to the Nations Ministries (WOLNM).

Ciudad de México, Mexico   Alejandro Peña-Ayala
April, 2023

The original version of the book has been revised. A correction to this book can be found at https://doi.org/10.1007/978-981-99-0026-8_8

Contents

Editor and Contributors

About the Editor

Alejandro Peña-Ayala is professor of Artificial Intelligence on Education and Cognition in the School of Electric and Mechanical Engineering of the National Polytechnic Institute of México. Dr. Peña-Ayala has published more than 50 scientific works and is author of three machine learning patents (two of them in progress to be authorized), including the role of guest-editor for six Springer Book Series and guest-editor for an Elsevier journal. He is fellow of the National Researchers System of Mexico, the Mexican Academy of Sciences, Academy of Engineering, and the Mexican Academy of Informatics. Professor Peña-Ayala was scientific visitor of the MIT in 2016, made his postdoc at the Osaka University 2010–2012 and earned with honors his PhD, MSc, and BSc in Computer Sciences, Artificial Intelligence, and Informatics, respectively.

Contributors (Reviewers)

Meital Amzalag Holon Institute of Technology, Holon, Israel

Mark Anthony R. Aribon Ateneo de Manila Universi, Manila, Philippines

Kent Levi A. Bonifacio Central Mindanao University, Maramag, Philippines

Leandro S. G. Carvalho Federal University of Amazonas, Manaus, Amazonas, Brazil

John Maurice Gayed Tokyo Institute of Technology, Tokyo, Japan

Ernani Gottardo Federal Institute of Education, Porto Alegre, RS, Brazil

Woojin Kim Korea University, Seoul, South Korea

ByeongJo Kong ALFA, MIT CSAIL, Cambridge, MA, USA

Yiwen Lin University of California, Irvine, CA, USA

José-Antonio Marín-Marín University of Granada, Granada, Spain

Imperial Joseph Marvin National University, Manila, Philippines
De la Salle University, Manila, Philippines

Alejandra Ruiz-Segura McGill University, Montreal, QC, Canada

Vinitra Swamy UC Berkeley, Berkeley, CA, USA

Christos Vaitsis Karolinska Institute, Solna, Sweden

Tao Wu South China Agricultural University, Guangzhou, China

Contributors

Sasipa Boonyubol Department of Transdisciplinary Science and Engineering, School of Environment and Society, Tokyo Institute of Technology, Ookayama, Meguro-ku, Tokyo, Japan

Rachel Van Campenhout VitalSource Technologies, Raleigh, NC, USA

May Kristine Jonson Carlon Department of Transdisciplinary Science and Engineering, School of Environment and Society, Tokyo Institute of Technology, Ookayama, Meguro-ku, Tokyo, Japan
Online Content Research and Development Section, Center for Innovative Teaching and Learning, Tokyo Institute of Technology, Ookayama, Meguro-ku, Tokyo, Japan

Jeffrey S. Cross Department of Transdisciplinary Science and Engineering, School of Environment and Society, Tokyo Institute of Technology, Ookayama, Meguro-ku, Tokyo, Japan
Online Content Research and Development Section, Center for Innovative Teaching and Learning, Tokyo Institute of Technology, Ookayama, Meguro-ku, Tokyo, Japan

Konomu Dobashi Aichi University, Nagoya, Aichi, Japan

Carrie Demmans Epp Department of Computing Science, University of Alberta, Edmonton, AB, Canada

Gunnar Friege Institute for Didactics of Mathematics and Physics, Leibniz University Hannover, Hannover, Germany

Bill Jerome VitalSource Technologies, Raleigh, NC, USA

Benny G. Johnson VitalSource Technologies, Raleigh, NC, USA

Nopphon Keerativoranan Department of Transdisciplinary Science and Engineering, School of Environment and Society, Tokyo Institute of Technology, Ookayama, Meguro-ku, Tokyo, Japan

Eirini Ntoutsi Research Institute Cyber Defence and Smart Data (CODE), Universität der Bundeswehr München, Munich, Germany

Luc Paquette The Department of Curriculum and Instruction, College of Education, University of Illinois at Urbana-Champaign, Champaign, IL, USA

Alejandro Peña-Ayala Artificial Intelligence in Education Lab, WOLNM & Sección de Estudios de Posgrado e Investigación, ESIME-Z, Instituto Politécnico Nacional, CDMX, Mexico

Tai Le Quy L3S Research Center, Leibniz University Hannover, Hannover, Germany

Jinnie Shin College of Education, University of Florida, Gainesville, FL, USA

Yingbin Zhang Institute of Artificial Intelligence in Education, South China Normal University, Guangdong, China

Part I
Logistic

Chapter 1
Engaging in Student-Centered Educational Data Science Through Learning Engineering

Rachel Van Campenhout, Bill Jerome, and Benny G. Johnson

Abstract As educational data science (EDS) evolves and its related fields continue to advance, it is imperative to employ EDS to solve real-world educational challenges. One such challenge is to research how students learn and study effectively in digital learning environments and apply those findings to better their learning resources. The volume of educational data collected by digital platforms is growing tremendously, so it is a pivotal moment for EDS to be applied with an ethical approach in which the best interests of the learner are kept at the forefront. Learning engineering provides a practice and process to engage in EDS with a student-centered approach. In this work, we exemplify how the learning engineering process (LEP) guided large-scale data analyses to advance learning science (i.e., the doer effect), developed new artificial intelligence (AI)–based learning tools, and scaled both effective learning methods in natural educational contexts and automated data analysis methods—all in the service of students. The examples of analyses in this chapter serve to showcase how EDS—applied as a part of learning engineering—can validate learning science theory and advance the state of the art in learning technology.

Keywords Educational data science · Learning engineering · Courseware · Learn by doing · Doer effect · Artificial intelligence · Automatic question generation

Abbreviations

AG   Automatically generated (questions)
AI   Artificial intelligence
AQG   Automatic question generation
CIS   Content improvement service

R. Van Campenhout (✉) · B. Jerome · B. G. Johnson
VitalSource Technologies, Raleigh, NC, USA
e-mail: rachel.vancampenhout@vitalsource.com

© The Author(s), under exclusive license to Springer Nature Singapore Pte Ltd. 2023
A. Peña-Ayala (ed.), *Educational Data Science: Essentials, Approaches, and Tendencies*, Big Data Management,
https://doi.org/10.1007/978-981-99-0026-8_1

EDS   Educational data science
HA   Human-authored (questions)
ICICLE   Industry Connections/Industry Consortium on Learning Engineering
ML   Machine learning
NLP   Natural language processing
OLI   Open Learning Initiative
SLE   Smart learning environment

1.1 Introduction

To help define the developing field of educational data science (EDS), several comprehensive reviews have been presented in recent years that have worked to clearly articulate what EDS is—and what it is not. The formalization of EDS was largely motivated by the emergence of novel data and big data from administrative and technology sources [1–3]. Educational data science requires three domains to converge: mathematics and statistics; computer science and programming; and knowledge about teaching, learning, and educational systems [3]. McFarland et al. [2] argue for a more expansive definition where EDS becomes "an umbrella for a range of new and often nontraditional quantitative methods (such as machine learning, network analysis, and natural language processing) applied to educational problems often using novel data" (p. 2). There is a drive to apply the learning sciences, learning analytics, educational data mining, and data science to explore and solve real-world educational challenges. In this work, we use the research and practices of an educational technology company, Acrobatiq by VitalSource, to expand on how these areas coalesce to reveal learning insights from a digital courseware platform.

Educational data science is the investigation and interpretation of educational data and, therefore, is inherently and fundamentally the result of (a) the creation of educational products that collect data and (b) the implementation of those products with students in natural learning environments. EDS does not exist in a vacuum, but rather as a situated step in a larger process of creation and implementation. EDS is also purpose-driven. The technologies of today produce enormous quantities of data simply as a byproduct of learners engaging in their learning environments. EDS seeks to make meaning from data—to better understand how learners learn and to optimize learning over time. Yet investigation of data is also shaped by the learning product that collected the data, the students who generated the data, and the learning theories and principles underlying the product, data, and analyses.

In this chapter, we propose that learning engineering is a practice and process that can be used to ground educational data science. Learning engineering combines the learning sciences, human-centered engineering design, and data-informed decision making in an iterative process that provides a contextualization and structure for EDS. Throughout this chapter, we will showcase how the learning engineering process (LEP) supported our application of EDS as a part of a larger goal of creating and evaluating educational products at Acrobatiq by VitalSource—from adaptive courseware to questions generated through artificial intelligence (AI) to an automated content improvement service (CIS). An overarching goal for each of these products was to support students and their learning process, which inherently creates a need to evaluate whether the product was successful. Educational data science was an integral part of each learning engineering process.

In this chapter, we will examine each learning engineering goal and how EDS helped answer the following questions:

- Does the data from courseware support existing learning science principles?
- How do questions generated from artificial intelligence perform for students?
- How can data analytics be applied at scale to evaluate questions in real time?

1.1.1 Chapter Overview

This chapter covers a series of related educational technology projects and therefore is broken up into topics and contextualized by the corresponding learning engineering process. Each section is summarized as follows:

"Learning Engineering." As more data is being collected and used than ever before and technology is advancing every day, it is also a critical time to consider how to engage in EDS for the study and creation of learning products. We propose utilizing learning engineering as both a practice and process for applying EDS. Learning engineering provides a student-centered process based on engineering systems design to make data-informed decisions [4]. Applying educational data science to make a meaningful impact in student learning is a complex process that—more often than not—also requires a diverse team. The process itself is useful when employing EDS and the many disciplines involved therein, but it also requires a student-centered approach that helps the research and development in EDS to maintain an ethical process. This helps ensure that in the complex interactions of various disciplines and teams, the best interests of students are kept at the heart of educational data science. In this section, learning engineering is defined, its relationship to EDS explained, and Acrobatiq's application of the learning engineering process outlined.

"Courseware in Context." Understanding context is an important component of learning engineering and should be accounted for in the learning engineering process. Context for courseware itself is also necessary in this chapter to better understand the related learning engineering challenges and solutions discussed and the learning science research relevant to courseware and its evaluation. Additionally, this section has the practical objective of defining courseware and discussing how data is gathered and used in the courseware environment—providing context for the following sections that provide examples of EDS from courseware-generated data.

"EDS to Investigate Learning by Doing." The purpose of Acrobatiq—originally a start-up from Carnegie Mellon University's Open Learning Initiative (OLI)—was to extend the learning science findings from OLI by making courseware more widely accessible to students in higher education. The primary learning method used in the courseware is a learn by doing approach—integrating formative practice questions with the text content—which generates the doer effect. The doer effect, first researched at OLI, is the learning science principle that doing practice while reading is not only more effective for learning than just reading but also causal to learning [5–7]. In this section, we address the initial learning engineering challenge of creating effective courseware and explore how the data gathered by the courseware platform can be used to answer the question: how can EDS be employed to identify effective learning methods in courseware?

"EDS for Evaluating AI-Generated Questions." The pursuit of educational research to confirm learning science findings is in itself a beneficial goal, but the natural next step is apply this knowledge. The practical application of educational data science is to support and create learning resources and environments intended to help students. In this section, we outline how the doer effect research confirmed in the courseware data analyses led to an entirely new learning engineering challenge: using artificial intelligence (AI) to generate questions. The creation of an automatic question generation (AQG) system itself is a task that requires learning science, data science, machine learning (ML), natural language processing (NLP), and more. However, the evaluation of these questions as a learning tool for students is paramount. In this section, we look at how data can help answer: how can EDS evaluate the performance of AI-generated questions using student data?

"EDS for Question Evaluation at Scale." The deployment of millions of AI-generated questions to thousands of students presents both a challenge and opportunity to analyze educational data on an enormous scale. To evaluate student interactions with these AI-generated questions manually would be near impossible, so an automated content improvement service (CIS) was developed to monitor student engagement with millions of questions in real time. This adaptive system uses EDS to evaluate each individual question to make determinations on quality and can also surface broader insights about students or content. The goal of this work is to use AI to automate the learn by doing method for digital textbooks to give more students the ability to harness the doer effect while also automating the means for educational data analysis at an enormous scale.

1.1.1.1 State of the Art

Let us consider each of these topics in relation to the *state of the art*. The learn by doing method of engaging students in practice while they learn is not a new or novel concept; teachers have been putting this into practice in many forms for students of all ages for decades. However, the advances in digital learning resources (such as courseware) allowed for the first research to confirm that this learn by doing approach was effective and causal to learning [6]. As noted by McFarland et al.

[2], "Not only are old research questions being analyzed in new ways but also new questions are emerging based on novel data and discoveries from EDS techniques" (p. 1). In this way, a long-standing pedagogical approach has sparked new research as digital resources generate data. The technology itself has revolutionized educational research with the data it gathers, as "questions that were either costly or even impossible to answer before these data sources were available can now be potentially addressed" ([1], p. 132).

Automatic question generation has become a popular area of research, yet it was noted by Kurdi et al. [8] in their systematic review that no "gold standard" yet exists for AQG, and very little research has focused on the application and evaluation of generated questions in natural learning contexts. Thus, the large-scale integration of automatically generated (AG) questions with digital textbooks is a novel advancement for both scaling the learn by doing method and advancing research on automatically generated questions. The content improvement service—developed to perform real-time analysis on the millions of generated questions and adapt the content as needed—is the first system of its kind to employ EDS for such a purpose. Lastly, learning engineering as a process and practice has only in recent years been formalized. Applying EDS in a learning engineering practice will continue to push many EDS practitioners forward to new states of the art for years to come.

1.2 Learning Engineering

In 1967, Nobel laureate and Carnegie Mellon University professor Herb Simon introduced the term *learning engineer*, proposing that this role was for professionals who could create learning environments to increase student learning [9]. Through leadership from Carnegie Mellon on the development of the learning engineer role, it spread to other institutions and organizations. The IEEE Industry Connections/ Industry Consortium on Learning Engineering (ICICLE) was formed to formalize learning engineering as a discipline and defines learning engineering as "a process and practice that applies the learning sciences using human-centered engineering design methodologies and data-informed decision making to support learners and their development" [4, 10]. As indicated in this definition, learning engineering is a multidisciplinary practice and therefore often utilizes a diverse team with unique areas of expertise to accomplish its goal [4]. Notably, *practice* and *process* are both used in this definition as well; learning engineering is a *practice* for both individuals and teams to engage in while also being a *process* for approaching a variety of challenges and projects.

"Learning engineering is also a repeatable *process* intended to iteratively design, test, adjust, and improve conditions for learning" [11]. Goodell and Thai [12] describe the learning engineering process and its benefits:

> The learning engineering process enables data-informed decision-making through development cycles that include learning sciences, design-based research, and learning analytics/ educational data mining. It leverages advances from different fields including learning

Fig. 1.1 The learning engineering process [11]. (*CC BY Aaron Kessler*)

sciences, design research, curriculum research, game design, data sciences, and computer science. It thus provides a social-technical infrastructure to support iterative learning engineering and practice-relevant theory for scaling learning sciences through design research, deep content analytics, and iterative product improvements (p. 563).

The ICICLE design special interest group, led by Aaron Kessler, developed a learning engineering process (LEP) model [13] that shows how a project moves from a context-driven problem, through iterations of design and instrumentation, to implementation in natural contexts to data analysis and further iterative improvements based on the results. This model was further iterated on, as shown in Fig. 1.1, to represent the cyclical nature of the LEP [11]. In this model, the learners, team, and context all shape the central learning engineering challenge for the LEP. The creation phase encompasses designing, building, data instrumentation, and implementation planning—all of which may involve many iterations and sub-challenges that could be occurring simultaneously [11]. This process was also designed to be broadly applicable, as learning engineering is practiced in increasingly diverse contexts.

1.2.1 Learning Engineering at Acrobatiq

It is notable that courseware and learning engineering have a shared history. Carnegie Mellon fostered and spread learning engineering [14], through such institutes as the Open Learning Initiative (OLI) where early courseware was developed and researched (described later in the chapter). Acrobatiq was founded from OLI, and the team who formed Acrobatiq intentionally organized the company and its development process around the learning engineering process, led by learning engineers [15]. The Acrobatiq team engaged in many different LEPs over the years. One primary challenge was the creation of courseware that was effective for student learning, as shown in Fig. 1.2. This broad primary challenge encompassed many smaller iterations within the design phase. It also led to a distinct goal in the investigation phase: how to evaluate if the learn by doing method of the courseware was effective.

This is the first example of educational data science applied within the LEP discussed in this chapter. In the courseware creation challenge shown in Fig. 1.2, we can map educational data science to the phases of the learning engineering process outlined. In the creation phase, instrumentation for data is completed, i.e.,

Fig. 1.2 The learning engineering process for initial courseware creation at Acrobatiq

developing the infrastructure for how to collect data as a part of the courseware platform. This included making sure unique student user accounts would be associated with clicking pages and answering questions—the actions that generate the foundational clickstream data. Instrumentation also included developing a learning objective architecture that tagged questions to objectives and skills to add layers of knowledge and allow for future analyses to make sense of what students are learning. The creation phase was critical for planning for data collection and how that would be utilized by the platform, either as predictive analytics or triggers for student adaptivity. The implementation phase is when the data required for educational data science is collected. In this case, implementation was when the courseware was used by students in their natural learning contexts. Knowing the context for the implementation is important for later making sense of the data. Finally, the investigation phase is often when EDS is applied most robustly, as this is when we would seek to investigate the data and answer questions like, was the courseware effective for student learning?

Beyond this initial LEP of creating the courseware itself, many other challenges arose that were solved with either iterations of the LEP or an entirely new LEP altogether. For example, in the investigation of the original courseware creation LEP, it was identified that the adaptive activities were being underutilized. This triggered an investigation that began a new iteration of that LEP to change the design and reimplement this portion of the courseware [16]. Sometimes an entirely different challenge was identified for a new LEP, such as how to scale context-specific instructor training for the courseware [17].

At Acrobatiq, learning engineering is engaged beyond the process itself, but as a practice as well. Learning engineering is student-centered, and this provides a guiding purpose to the products and learning environments created. Each decision and product created should be in service to students and their learning process. This student-centered purpose also drives the investigation phase of the learning engineering process because evaluating how the product helps students learn is critical to ensuring that it is serving students. In this way, educational data science is purpose-driven as a part of the learning engineering process and practice. Data science–driven investigation either proves success or highlights improvements and necessary iterations on the product—and either outcome provides transparency and accountability for both the team developing the products and the students and instructors using them.

As previously mentioned, this chapter discusses how educational data science, as a part of the LEP, guided the development and evaluation of an evolution of learning resources for students. We can see this as three distinct LEPs (Fig. 1.3), each instigated by the previous process and the need to solve a new challenge and evaluate the effectiveness of the solution:

- First, the initial challenge was to create courseware that would help students learn effectively. This LEP required EDS in the investigation phase to determine if the learning by doing method of the courseware helped students get the learning benefits of the doer effect learning science principle.

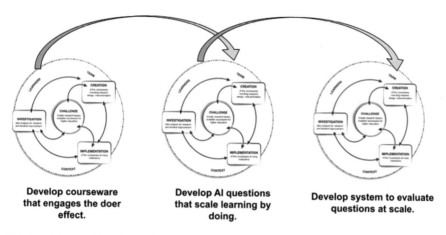

Develop courseware that engages the doer effect.

Develop AI questions that scale learning by doing.

Develop system to evaluate questions at scale.

Fig. 1.3 The three learning engineering processes described

- Second, the next challenge was to scale the learning by doing method by using artificial intelligence to generate courseware and questions. This LEP required EDS to evaluate the performance of these automatically generated questions using student data.
- Third, the final challenge was to evaluate (and intervene with) the millions of automatically generated questions. This required EDS to automate a system to perform this data analysis at scale, continuously.

1.3   Courseware in Context

Context is a key aspect of engaging in learning engineering, both as a practice and process. The purpose of this section is to provide context for courseware as a learning resource, as well as the learning engineering processes and EDS investigations discussed in the following sections. In this work, research and learning science not only informed the design and features of this courseware but also provided context for the learning engineering challenges and EDS investigations outlined in detail in later sections.

Courseware itself requires a definition, as courseware has evolved over time and there are varying ways of creating a learning resource with the set of features that constitutes "courseware." The Courseware in Context (CWiC) Framework was developed with support of the Open Learning Consortium, the Bill and Melinda Gates Foundation, and Tyton Partners [18]. This framework was developed for postsecondary decision makers to define and navigate courseware selection. They define courseware as "instructional content that is scoped and sequenced to support delivery of an entire course through software built specifically for educational purposes. It includes assessment to inform personalization of instruction and is equipped for adoption across a range of institutional types and learning environments." The need to define courseware for stakeholders and practitioners in education also showcases (a) the prevalence of courseware as a resource and (b) the need to define and provide context when discussing courseware in a research context.

Courseware, as developed at Acrobatiq, is a comprehensive learning environment that combines learning content with frequent formative practice in short, topical lessons aligned to learning objectives that are grouped into modules and units and followed by adaptive activities and summative assessments [15]. Students receive all the learning content, practice, and assessments needed for a semester-long university course in one unified learning environment.

Courseware is also *not* several categories of student learning environments, such as a cognitive tutor or intelligent tutor, where several additional sub-steps of student feedback are provided [19]. While determining a classification system for these various systems is not a goal of this work, identifying a relationship among these tools provides a useful conceptualization for how these learning resources are categorized. Hwang [20] described a smart learning environment (SLE) as one in which learners could "access digital resources and interact with learning systems in any place and at any time" while also giving students "the necessary learning guidance, hints, supportive tools or learning suggestions" (p. 2). Given this definition of an SLE, we think of the success of intelligent tutoring systems. Yet as noted by Baker [21], the complexities of such systems may have limited their ability to scale. Kinshuk et al. [22] describe three features of SLE development: full context awareness, big data and learning analytics, and autonomous decision making and dynamic adaptive learning. These foundational features are applicable to intelligent tutoring systems as well as the courseware environment. While not completing the same level of sub-step support of the intelligent tutoring system, courseware also fulfills the key features of SLEs, as outlined by Kinshuk et al. [22].

When considering the nature of SLEs as one in which students engage with learning activities and data are produced and made available for many forms of intervention, we can quickly see how SLEs are becoming available in nearly all educational ecosystems. However, it is not enough to design and deliver a system that gathers data and adapts based on that data. SLEs must also be designed, developed, and iterated on based on proven learning science research. What benefit is a data-driven SLE if it is not designed and proven to help students learn? Therefore, we propose that a smart learning environment—such as courseware—should be based in research as well as be data-driven.

As previously noted, courseware has been developed and researched for decades at places such as Carnegie Mellon's OLI, where learning engineering was also put in practice. Regarding the design of OLI courseware in a research study, Lovett et al. [23] stated: "The OLI-Statistics course was designed to make clear the structure of statistical knowledge, include multiple practice opportunities for each of the skills students needed to learn, to give students tailored and targeted feedback on their performance, and to effectively manage the cognitive load students must maintain while learning. All of these principles would be predicted to foster better, deeper learning, and our results across all three studies support that prediction" (p. 15).

Their results found that the OLI statistics courseware could help students achieve similar grades asynchronously compared with a traditional face-to-face course, improve grades when used in a hybrid format, accelerate learning compared with a traditional course, and increase knowledge retention.

Formative practice has long been known to be an effective technique for learning [24]. A key feature of courseware is the integration of formative practice questions with learning content in a learn by doing method. A cognitivist approach, this learn by doing method aims to engage students to be active in the learning process rather than passively reading. While there are many ways of engaging students in active learning (e.g., interactive labs, group discussion, simulation, etc.), this is a more specific approach akin to the testing effect, wherein practice offers immediate feedback [6]. Formative practice acts as a no-stakes practice testing [24] while also offering immediate feedback. Questions with feedback have been shown to benefit student learning over questions without [24, 54]. Formative practice is an established technique that benefits learning for all students in all subjects, and studies have shown that while formative practice benefits all students, it benefits low-performing students most of all [25].

The integration of formative practice in a courseware learning environment has specifically been proven to create the doer effect—the learning science principle researched at OLI that students who do practice while reading have higher learning gains than those who only read [6]. Research found doing practice had a median of six times the effect on learning than simply reading [5], and follow-up research found this relationship to be causal [6]. These causal findings have been replicated on courseware, extending the external validity of the doer effect [26]. The doer effect has also been studied controlling for student demographic characteristics, and the learning benefit of this learn by doing method remained relatively unchanged [5, 27]. Proven learning science research such as the doer effect research is a critical component of what makes a courseware an SLE.

1.3.1 Acrobatiq Courseware: Data-Driven

1.3.1.1 Learn by Doing Data

The primary learning feature of this courseware is the integration of formative practice with the textbook content in a learn by doing approach. There are many types of questions (multiple-choice (MC), fill-in-the-blank (FITB), drag-and-drop (D&D), free response, etc.), and each question provides immediate feedback and unlimited opportunities to answer. Each student-question interaction generates data. This includes

- The time the student interacted with the question
- Whether a correct or incorrect answer was selected or input

- Additional attempts at the answer, with the time of each additional attempt
- Help or hints requested

This clickstream data is stored in its raw format, and Fischer et al. [1] categorizes this as microlevel data. A single class answering formative questions in a single course can easily generate hundreds of thousands of data events. On its own, this large data pool does not reveal insights, but once the right questions are asked and data science/learning analytics has been applied, it becomes very useful for investigating how students use the practice and how that relates to their learning.

In Acrobatiq courseware, all formative practice is tagged to student-centered, measurable learning objectives. These learning objectives align the content and practice on a lesson page to make it clear to students what they should be learning, but the learning objectives also provide another level of data to analyze. This combination of formative practice data tagged to learning objectives generates insights used for both adaptivity and instructor dashboards.

1.3.1.2 Data-Driven Adaptivity

Vandewaetere et al. [28] developed a framework for classifying adaptive systems according to the source of adaptation (what determines adaptation), the target of adaptation (what is being adapted), and the pathway of adaptation (how it is adapted). The most common adaptive systems use student knowledge or action as the source of adaptation, the learning path as the target of adaptation, and some predetermined rules as the pathway of adaptation. That is generally true of the adaptive activities described here. In Acrobatiq courseware, the learn by doing method is also what drives adaptation, as these questions provide data to the predictive learning model, which in turns drives adaptive activities personalized to each student (Fig. 1.4). The adaptive activities rely on a data architecture of practice questions and learning objectives. The learning objectives are located at the top of each lesson to help guide students, but also serve as the objectives to which all questions are aligned and tagged. As students answer questions, those data are collected and fed to the analytics engine, which generates a predictive learning estimate. The adaptive activities are designed to include a set of low-, medium-, and high-difficulty questions for each learning objective. Scaffolded questions and content in this way align with Vygotsky's zone of proximal development—structuring content and interactions in such a way as to meet the learner at their level of understanding and build upon it [29]. Providing struggling students with scaffolds and immediate feedback helps reduce cognitive load in a similar way as worked examples [30]. As the student begins the adaptive activity, the learning estimate determines the level of scaffolding the student receives for each learning objective in the activity. Research has shown that students who increase their learning estimate in the adaptive activities also increase their scores on the summative assessments [15].

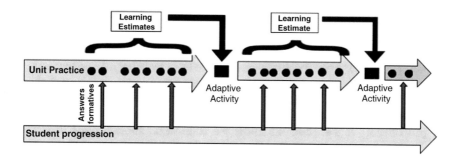

Fig. 1.4 The feedback loop of student data from student interactions with formative practice to the generation of predictive learning estimates to the adaptive activities

1.3.1.3 Data for Instructor Feedback Loops

In addition to being used for real-time course personalization for each individual student, data are also aggregated and surfaced to instructors to provide meaningful insights into both student and course performances. This allows instructors to engage in data-driven feedback loops with students in real time. For example, student responses to formative questions produce predictive learning estimates. While a student's learning estimate may vary from objective to objective, aggregated learning estimates can very quickly identify students who are struggling and may be at risk of failure. When the instructor sees this—often within the first weeks of a course—they can take action to provide early intervention and remediation.

Data are also aggregated from all students to provide insight into the entire class's performance on each learning objective. This surfaces issues at the learning objective level, indicating a need to investigate an entire topic of the course. It could be that an objective is particularly difficult to master and the class needs additional remediation from the instructor or that there could be additional content or questions needed to better teach the topic.

The importance of data for instructor feedback loops should not be overlooked. Instructor use of and satisfaction with the courseware is imperative to the success of this SLE for students. Implementation is a complex phase of the learning engineering process for the student/technology/instructor interactions and the many external factors that can influence each SLE implementation context. Research by Kessler et al. [31] using a mathematics cognitive tutor similarly found several different interaction models between students, instructors, and the cognitive tutor that affect and support student learning in different ways. This indicates that even in intelligent tutoring systems—considered the most complex computer-based instruction with learning gains on par with one-on-one human tutoring [19]—instructors can still be highly influential during implementation. An analysis of the same courseware implemented with multiple instructors at multiple institutions found that student engagement was ultimately highly impacted by instructor implementation and course policy choices [32].

Baker [33] suggested that perhaps instead of attempting to scale complex intelligent tutoring systems, we focus on tutoring systems "that are *designed intelligently, and that leverage human intelligence*" (p. 603). The courseware described herein does not have the same type of step-level feedback as is common in intelligent tutoring systems [19], yet focuses on an effective learn by doing method that collects and surfaces meaningful data to humans. Who better to make use of this information than the instructor who understands the needs of their students and complexities of their learning context? Qualitative reports from faculty using AI-based courseware as part of their courses described how the courseware allowed them to change teaching models (to a flipped blended model), focus energy where students needed it most, and improve their own satisfaction in teaching [34]. Furthermore, by evaluating the data provided by the courseware environment, instructors are able to iterate on their own course policies to increase student engagement and ultimately increase student outcomes [35]. This calls for a renewed focus on educators as part of smart learning environments and the learning engineering process, for their role in implementation can impact student outcomes and the data that drives both learning and research.

1.4 EDS to Investigate Learning by Doing

Digital learning platforms, such as the courseware described herein, generate enormous quantities of micro-data in the form of clickstream data from every interaction students have with the courseware. A single semester-long course of 30 students can easily surpass a million data events. Yet that volume of micro-data does not reveal insights on its own. First, a question must be asked, such as, can this data be used to replicate the doer effect analyses developed at Carnegie Mellon [6], showing the same causal findings between doing practice while reading and increased learning outcomes? Therefore, the central challenge for this learning engineering process was to develop courseware with a learning by doing method that would help students learn more effectively, as shown in Fig. 1.5. These analyses require a combination of data science, statistics, and knowledge of learning processes and principles in order to understand a simple and profound question: how do we help students learn better?

1.4.1 The Doer Effect

Let us consider the question more formally: "how do we help students learn better?" The word *better* can mean many things. In analyzing current doer effect research, we are comparing the relative benefit of the time a student spends between reading text, watching video (if applicable), and doing practice. In this sense, "better" implies a more beneficial use of the time of the student to further their learning.

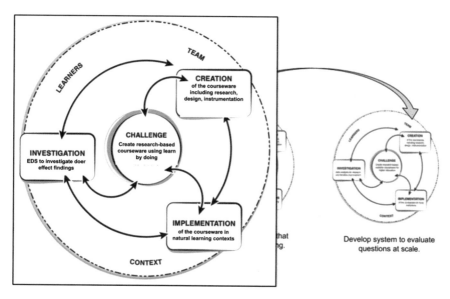

Fig. 1.5 The LEP diagram with the central challenge of creating learn by doing courseware

The particular data science applied to this question is detailed below, but the concept is in some ways easy to understand, and indeed it is expected. If a learner is somewhat frequently asked to pause and reflect on their recent learning, there is a benefit. Any of us might read a book or a paper or a chapter about learning and realize we stopped paying attention at the last paragraph about the doer effect. Asking a learner to pause and reflect not only gives them a chance to check their learning but might snap them out of the "trance" of skimming along without internalizing what they are reading. Keeping the learning at the center of this investigation is critical and a key component of the learning engineering process. Let us examine then how this effect was proven and how the results were replicated.

Koedinger et al. [5–7] used student data from natural learning contexts to investigate the relationship between the practice in OLI courseware and outcomes on summative assessments. It is well known that practice generally correlates positively with outcomes. However, in order to recommend this learn by doing approach with high confidence, it is necessary to know that there is a causal relationship between practice and better learning. For example, a highly motivated "go-getter" student may do more practice and also obtain better learning outcomes, but this would not be able to tell us whether the better outcomes were caused by doing the practice. Koedinger et al. used causal discovery methods [5, 7] and mixed-effects regression [6] to determine that the doer effect was in fact causal. There is no better explanation of the importance of causal relationships than stated in Koedinger et al. [6]: "It should be clear that determining causal relationships is important for scientific and practical reasons because causal relationships provide a path toward explanatory theory and a path toward reliable and replicable practical application."

The importance of analysis in a wide variety of natural learning contexts, e.g., real courses, should not be overlooked. This is key to ensuring the external validity of the method, i.e., confidence that it is generalizable to other contexts in which it was not analyzed. As Koedinger et al. [6] said, "If we can be increasingly certain a learning method is causally related to more optimal learning across a wide variety of contexts and content, then that method should be used to guide course design and students should be encouraged to use it."

This goes hand in hand with replication research, which is critical in the learning sciences to provide additional evidence to support—or refute—claims made about effective learning practices. A large fraction of published research in the social sciences has not been replicated, and studies that cannot be reproduced are cited more frequently than those that can [36]. By replicating and sharing the data analysis and findings as part of the LEP, researchers and developers maintain transparency and accountability to the learner [16]. Furthermore, replicating findings that are based on large-scale EDS studies provides valuable verification of the results, as the volume and type of data analyzed can be difficult to obtain. Replicating these causal doer effect studies helps support a practical recommendation that students can improve their learning outcomes by increasing the amount of formative practice they do.

Using the mixed-effects regression approach of Koedinger et al. [6], we first replicated the causal doer effect study using final exam data for 3120 students in a competency-based Macroeconomics course at Western Governors University [37]. The regression model analyzed the relationship of student reading and doing in each unit (competency) of course content to scores on that unit's portion on the final exam. The key innovation in the model was to control for the total amounts of reading and doing in *other* units of the course. Reading and doing outside a unit can act as a proxy for a third variable like motivation that can lead to correlation between level of effort and outcomes. In this way, if the doer effect is causal, then the amount of doing within a unit should be predictive of the student's score on that unit's content even when accounting for doing outside that unit.

The findings were consistent with those of the original study replicated. Both within-unit doing and outside-unit doing were strongly, positively significant, indicative of a causal doer effect. Replicating these results using courseware designed with the same learning science principles but in a different domain and at a different higher education institution adds support for the external validity of the doer effect. In further work [27], several correlational and causal doer effect analyses were conducted on multiple types of summative outcomes for this. Not only were causal doer effects documented, but another key finding was also that the doer effect was not accounted for by student demographic characteristics, consistent with Koedinger et al. [5].

1.4.2 Summary

A critical role of educational data science is to validate or invalidate ideas we hold as common. In the field of education, it was a long-held belief that "doing is good for learning," but only recent data analysis has proven that the link between doing and learning is both *measurable* and *causal*. It might seem obvious as an idea, but using data science to prove this enables us to hone in on those features of learning environments that most help students learn. In this case, uncovering the causal link between doing and learning allows us to further improve learning experiences for students and recommend this practice at scale. Likewise, any new learning interventions can be evaluated to a similar metric going forward.

In this learning engineering process, EDS was needed to investigate an established learning science principle, and so the research and courseware context were particularly important for successful analysis. The contextual factors and prior research accounted for in the LEP were necessary for a successful investigation of the learn by doing method.

1.5 EDS for Evaluating AI-Generated Questions

With the doer effect results replicated using data from courseware, a new learning engineering challenge was identified: how to scale this learning by doing method. This section describes the methods used for three major phases of the LEP: the development of AI-generated courseware and its additional enhancements, the implementation of the courseware, and the types of data analyses that were done to evaluate the performance of this AI-generated courseware. Unlike the doer effect investigation where the question data was used to identify student learning, this analysis used that same formative question data but to investigate the performance of the questions themselves (Fig. 1.6).

1.5.1 AI-Generated Courseware

The courseware used to illustrate this LEP was generated using artificial intelligence, through the SmartStart process on Acrobatiq's platform [38]. SmartStart was designed to apply learning science principles to static content and complete certain instructional design tasks as an expert would. There are three main tasks that SmartStart does to transform an online textbook into courseware: create short lessons of content, align learning objectives to lessons, and automatic question generation (AQG) [52]. These tasks are designed to be content agnostic so as to work across subject domains. SmartStart uses supervised and unsupervised machine learning (ML) models to perform content-related tasks (such as extracting learning

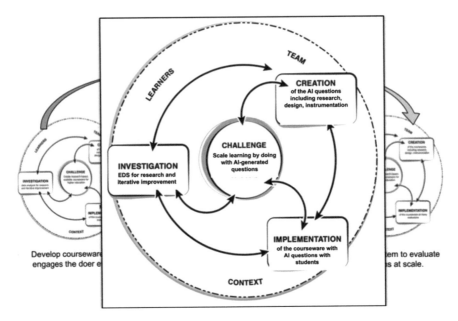

Fig. 1.6 The LEP diagram with the central challenge of creating AI questions to scale learning by doing

objectives from textbook chapters and aligning lesson content to them) and a variety of syntactic and semantic natural language processing (NLP) methods to understand textbook content in more detail (such as identifying content that is important for questions and then automatically generating those questions and placing them strategically throughout the course as formative practice).

The first step is the course structure determination. Textbooks provide high-quality content in contiguous blocks typically organized in chapters. Students frequently do not read textbook content as intended [39, 40], and these long sections of content are not beneficial for content retention and cognitive load. Shorter lessons on a single topic provide students with more manageably sized content sections, which help them construct better mental models.

SmartStart analyzes the textbook structure and identifies how the content can be chunked into smaller lessons that revolve around a single topic or learning objective. Structural features of the textbook are identified to help determine how the content should be sectioned, and expert rules for content design were used to guide the application's process. Instructional design best practices combined with data analysis from dozens of courseware environments used by thousands of students provided insights into optimal lesson length.

The next step is learning objective alignment. Learning objectives are placed at the top of each lesson page and provide students with guidance on what they are about to learn and how they will be evaluated on that content. In addition to providing context for the student, objectives also will be tagged on all formative

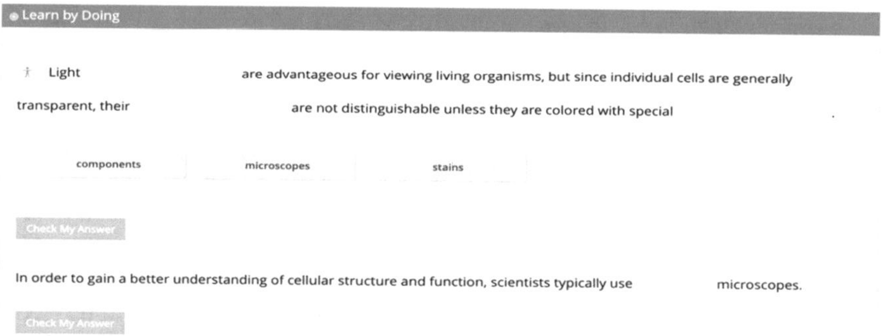

● Learn by Doing

} Light   are advantageous for viewing living organisms, but since individual cells are generally

transparent, their   are not distinguishable unless they are colored with special   .

components   microscopes   stains

Check My Answer

In order to gain a better understanding of cellular structure and function, scientists typically use   microscopes.

Check My Answer

Fig. 1.7 Examples of AG matching and FITB questions

practice aligned to that lesson for organizing instructor dashboards, optional adaptive enhancements, and post hoc data analysis.

SmartStart has two distinct tasks related to learning objectives. The first task is to identify learning objectives in the textbook. Upon initial inspection of only a few dozen online textbooks, there was not sufficient consistency in terminology usage, HTML markup, or placement of learning objectives that would allow for development of a simple rule set to apply with sufficient accuracy to this small set, let alone generalize to thousands of titles. Therefore, a supervised machine learning model was developed for learning objective identification. The model uses a variety of features including specific identifiers, placement characteristics, and Bloom's Taxonomy verbs [41]. The learning objective identification model and other ML models used in SmartStart were developed using the scikit-learn library [42].

Once the learning objectives have been identified (if present in the textbook), the second task is to place them with the appropriate lessons. Alignment of learning objectives to content is a critical task. SmartStart accomplishes this task by using a statistical model developed to evaluate the content of the objective and the content of the lessons to propose the best placement.

The final step is the automatic question generation. As formative practice questions are critical for a more effective learning environment, AQG is one of the most important processes in SmartStart. Two types of questions, fill-in-the-blank (FITB) and matching, are currently generated to accompany the content, with expanded question types to be released in 2022. The AG FITB questions require students to read a sentence and type in a missing term, making this type a recall question. The AG matching questions require students to evaluate three terms and drag-and-drop them into the correct locations in a sentence, making this type a recognition question. Both are cloze question types constructed from important sentences in the textbook content and purposefully target the *remembering* cognitive process dimension in the revised Bloom's Taxonomy [41]. The purpose of the formative questions is to help students actively engage in the learning process, so foundational Bloom's level questions are not only necessary but appropriate (Fig. 1.7).

Automatic question generation is an active research field that is growing rapidly because of increasing improvements in NLP and ML. The recent systematic review by Kurdi et al. [8] provides a categorization system for AQG that includes level(s) of understanding and procedure(s) for transformation. For level of understanding, SmartStart AQG uses both syntactic and semantic knowledge to perform two main tasks: selecting the content sentences to be used for question generation and selecting the terms to be used as the answers to create cloze questions. Syntactic information, such as part-of-speech tagging and dependency parsing, was used both in sentence selection and to identify terms suitable as answer candidates. Semantic knowledge was also used in detecting important sentences. The various NLP analyses employed in AQG were carried out using the spaCy library [43].

Rule-based selection is the primary method for the procedure of transformation. These rules use both the syntactic and semantic knowledge of the content derived through NLP to identify the best sentences and then the best answer terms within those sentences for creating the cloze questions. For example, one rule for sentence selection uses the dependency tree to identify and reject sentence fragments. Another example rule uses syntactic knowledge to ensure answer terms for matching questions agree in number (for nouns) or degree (for adjectives), to help prevent answers from being guessable without needing to understand the content.

The processing of the textbook corpus and subsequent application of the selection rules generates a large set of questions to cover the chunked lessons in the course. Not all questions make it into the courseware, however. AQG uses an overgenerate-and-rank approach [44], which selects only the top four questions of each type (FITB and matching) generated from each lesson page's content to appear on that page.

1.5.2   Courseware Adaptive Development

The SmartStart AI-generated courseware provides a clearly organized structure and learning by doing at a lesson level, but does not currently create the adaptive elements. The adaptive activities require learning estimates in order to determine which questions to deliver to each student. Learning estimates are a predictive measure generated by an analytics engine that is generated for each student on each learning objective to estimate how well that student might perform on assessment items on those learning objectives. The machine learning model that generates the learning estimate is based on item response theory (IRT) [33, 45] to construct an estimate of the student's ability on a learning objective. That ability estimate is then used to predict how well a student would perform on that objective in a summative assessment. The IRT model is able to take certain psychometric properties of the formative practice into account when constructing the ability estimate. A two-parameter logistic regression is used that models difficulty and discrimination for each question. A Bayesian approach [46] is used to estimate the posterior distributions of the IRT question parameters from the data, as well as the student ability posterior distributions from the formative questions they answered. A numerical learning estimate between 0 and 1 is derived from the ability posterior distribution, where higher values indicate better expected performance. When the model has sufficient confidence in its learning estimate based on the available data, a learning estimate category of low, medium, or high is assigned.

The Neuroscience courseware initially created via the SmartStart process was then further enhanced by the textbook publisher to include additional formative questions, adaptive activities, and summative assessments. The goal for the team adding these features was to take full advantage of the Acrobatiq courseware platform's adaptive and assessment features. The SmartStart process produced formative practice questions; however, the platform had a minimum requirement for questions per learning objective in order to produce the predictive analytics. Therefore, the development team added additional formative items to fulfill these requirements for each learning objective when necessary. Questions were repurposed from end-of-chapter materials or ancillary materials written to accompany the textbook. The questions were originally written by subject matter experts to align to the textbook, and editors and instructional designers ensured the additional questions were placed with the appropriate lessons. When there were still an insufficient number of practice items after repurposing existing questions, subject matter experts wrote new items for the content until the threshold was met.

The adaptive activities were also created by the publisher team, according to specific requirements. A minimum of one scaffolding question for the low and medium levels was required per learning objective. While content such as scenarios and images was taken from ancillary materials, the majority of these questions were written by subject matter experts.

1.5.3 Courseware Implementation

This AI-generated courseware was enhanced by the publisher team and then adopted by schools across the United States and Canada. While the Neuroscience textbook [47] used as the basis for courseware generation is a higher-level text than would be used in an introductory college course and would likely be used by students in a science major, no student population characteristics can be drawn from the data available from the courseware platform (no student demographic information is collected, and no personal identifiable information is included in the data for analysis).

With course sections being run at over 40 institutions, there are many instructor implementation factors that cannot be accounted for. As discovered in previous research, student engagement with the courseware varies widely based on instructor practices [32]. However, in this instance, there was a recommendation by the publisher to assign completion points for the formative practice. Assigning points, even a small percentage of student grades, has been found to increase student engagement [34]. When aggregating the data for all course sections, we can see the effect of this practice in a data visualization called an engagement graph.

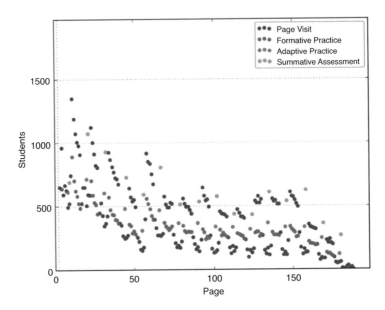

Fig. 1.8 An engagement graph showing student reading, doing, and assessment activity for the Neuroscience courseware

Engagement graphs show the total number of students who were enrolled in the course on the y-axis and the pages of the course on the x-axis. The colored dots represent what students did on each page of the course. The green dots indicate completion of summative assessments. The blue dots show the total number of students who visited each page (a proxy for reading), and the red dots show the total students who did formative practice. It is common for there to be more students who read than do practice (called the "reading-doing gap"), and it is also common to see student attrition as the course (and semester) progresses [32].

The engagement graph for this course (Fig. 1.8)—aggregated for all course sections—shows some interesting trends that can be studied to answer our first research question on how instructors can influence student engagement trends. The green dots—showing summative assessment completion—are at the top, indicating nearly all students did them, even if they did nothing else. This indicates they were likely assigned in all course sections. Next, we see a high line of blue dots for student reading and a lower line of red dots for students who did practice, as expected. The very low blue dots were end-of-chapter resources, which were not assigned. There is also a downward angle in the first 50 pages, which is also typical as some students drop the course and others stop reading the textbook altogether. An unusual feature is the very horizontal red doing trend that becomes apparent from page 75 onward. There are approximately 300 students who consistently did the practice until the end of the course content. This is not a typical pattern observed in prior research, as the attrition usually gets stronger at the very end. The presumptive cause of this pattern is the instructors who assigned completion points for formative practice.

1.5.4 Description of Analyses

In order to evaluate how this courseware is performing, there are two types of analyses to do. The first is to do a comparison of the automatically generated questions and human-authored (HA) questions in this AI-generated courseware. The purpose of this analysis is to check performance metrics of AG questions against their traditional HA counterparts. Should these performance metrics reveal meaningful differences in how students chose to engage with the AG questions or in the difficulty of the AG questions compared with the HA questions, that could indicate students perceived a meaningful difference in the quality of AG questions.

To compare the AG and HA questions, we created a data set that consisted of all opportunities students had to engage with the formative practice in the course. Engagement opportunities were defined as all student-question pairs on course pages the student visited (page visits under 5 s were excluded). A mixed-effects logistic regression model was used to model the probability of a student answering a question as a function of question type. The next comparison is question difficulty, and this analysis is done in a similar manner. The data set for difficulty was a subset of the engagement data that included all questions answered by students. The same mixed-effects logistic regression model was used for difficulty. The final analysis is persistence, resulting from when students answer questions incorrectly on their first attempt.

Once the AG questions have been evaluated, the next analysis is to determine how well the courseware is performing as an adaptive system. To do this, we created a data set that included learning objectives with formative, adaptive, and summative questions tagged to them and all student responses. We also identified the predictive learning estimates before and after the adaptive activity for each student. This allows us to group students by learning estimate category and evaluate the impact of the adaptive activities for each group.

1.5.5 Evaluation of AG Questions

1.5.5.1 Engagement

Students need to do the formative practice in order to gain the benefits of the doer effect. Therefore, the first analysis of the AG questions should be on how students choose to engage with them. It was previously discussed that many instructors chose to assign the formative practice for completion points, and we saw the results of that in Fig. 1.8. However, there were still a number of sections where this was not the case, and it is still reasonable that student use of some question types could have decreased and still remained within the instructor's completion parameters.

As described previously, the SmartStart process produced AG FITB and matching question types. The course also contains seven additional HA question

Table 1.1 Engagement regression results for the Neuroscience courseware

Fixed effects	Observations	Mean	Significance	Estimate	p
Intercept	–	–	***	−1.73130	<2e−16
Course page	–	–	***	−0.46964	<2e−16
Module page	–	–	***	−0.27163	<2e−16
Page question	–	–	***	−0.11295	2.47e−12
HA D&D image	468	60.5		0.47962	0.22129
HA D&D table	924	64.1	**	0.84878	0.00565
HA pulldown	109,509	54.3	***	0.53040	<2e−16
AG matching	117,411	54.9	***	0.44413	<2e−16
HA MC	187,107	54.8	***	0.57135	<2e−16
HA MCMS	88,303	54.3	***	0.51305	<2e−16
HA passage selection	1090	48.3	**	−1.01289	0.00132
AG FITB	100,487	50.7	–	–	–
HA FITB	129,901	49.0	.	−0.08670	0.08572
HA numeric input	12,878	52.3		0.05510	0.64440

types. Two are recall question types (HA FITB and HA numeric input), and five are recognition types (HA drag-and-drop (D&D) table, HA D&D image, HA pulldown, HA multiple-choice (MC), and HA multiple-choice multiple-select (MCMS)).

The engagement data set for the Neuroscience course consists of 748,078 unique student-question observations from 1389 students on 1244 questions. For every opportunity a student had to complete a question, an answered question was recorded as 1, and an unanswered question was recorded as 0. A logistic regression is used to model the probability that a student will answer a question as a function of question type. Why not simply use the mean total engagement? While mean engagement is valuable (and recorded in Table 1.1), we know that student engagement declines over the semester and even within units and lesson pages [32]. Therefore, we take the location of the question into account as covariates in the model. Furthermore, there are multiple observations per student and per question, so the observations are not independent, and a mixed-effects model is required. The AG FITB question type is used as the baseline for the question type categorical variable. The R logistic regression model is

```
glmer(answered ~ course_page_number + module_page_number
        + page_question_number + question_type + (1|student)
        + (1|question), family = binomial(link = logit),
        data = df)
```

When we consider only mean engagement for each question type, we see some clusters in engagement. The HA D&D questions were both above 60%; the recognition types AG matching, HA pulldown, HA MC, and HA MCMS were all around 54%; and the recall types AG and HA FITB were both close to 50%. The results of the regression model show that all the location variables were significant ($p < 0.001$) and negative, confirming that students are less likely to answer questions at the end of a page, module, and course. After controlling for the location of the questions, we can examine how engagement differs by question type. We can see the difference in the results of the model when we look at the HA D&D types, which both had the highest mean engagement: the HA D&D table was more likely to be engaged with than the AG FITB ($p < 0.01$), but the HA D&D image was not statistically different than the AG FITB. The four recognition types that were similar in mean score are all more likely to be engaged with than AG FITB ($p < 0.001$), though there is a larger spread in the estimates than there was in mean engagement. AG FITB is not statistically different in engagement than either HA FITB or HA numeric input—all recall types.

1.5.5.2   Difficulty

When students choose to engage with questions, we can use their first attempt accuracy to assess question difficulty. A correct response was recorded as 1, and an incorrect response was recorded as 0. There were a total of 397,143 observations from 1197 students on 1243 questions in this data set. The same regression model was used with this data set, using correctness instead of engagement as the outcome. Table 1.2 shows the number of observations, mean difficulty, and the regression results. Only the question types are presented, as the location was not significant.

The mean scores show a wide range across question types. The easiest question types were AG FITB and HA D&D image with means above 70%. Five question types (of both recognition and recall types) were in the mid- to high 60% range, while two HA types were below 40%. Yet when we review the results from the model, the most notable finding is how few question types are significantly different in difficulty than the AG FITB. AG matching is easier, while HA MCMS and HA passage selection are harder ($p < 0.001$). Given the mean difficulties, this is an expected outcome. However, all other HA questions were not significantly different than the AG FITB, which include both the HA recall and the HA recognition types.

Table 1.2 Difficulty regression results for the Neuroscience courseware

Fixed effects	Observations	Mean	Significance	Estimate	p
HA D&D image	283	71.0		0.296868	0.615061
HA D&D table	592	49.5		−0.734414	0.113760
HA pulldown	59,508	66.0		0.075244	0.417360
AG matching	64,458	74.9	***	0.618933	7.38e−12
HA MC	102,452	64.2		−0.001336	0.987318
HA MCMS	47,928	39.0	***	−1.312614	< 2e−16
HA passage selection	527	30.9	***	−1.725002	0.000624
AG FITB	50,991	65.7	−	−	−
HA FITB	63,671	68.9		0.140471	0.113482
HA numeric input	6733	68.4		0.191800	0.360261

1.5.5.3 Persistence

The final performance metric to evaluate is persistence. Because these questions are formative in nature, students receive immediate feedback and are allowed to try again if they get the question wrong. Persistence is the rate at which students continue to answer a question until they choose the correct response. Therefore, persistence is the smallest subset of data, as it is only the initial incorrect student-question pairs. This data set includes 142,439 observations from 1171 students on 1243 questions. If the student persisted until correct, the response was recorded as 1, and if they did not persist, it was recorded as 0. The mean persistence and results of the regression model are in Table 1.3.

The mean persistence rates are generally very high across most question types. AG matching is above 97% along with the majority of the HA recognition types. AG FITB is below this group at 89.1% and yet is higher than the HA recall types as well as the lowest outlier—HA passage selection. The model shows that every question type is significantly different from AG FITB, with the exception of HA D&D image (which may be a statistical power issue). As anticipated based on the mean persistence values, the recognition types have higher persistence, while the HA recall types have lower persistence.

To date, this is the largest known evaluation of AG questions in natural learning contexts from a single course. The analyses of the AG questions revealed no meaningful differences in how students engaged with AG versus HA questions or their difficulty or persistence. Instead, we see trends grouped by the cognitive process dimension of the question, i.e., recognition or recall. This finding is consistent with prior evaluations of AG questions in similar contexts [48]. This is the first step in confirming that the AG questions would be equally as useful for the predictive analytics model as the HA questions.

Table 1.3 Persistence regression results for the Neuroscience courseware

Fixed effects	Observations	Mean	Significance	Estimate	p
HA D&D image	82	93.9		0.801627	0.369766
HA D&D table	299	98.7	***	3.245109	0.000135
HA pulldown	20,233	98.4	***	3.069654	<2e−16
AG matching	16,152	97.5	***	2.412853	<2e−16
HA MC	36,695	98.3	***	2.804181	<2e−16
HA MCMS	29,215	97.6	***	2.257901	<2e−16
HA passage selection	364	44.0	***	−4.141411	<2e−16
AG FITB	17,510	89.1	−	−	−
HA FITB	19,814	83.2	***	−0.880597	<2e−16
HA numeric input	2129	79.4	***	−0.971798	2.76e−07

1.5.5.4 Adaptive Analysis

The goal of the adaptive analysis is to evaluate the performance of the adaptive activities, which rely on an interconnected system of formative practice and predictive learning estimates in order to function. In this analysis, the data set consists of learning objectives, formative practice, adaptive practice, and summative assessments, as well as the learning estimate values for each student on each learning objective before and after completing the adaptive activity. In total there are 1496 students, 82 learning objectives, and 59,911 data points included in this data set. The goal of the adaptive activities is to use a student's learning estimate on each learning objective in the activity to provide the right level of scaffolding to that student. The goal is that for struggling students with low learning estimates, providing scaffolded questions will also help scaffold their understanding, resulting in higher learning estimates and higher scores on summative assessments.

As noted previously, the learning estimate itself is a value between 0 and 1, with 1 being a higher prediction of a student's ability. Comparing the raw increase or decrease in learning estimates after an adaptive activity, we found a net positive mean increase in learning estimate values. To look at these learning estimate changes in more detail, we can use the learning estimate categories to determine the changes between categories from before to after the adaptive activities. Table 1.4 shows the number of student–learning estimate pair changes between categories.

There are several observations of note from the number of learning estimate observations that moved categories after the adaptive activities. First, most learning estimates stay in their category. This is not unexpected, as it would take a dramatic increase or decrease of performance on adaptive questions to move to a new category. Next, we see that more students increase their learning estimate category for learning objectives than decrease. This shows the net positive increases as they also change a student's category.

As the adaptive activities are created to help better prepare students for the summative assessment, it is necessary to look at student mean summative scores within their learning estimate categories. Table 1.5 shows mean scores for each category after the adaptive activities. As we would expect, students who have high learning estimates have the highest mean scores on the summative assessment. What is meaningful, however, is that within the "before adaptive" learning estimate categories, students who increased their learning estimate after the adaptive activities also performed better on the corresponding summative assessments than their peers

Table 1.4 The number of instances of learning estimate changes after completing adaptive questions, grouped by learning estimate category

Learning estimate category	High (after adaptive)	Medium (after adaptive)	Low (after adaptive)
High (before adaptive)	8048	285	1
Medium (before adaptive)	609	2800	426
Low (before adaptive)	11	521	3109

Table 1.5 Mean summative scores by learning estimate category before and after the adaptive activity questions

Learning estimate category	High (after adaptive)	Medium (after adaptive)	Low (after adaptive)
High (before adaptive)	0.820	0.737	0.667
Medium (before adaptive)	0.768	0.709	0.670
Low (before adaptive)	0.805	0.693	0.589

who did not. Results of a Kruskal-Wallis test indicated groups were statistically different from each other, with a Mann-Whitney post hoc test for the medium learning estimate category indicating significance at $p < 0.01$. This is especially important for the students who were able to move from low to medium or medium to high, as those increases in score could potentially translate to crossing a grade threshold in their course. The additional scaffolded questions those students received likely benefited those students as intended.

These results align with previous adaptive courseware analyses from courseware written entirely by subject matter experts [15]. This shows that AI-generated courseware that has AG questions as formative practice serves students in a similar way as a course with only HA questions.

1.5.6   Summary

This learning engineering process was focused on a central challenge of creating AI-generated questions that would serve as an effective solution to scaling the learn by doing benefits of courseware. The investigation phase used data from the Neuroscience courseware used by students in a natural learning context to evaluate how these questions (a) performed in comparison with other human-authored questions and (b) contributed to the predictive analytics that drove the adaptive activities. The same type of clickstream data that was used to study the doer effect was used to evaluate the performance of these AG questions. The results of the analyses revealed that the AG questions performed similarly to human-authored questions on engagement, difficulty, and persistence. The AG questions also contributed to the predictive analytics that triggered the scaffolding in the adaptive activities, which performed on par with those from courses with only human-authored questions. These analyses confirmed that the AG questions in courseware performed similarly to human-authored questions.

Unlike the doer effect research that was focused on replicating proven learning science principles, this analysis was focused on the effectiveness of this specific type of question as a learning tool, not a learning method. When artificial intelligence becomes a part of student learning environments (especially smart learning environments), there is an increased level of scrutiny that naturally accompanies it. The original learning engineering challenge inherently included the investigation phase where the validity of the questions could be determined. These analyses were not required or even requested. There are few people who would even inquire how the AI questions would impact predictive analytics and, subsequently, the functionality of the adaptive activities. Yet this thorough level of EDS investigation is established as part of the LEP and, over time, becomes a natural part of a team's expectations and planning. In this way, the LEP helps teams establish responsible, student-centered practices.

1.6   EDS for Question Evaluation at Scale

1.6.1   The Content Improvement Service

The data science investigation from the previous LEP confirmed that the automatically generated questions performed as well as their human-authored counterparts. Yet the solution to scaling the learn by doing approach—AI-generated questions—presents a new challenge altogether. Millions of generated questions cannot be managed through human review. An automated method was needed to monitor student data for each question and evaluate that question's performance. In this LEP, the central challenge was one of educational data science: how to apply EDS at scale. Unlike the previous LEPs where EDS was primarily employed in the investigation phase, EDS was the primary focus of the creation and implementation phase in this LEP as well (Fig. 1.9).

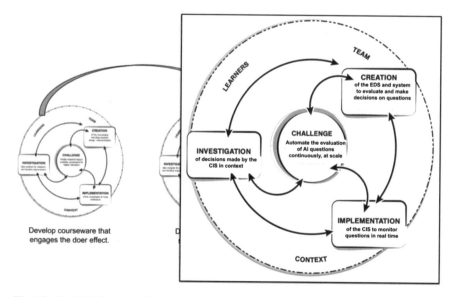

Fig. 1.9 The LEP diagram with the central challenge of automating AI question evaluation

The content improvement service (CIS) is an adaptive system that works as a part of the platform architecture to monitor all AI-generated questions, evaluate their performance, and make a determination on the removal and replacement of sub-par questions, in real time [49]. The fundamental purpose of the CIS is not new, even to Acrobatiq. Historically, using student data to evaluate the performance of questions, assessments, and courseware content was a task assigned to a learning engineer. However, this type of human review only happened at the end of a course when all students had completed the questions and was typically labor and cost intensive to act upon. To do this process at scale was simply not feasible. Given the scale of millions of questions spread over thousands of textbooks, human review could not be applied in this situation. To solve this problem, the CIS was developed to make use of the big data accumulated from questions and the existing knowledge of AG question performance [48, 50].

The CIS is also fundamentally an adaptive system. If we recall Vandewaetere et al.'s [28] framework for classifying adaptive systems, we can map the source of adaptation (what determines adaptation), the target of adaptation (what is being adapted), and the pathway of adaptation (how it is adapted). In the case of the CIS, the question itself is both the source of adaptation and the target of adaption, with the methods of the CIS being the pathway of adaptation. Put another way, the student data collected on the question is used to decide if that question will remain in the course. In this way, the CIS is continuously monitoring all AI questions in the platform simultaneously and making micro-changes on a question-by-question basis to optimize the content for all students.

The CIS is specifically designed to be an accountable system in two ways. First, the CIS has a guiding philosophy of recall over precision. The precision-recall tradeoff is the tension between having a high degree of certainty that a question is unsatisfactory before removing it and that as many unsatisfactory questions are identified and removed as possible. As the AG questions are part of a student's learning resource, it is much more important to ensure that students have the best questions possible. Therefore, it is necessary to have a recall philosophy when evaluating questions. Second, the CIS does not make decisions with a black-box approach wherein the method for determining whether or not a question stays is unexplainable. Instead, the CIS utilizes a gray-box approach, described by Sharma et al. [51] as a method

> ...where the input features can be informed from the context and the theory/relevant research, the data fusion is driven by the limitations of the resources and contexts (e.g., ubiquitous, low-cost, high precision, different experimental settings), and the [machine learning] method is chosen in an informed manner, rather than just as a way to obtain the optimal prediction/classification accuracy. In other words, this contribution aims to invite researchers to shift from the optimal ends (outputs) to the optimal means (paths) (p. 3007).

The gray-box approach is also well aligned to the practice of learning engineering. There is an emphasis on considering context and relevant theory and research when developing the system. The methods are chosen in an informed manner. This supports the student-centered approach of learning engineering where the decisions made in a student's learning environment should be explainable and made with clear, informed methods. The CIS must be able to operate autonomously in order to continuously monitor the millions of questions it is responsible for, and yet the methods of the CIS were designed based on expert rules and decisions so that the function of the CIS is, in this sense, supervised by the experts who designed it.

The CIS can add "tools" to its evaluation decision making, allowing it to change over time. One tool that it currently uses for question evaluation is simple student helpfulness ratings where it uses student feedback on the question to make a decision [49]. The CIS also considers question quality as determined from prior research on the performance metrics of AG questions. For example, the research findings have suggested that questions that fall below a certain difficulty index boundary (have a low mean score) can be too challenging and diminish student persistence [48, 53].

If a question's mean score is not acceptable, we wish to learn this as quickly as possible from the observed student data in order to minimize exposure of poorly performing questions. A Bayesian approach allows the CIS to assess question difficulty from a small sample of student data. A Bayesian method combines prior knowledge about a model's parameter distributions with the likelihood of the observed data under the model to determine the joint posterior distribution of the model parameters. A simple approach is to model each question independently, treating students' answers to the question as Bernoulli trials, with each question having a single parameter p, the probability of success, representing the question's mean score. While a more complex model like item response theory [46]—that takes student ability into account—would be more accurate, it would also require collecting much more data to make the assessment, which is at odds with our requirements. Furthermore, when the mean score of an AG question is very low, it is sometimes indicative of an error in the generation process, making individual student abilities less relevant.

A Bayesian approach can help us arrive at better-quality decisions more quickly by allowing us to incorporate what is known about question mean scores from prior experience. For example, the shape of the mean score's prior distribution can be determined by fitting the mean and variance of an empirically observed set of question mean scores. This gives a so-called "informed prior." A beta distribution fit in this manner to a large data set from a previous study of difficulty of AG questions [48] is shown in Fig. 1.10.

To illustrate the analysis, we can use a real example from the Neuroscience textbook. Suppose we want to remove a question if it is at least 95% likely that its mean score is less than 0.5 (more students answer incorrectly than correctly). For the example question, 2 students answered correctly, and 13 students answered incorrectly. Should this question be removed? To decide, we must construct the posterior distribution of the question's mean score from the prior and observed data and then evaluate the decision rule with it. For Bernoulli trials and a beta prior distribution for p, the posterior is also a beta distribution with shape parameters updated to take the observed number of correct and incorrect answers into account. The probability that the mean score is less than 0.5 is then simply the posterior's cumulative distribution function evaluated at 0.5, the shaded area in Fig. 1.11. Note that the posterior has been shifted significantly to the left of the prior based on the observed data. The

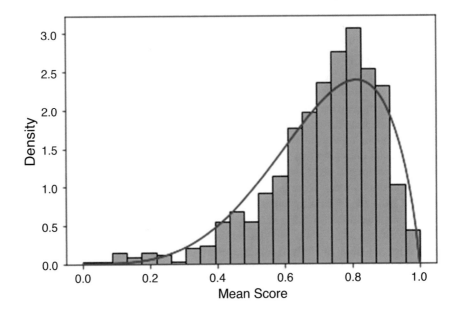

Fig. 1.10 Informed prior distribution for mean score obtained from AG question mean score data set

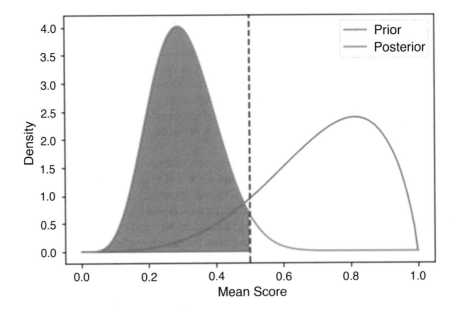

Fig. 1.11 Posterior distribution for mean score, shown together with the prior distribution from Fig. 1.10 for comparison

shaded area is 0.967, or 96.7%, which is greater than the 95% threshold, so the question should be removed.

So this question was ultimately removed. But what was wrong with it? While the CIS would not diagnose the problem with the question once it determined the question should be removed and replaced, we often look at removed questions for our own edification and possible systematic iterative improvements (i.e., engaging in the investigation step of the LEP in order to determine if the AG questions could be iterated on). The removed question was:

```
Arranged in networks that vary enormously in complexity, _____
of neurons define and control all of our abilities and behaviors,
from the simplest reflexes to the most complex intellectual
processes.
```

The correct response is "circuits." Reading that answer in the sentence, it makes sense. So why did so many students get the question incorrect? It was not because students were spelling it incorrectly (the AG questions come with a spelling assistance feature specifically to avoid this issue). Some other student answers (and their frequencies) were "networks" (3), "axons" (2), "billions" (1), and "ganglia" (1). If we reread the question with "networks" as the input, that also sounds like a reasonable choice. In this case, there were other non-synonym terms that seemed plausible for completing this sentence. At first glance, this question might not have seemed problematic during a human review. However, student data shows that this question was more challenging than expected because of context. In this case, the CIS was able to identify a problem with only 15 responses and removed the question. No other students will ever see this question.

1.6.2 Summary

This final learning engineering challenge existed because of the success of the previous challenge—creating (and validating) AI-generated questions. The central goal was to create an automated system to evaluate and determine appropriate actions on all AI questions on the platform, continuously. The CIS is essentially automated educational data science. This system adapts the questions in the learning environment and therefore optimizes the learn by doing method for students. As seen in the Neuroscience example, the CIS can identify questions that may not have been flagged by human review and remove them with very few students having contact with them. These questions are a learning tool and as such should meet rigorous standards. The CIS is a system designed to do what humans could not—monitor all questions in real time. Designed as a gray-box system, the CIS maintains transparency with its actions and accountability to the learner.

1.7   Conclusion

In this chapter, we shared our experience at Acrobatiq and VitalSource to showcase how educational data science was applied as a part of the learning engineering process to solve educational challenges. This included applying statistical analysis to data to investigate if the doer effect findings could be replicated—did the learn by doing method of the courseware improve student outcomes? The investigation of this data confirmed that, yes, this learning science principle was evident in the courseware. This finding led directly to the next learning engineering challenge—how to scale courseware and the learn by doing method that is so effective for student learning. Using artificial intelligence to generate questions solved the practical issue of how to scale learning by doing, but the evaluation of their performance for students in natural learning contexts was vital to confirming the validity of these questions as a learning tool. In this way, the EDS involved in the investigation of these questions was as important as the creation of the questions themselves. As seen in the Neuroscience course used as an example, the automatically generated questions performed as well as their human-authored counterparts. These AG questions also supported the predictive analytics that drove the adaptive activities, and that analysis also revealed the activities performed on par with courses that used only human-authored questions. Given the conclusion that AI-generated questions are acceptable for the learn by doing approach, the last learning engineering challenge examined was how to monitor millions of AI-generated questions. In this challenge, EDS was the focus of the creation phase, as the content improvement service developed is itself an automated EDS system. The CIS is able to monitor all questions in a continuous manner and make decisions on the quality of the questions using very little data. This gray-box system was designed to execute data analyses, which were developed with learning science research as a base. The CIS adapts the questions across the entire platform to optimize the learning experience for students in real time. The CIS is truly a culmination of the learning engineering investigation phase and continuous iterative improvement that could only have been achieved through the path of the previous learning engineering challenges and EDS-supported investigations.

These learning engineering examples also showcase how to develop educational technology with an ethical approach. Learning engineering is a human-centered practice, and in this case, each of these learning engineering processes and the data science–driven investigations was focused on the student. Does this learn by doing method help students learn? Are these AI-generated questions as useful to students as human-authored questions? How do we ensure all AG questions are beneficial for students when the scale is too large to monitor through human effort? A student-centered approach drove each of these product development cycles and helped a diverse learning engineering team maintain the learner at the heart of the project and make research and data-based decisions in their best interests. In this way, our practice of learning engineering uses educational data science in an ethical, student-centered approach that maintains transparency and accountability to the learner.

Educational data science is a field capable of discovery and innovation, as the tools and methods of data science are being used to investigate and interpret vast quantities of educational data being produced from current digital learning environments. Yet the processes of EDS are still in need of context and purpose in order to go about the process of sense-making from raw data. Learning engineering—as both a practice and process—can provide a contextualized, challenge-oriented process for EDS to work within, as well as a student-centered approach. It is certainly clear how the learning engineering process contains space for EDS as a part of the investigation phase, as the goal of EDS is to investigate and make meaning of educational data. However, the LEP also *plans* for EDS as a part of the process. The context of the LEP takes into consideration the nature of the team, development/product constraints, the learners, the learning environment, and the relevant research. EDS needs this contextualization to know the right questions to ask for data analysis and to properly interpret data. The creation phase of the LEP plans the instrumentation of the learning resource. Including EDS in the planning and creation of educational products helps ensure all the data that would be necessary to evaluate the product or investigate learning theory is collected. This inclusion of EDS in the LEP also reaffirms the team aspect of learning engineering. To solve educational challenges, a team with diverse expertise is needed, and this includes expertise in data science.

Specific learning engineering challenges were outlined, and their EDS details were shared to provide an example of how learning engineering and EDS can work together to accomplish real-world educational applications. These are a small number of context-specific examples. The challenge for others working in educational data science and educational technology at large is to evaluate their own context and explore how learning engineering could provide a beneficial process and practice for their own circumstances. What processes or practices from the examples in this chapter resonated with you? What do you want to learn more about? The benefits of applying EDS within the learning engineering process will continue to become clear as others engage in the practice and share the results.

Acknowledgments We would like to acknowledge and thank the extended Acrobatiq and VitalSource team members who have contributed to the creation of these learning environments. Developing solutions to these challenges was only possible with the diverse expertise, collaboration, and commitment of all involved. We also thank the ICICLE community for the continued evolution and dissemination of learning engineering and for the support of this passionate community of practice. Lastly, we thank Michelle Clark for her thoughtful comments on this manuscript.

References

1. Fischer, C., Pardos, Z.A., Baker, R.S., Williams, J.J., Smyth, P., Yu, R., Slater, S., Baker, R., Warschauer, M.: Mining big data in education: affordances and challenges. Rev. Res. Educ. **44**(1), 130–160 (2020). https://doi.org/10.3102/0091732X20903304
2. McFarland, D.A., Khanna, S., Domingue, B.W., Pardos, Z.A.: Education data science: past, present, future. AERA Open. **7**(1), 1–12 (2021). https://doi.org/10.1177/23328584211052055
3. Rosenberg, J.M., Lawson, M., Anderson, D.J., Jones, R.S., Rutherford, T.: Making data science count in and for education. In: Research Methods in Learning Design and Technology, pp. 94–110 (2020)
4. Goodell, J.: What is learning engineering? In: Goodell, J., Kolodner, J. (eds.) Learning Engineering Toolkit: Evidence-Based Practices from the Learning Sciences, Instructional Design, and Beyond. Routledge, New York (2022)
5. Koedinger, K., Kim, J., Jia, J., McLaughlin, E., Bier, N.: Learning is not a spectator sport: doing is better than watching for learning from a MOOC. In: Proceedings of the Second ACM Conference on Learning@Scale, pp. 111–120 (Mar 2015). https://doi.org/10.1145/2724660.2724681
6. Koedinger, K.R., McLaughlin, E.A., Jia, J.Z., Bier, N.L.: Is the doer effect a causal relationship? How can we tell and why it's important. In: Proceedings of the Sixth International Conference on Learning Analytics & Knowledge, Edinburgh, UK, pp. 388–397 (Apr 2016). https://doi.org/10.1145/2883851.2883957
7. Koedinger, K.R., Scheines, R., Schaldenbrand, P.: Is the doer effect robust across multiple data sets? In: Proceedings of the 11th International Conference on Educational Data Mining, pp. 369–375 (2018)
8. Kurdi, G., Leo, J., Parsia, B., Sattler, U., Al-Emari, S.: A systematic review of automatic question generation for educational purposes. Int. J. Artif. Intell. Educ. **30**(1), 121–204 (2020). https://doi.org/10.1007/s40593-019-00186-y
9. Simon, H.A.: The job of a college president. Educ. Rec. **48**, 68–78 (1967)
10. ICICLE: What is learning engineering? https://sagroups.ieee.org/icicle/ (2020)
11. Kessler, A., Craig, S., Goodell, J., Kurzweil, D., Greenwald, S.: Learning engineering is a process. In: Goodell, J., Kolodner, J. (eds.) Learning Engineering Toolkit: Evidence-Based Practices from the Learning Sciences, Instructional Design, and Beyond. Routledge, New York (2022)
12. Goodell, J., Thai, K.-P.: A learning engineering model for learner-centered adaptive systems. In: Stephanidis, C., et al. (eds.) HCII 2020. LNCS, vol. 12425, pp. 557–573. Springer, Cham (2020). https://doi.org/10.1007/978-3-030-60128-7
13. Kessler, A., Design SIG Colleagues. Learning Engineering Process Strong Person. https://sagroups.ieee.org/icicle/learning-engineering-process/ (2020)
14. Lieberman, M.: Learning engineers inch toward the spotlight. Inside Higher Ed. https://www.insidehighered.com/digital-learning/article/2018/09/26/learning-engineers-pose-challenges-and-opportunities-improving (2018). Accessed 1 Nov 2021
15. Van Campenhout, R., Jerome, B., Johnson, B.G.: The impact of adaptive activities in Acrobatiq courseware: investigating the efficacy of formative adaptive activities on learning estimates and summative assessment scores. In: Sottilare, R.A., Schwarz, J. (eds.) Adaptive Instructional Systems, HCII 2020 Lecture Notes in Computer Science, vol. 12214, pp. 543–554. Springer, Cham (2020). https://doi.org/10.1007/978-3-030-50788-6_40
16. Van Campenhout, R.: Learning engineering as an ethical framework. In: Sottilare, R.A., Schwarz, J. (eds.) Adaptive Instructional Systems. Design and Evaluation. HCII 2020 Lecture Notes in Computer Science, vol. 12792, pp. 105–119. Springer, Cham (2021). https://doi.org/10.1007/978-3-030-77857-6_7
17. Van Campenhout, R., Kessler, A.: Developing instructor training for diverse & scaled contexts: a learning engineering challenge. In: Proceedings of eLmL 2022: The Fourteenth International

Conference on Mobile, Hybrid, and On-Line Learning, pp. 29–34. ISSN: 2308–4367. https://www.thinkmind.org/index.php?view=article&articleid=elml_2022_2_40_58010 (2022)

18. CWiC: http://coursewareincontext.org/defining-digital-courseware/ (n.d.)
19. VanLehn, K.: The relative effectiveness of human tutoring, intelligent tutoring systems, and other tutoring systems. Educ. Psychol. 46(4), 197–221 (2011)
20. Hwang, G.-J.: Definition, framework and research issues of smart learning environments—a context—aware ubiquitous learning perspective. Smart Learn. Environ. 1(1), 4 (2014). https://doi.org/10.1186/s40561-014-0004-5
21. Baker, R.S.: Stupid tutoring systems, intelligent humans. Int. J. Artif. Intell. Educ. 26(2), 600–614 (2016). https://doi.org/10.1007/s40593-016-0105-0
22. Kinshuk, Chen, N.S., Cheng, I.L., Chew, S.W.: Evolution is not enough: revolutionizing current learning environments to smart learning environments. Int. J. Artif. Intell. Educ. 26(2), 561–581 (2016). https://doi.org/10.1007/s40593-016-0108-x
23. Lovett, M., Meyer, O., Thille, C.: The open learning initiative: measuring the effectiveness of the OLI statistics course in accelerating student learning. J. Interact. Media Educ. 1, 1–16 (2008). https://doi.org/10.5334/2008-14
24. Dunlosky, J., Rawson, K., Marsh, E., Nathan, M., Willingham, D.: Improving students' learning with effective learning techniques: promising directions from cognitive and educational psychology. Psychol. Sci. Public Interest. 14(1), 4–58 (2013). https://doi.org/10.1177/1529100612453266
25. Black, P., William, D.: Inside the black box: raising standards through classroom assessment. Phi Delta Kappan. 92(1), 81–90 (2010). https://doi.org/10.1177/003172171009200119
26. Van Campenhout, R., Johnson, B.G., Olsen, J.A.: The doer effect: replicating findings that doing causes learning. In: Proceedings of eLmL 2021: The Thirteenth International Conference on Mobile, Hybrid, and On-Line Learning, pp. 1–6. ISSN: 2308-4367. https://www.thinkmind.org/index.php?view=article&articleid=elml_2021_1_10_58001 (2021)
27. Van Campenhout, R., Johnson, B.G., Olsen, J.A.: The doer effect: replication and comparison of correlational and causal analyses of learning. Int. J. Adv. Syst. Measure. 15(1&2) (2022). ISSN: 1942-261x. http://www.iariajournals.org/systems_and_measurements/tocv15n12.html
28. Vandewaetere, M., Desmet, P., Clarebout, G.: The contribution of learner characteristics in the development of computer-based adaptive learning environments. Comput. Hum. Behav. 27(1), 118–130 (2011). https://doi.org/10.1016/j.chb.2010.07.038
29. Sanders, D., Welk, D.: Strategies to scaffold student learning: applying Vygotsky's zone of proximal development. Nurse Educ. 30(5), 203–204 (2005)
30. Sweller, J.: The worked example effect and human cognition. Learn. Instruct. 16(2), 165–169 (2006). https://doi.org/10.1016/j.learninstruc.2006.02.005
31. Kessler, A., Boston, M., Stein, M.K.: Exploring how teachers support students' mathematical learning in computer-directed learning environments. Inform. Learn. Sci. 121(1–2), 52–78 (2019). https://doi.org/10.1108/ILS-07-2019-0075
32. Van Campenhout, R., Kimball, M.: At the intersection of technology and teaching: the critical role of educators in implementing technology solutions. In: Proceedings of the IAFOR International Conference on Education in Hawaii 2021, pp. 151–161. ISSN: 2189-1036 (2021). https://doi.org/10.22492/issn.2189-1036.2021.11
33. Baker, F.B.: The Basics of Item Response Theory, 2nd edn. ERIC Clearinghouse on Assessment and Evaluation. http://echo.edres.org:8080/irt/baker/
34. Schroeder, K., Hubertz, M., Johnson, B.G., Van Campenhout, R.: Courseware at Scale: Using Artificial Intelligence to Create Learning by Doing from Textbooks [Conference Session]. OLC Accelerate, Washington, DC (2021)
35. Hubertz, M., Van Campenhout, R.:. Teaching and iterative improvement: the impact of instructor implementation of courseware on student outcomes. In: Proceedings of the IAFOR International Conference on Education in Hawaii 2022. ISSN: 2189-1036 (2022). https://doi.org/10.22492/issn.2189-1036.2022.19

36. Serra-Garcia, M., Gneezy, U.: Nonreplicable publications are cited more than replicable ones. Sci. Adv. **7**, 1–7 (2021). https://doi.org/10.1126/sciadv.abd1705

37. Van Campenhout, R., Johnson, B.G., Olsen, J.A.: The doer effect: replication and comparison of correlational and causal analyses of learning. Int. J. Adv. Syst. Measure. **14**(1&2) (2021)

38. Dittel, J.S., Jerome, B., Brown, N., Benton, R., Van Campenhout, R., Kimball, M.M., Profitko, C., Johnson, B.G.: SmartStart: Artificial Intelligence Technology for Automated Textbook-to-Courseware Transformation (Version 1.0). VitalSource Technologies, Raleigh, NC (2019)

39. Fitzpatrick, L., McConnell, C.: Student reading strategies and textbook use: an inquiry into economics and accounting courses. Res. High. Educ. J., 1–10 (2008)

40. Phillips, B.J., Phillips, F.: Sink or skim: textbook Reading behaviors of introductory accounting students. Issues Account. Educ. **22**(1), 21–44 (2007). https://doi.org/10.2308/iace.2007.22.1.21

41. Anderson, L.W., Krathwohl, D.R., Airasian, P.W., Cruikshank, K.A., Mayer, R E., Pintrich, P. R., Raths, J., Wittrock, M.C.: A Taxonomy for Learning, Teaching, and Assessing: A Revision of Bloom's Taxonomy of Educational Objectives, Complete edn. Longman, New York (2001)

42. Pedregosa, F., Varoquaux, G., Gramfort, A., Michel, V., Thirion, B., Grisel, O., Blondel, M., Prettenhofer, P., Weiss, R., Dubourg, V., Vanderplas, J., Passos, A., Cournapeau, D., Brucher, M., Perrot, M., Duchesnay, E.: Scikit-learn: machine learning in python. J. Mach. Learn. Res. **12**(85), 2825–2830 (2011). http://scikit-learn.sourceforge.net

43. Honnibal, M., Montani, I., Van Landeghem, S., Boyd, A.: spaCy: Industrial-Strength Natural Language Processing in Python (2020). https://doi.org/10.5281/zenodo.1212303

44. Heilman, M., Smith, N.A.: Question Generation via Overgenerating Transformations and Ranking. www.lti.cs.cmu.edu (2009)

45. Embretson, S., Reise, S.: Item Response Theory for Psychologists. Erlbaum, Mahwah (2000)

46. Fox, J.: Bayesian Item Response Modeling: Theory and Applications. Springer, Heidelberg (2010). https://doi.org/10.1007/978-1-4419-0742-4

47. Watson, N.V., Breedlove, S.M.: The Mind's Machine: Foundations of Brain and Behavior, 4th edn. Sinauer Associates, Sunderland, MA (2021)

48. Van Campenhout, R., Dittel, J.S., Jerome, B., Johnson, B.G.: Transforming textbooks into learning by doing environments: an evaluation of textbook-based automatic question generation. In: Third Workshop on Intelligent Textbooks at the 22nd International Conference on Artificial Intelligence in Education. CEUR Workshop Proceedings, pp. 60–73. ISSN: 1613-0073. http://ceur-ws.org/Vol-2895/paper06.pdf (2021)

49. Jerome, B., Van Campenhout, R., Dittel, J.S., Benton, R., Greenberg, S., Johnson, B.G.: The content improvement service: an adaptive system for continuous improvement at scale. In: Meiselwitz, et al. (eds.) Interaction in New Media, Learning and Games. HCII 2022 Lecture Notes in Computer Science, vol. 13517, pp. 286–296. Springer, Cham (2022). https://doi.org/10.1007/978-3-031-22131-6_22

50. Johnson, B.G., Dittel, J.S., Van Campenhout, R., Jerome, B.: Discrimination of automatically generated questions used as formative practice. In: Proceedings of the Ninth ACM Conference on Learning@Scale, pp. 325–329 (2022). https://doi.org/10.1145/3491140.3528323

51. Sharma, K., Papamitsiou, Z., Giannakos, M.: Building pipelines for educational data using AI and multimodal analytics: a "grey-box" approach. Br. J. Educ. Technol. **50**(6), 3004–3031 (2019). https://doi.org/10.1111/bjet.12854

52. Jerome, B., Van Campenhout, R., Johnson, B.G.: Automatic question generation and the SmartStart application. In: Learning at Scale. (2021). https://doi.org/10.1145/3430895.3460878

53. Moeyaert, M., Wauters, K., Desmet, P., Van den Noortgate, W.: When easy becomes boring and difficult becomes frustrating: disentangling the effects of item difficulty level and person proficiency on learning and motivation systems. **4**(1), 14 (2016). https://doi.org/10.3390/systems4010014

54. Schaeffer, L.M., Margulieux, L.E., Chen, D., Catrambone, R.: Feedback via educational technologies. In: Educational Technologies: Challenges, Applications, and Learning Outcomes, pp. 59–72 (2016)

Part II
Reviews

Chapter 2
A Review of Clustering Models in Educational Data Science Toward Fairness-Aware Learning

Tai Le Quy ⓘ, Gunnar Friege ⓘ, and Eirini Ntoutsi ⓘ

Abstract Ensuring fair access to quality education is essential for every education system to fully realize every student's potential. Nowadays, machine learning (ML) is transforming education by enabling educators to develop personalized learning strategies for the students, providing important information on student progression and early identification of potential points of struggle, developing more efficient grading systems, etc. The role of the Educational Data Science (EDS) domain in educational activities for both teachers and learners is becoming therefore increasingly important. However, ML-driven decision-making can be biased, resulting in underperforming ML models and/or ML models that discriminate against individuals or groups of students based on protected attributes like gender or race. Mitigating bias and discrimination in ML is of paramount importance. In this work, we focus on one of the most effective ML tasks, clustering, which is widely used in EDS as an exploratory tool to understand student characteristics and behavior but also as a stand-alone tool for, e.g., group assignments. Traditionally, clustering algorithms focus on finding groups or clusters of similar students and ignore aspects of fairness and discrimination. However, both cluster quality and fairness of the resulting clusters are needed. This chapter provides a comprehensive review of different clustering models in EDS, with greater emphasis on fair clustering models. Among the fair clustering models, we mainly focus on models that have been proposed and/or applied in educational activities to ensure their usefulness and applicability for fairness-aware EDS.

T. Le Quy (✉)
L3S Research Center, Leibniz University Hannover, Hannover, Germany
e-mail: tai@l3s.de

G. Friege
Institute for Didactics of Mathematics and Physics, Leibniz University Hannover, Hannover, Germany
e-mail: friege@idmp.uni-hannover.de

E. Ntoutsi
Research Institute CODE, Bundeswehr University Munich, Munich, Germany
e-mail: eirini.ntoutsi@unibw.de

43

Keywords Bias · Fairness · Clustering · Educational datasets · Machine learning

Abbreviations

ACC	Clustering accuracy
AI	Artificial intelligence
AIED	Artificial intelligence in education
ANOVA	Analysis of variance
ARI	Adjusted rand index
BIRCH	Balanced iterative reducing and clustering using hierarchies
BMI	Body mass index
BMU	Best matching unit
CFSFDP-HD	Clustering by fast search and finding of density peaks via heat diffusion
CHI	Calinski–Harabasz index
CLARA	Clustering in LARge Applications
CLARANS	Clustering Large Applications based on RANdomized Search
CORE	Computing Research and Education Association of Australasia
DBI	Davies–Bouldin index
DBLP	Database systems and logic programming
DBSCAN	Density-based spatial clustering of applications with noise
DI	Dunn index
DP	Dirichlet process
EDM	Educational data mining
EDS	Educational data science
EM	Expectation–maximization
EMT	Ensemble meta-based tree
FCM	Fuzzy c-means
FIE	Frontiers in education
ICALT	International Conference on Advanced Learning Technologies
ITS	Intelligent tutoring system
KPCA	Kernel-based principal component analysis
LA	Learning analytics
LD	Learning design
LMS	Learning management system
MIT	Massachusetts Institute of Technology
ML	Machine learning
MOOC	Massive open online course
NMI	Normalized mutual information
OPTICS	Ordering points to identify the clustering structure
OULAD	Open University Learning Analytics
PAM	Partition around medoids
PISA	Program for International Student Assessment

RQ	Research question
SJR	Scimago Journal & Country Rank
SOM	Self-organizing map
SSE	Sum of squared error
SVM	Support vector machine

2.1   Introduction

Fairness is a fundamental concept for every education system to ensure that all students have equal opportunities in studying or are being treated fairly regardless of, e.g., their gender, race, household income, assets, knowledge, or domain-specific abilities. Fairness in the education system plays an important role in a wide range of education-related activities, such as assessment and measurement [1, 2], students' group work and group assignment [3–5], graduate school admission [6], and predicting student performance [7]. One of the kernel demands for justice is education; therefore, having a fair education system is crucial to achieving justice in society [8].

Nowadays, many of the decisions in the educational domain are supported via machine learning (ML) and (big) data; examples include grading, assignments, group formation, etc. The domain of Educational Data Science (EDS) is conceptualized as "the application of tools and perspective" from statistics, probability, machine learning, data mining, knowledge discovery in databases, knowledge-based systems, and analytics to educational phenomena and problems, which "works with data gathered from educational environments/settings to solve educational problems" [9, 10]. EDS begins to play an important role in educational activities by assisting teachers, giving feedback to learners, and analyzing student data [9]. In addition, analyzing student data in many aspects, including classroom assessment, cognitive, psychological, and pedagogical, is an essential task of EDS.

From a different perspective, researchers consider EDS as four communities: learning analytics (LA)/educational data mining (EDM), learner analytics/personalization and educational recommender systems, academic/institutional analytics, and systemic/instructional improvement (or data-driven decision-making) [10]. Either way, the growth of EDS is closely related to the advances in ML. ML has been used in a wide variety of decision-making tasks in education [9, 11–13], for example, student dropout prediction [14, 15], education admission decisions [6], or forecasting on-time graduation of students [16, 17]. Therefore, the results of ML models are the basis for building applications in EDS, such as student performance analysis and learning support.

However, along with the benefits of ML technology also come negative consequences like discrimination and lack of transparency. The discriminative impact of ML-driven decision-making on certain population subgroups defined based on protected attributes like gender or race has already been observed in a variety of cases, including education [18, 19]. The domain of fairness-aware machine learning focuses on methods and algorithms for understanding, mitigating, and accounting for bias in ML models [20]. Research on fairness-aware ML has been carried out in various domains such as finance, education, healthcare, criminology, and social issues [21, 22]. Fairness is an essential requirement for educational systems to fully realize every student's potential. This need is reflected in the steadily growing body of work on fairness-aware EDS. The majority of these works focus on supervised learning models, i.e., classification and regression/predictive models [23–25]. Recently, there have also been several surveys on algorithmic bias and fairness in education, covering different definitions of fairness and supervised learning models [26, 27].

In this chapter, we focus on unsupervised learning methods and, in particular, on clustering models. Clustering methods are widely employed in EDS [9, 11]. The main goal of clustering is to group/cluster instances or students into groups of similar students; such grouping allows for gaining insight, understanding student achievement [28], characterizing students' learning behavior [29], etc. However, only a few works consider the aspect of fairness and discrimination in clustering models in education [30]. Furthermore, an overview of clustering models and fairness-related aspects for EDS is still missing. Hence, the goal of our chapter is to provide a review of commonly used clustering models in EDS tasks and associated fairness aspects. We believe that the review will comprise a useful resource for EDS researchers and practitioners, for example, to raise awareness regarding bias and discrimination of clustering models, to help select the most appropriate clustering model for the problem at hand, and to initiate further research in the area of fair clustering for EDS.

The survey is organized around the following key research questions (*RQs*):

- RQ_1. Why is clustering an important task for EDS? What clustering models and for what purposes have been applied in EDS?
- RQ_2. What is fairness in clustering? How to achieve fair clustering results? How to define and achieve fairness in clustering in EDS?
- RQ_3. How to evaluate the quality of (fair) clustering in EDS? What is the proper evaluation setup (datasets, methods, quality measures, etc.) performed? Are there benchmark datasets?
- RQ_4. What other issues/requirements, beyond fairness, should be considered for clustering in EDS?

The rest of the chapter is organized as follows: The most commonly used clustering models in EDS are presented in Sect. 2.2. An overview of fair clustering models and their potential for EDS is presented in Sect. 2.3. A detailed discussion on evaluation aspects for fair clustering in EDS is presented in Sect. 2.4. Requirements for clustering in EDS, beyond the quality of clustering and fairness of the resulting clusters, are covered in Sect. 2.5. Finally, Sect. 2.6 summarizes this chapter and provides an outlook for future work.

Table 2.1 Set of keywords used for literature review w.r.t. RQ_1

Keywords	Synonyms	Antonyms
Bias	Discrimination	Fair, unbiased
Fairness	Equity, justice	
Clustering	Grouping	
Learning analytics		
Educational data mining		
Educational datasets		

2.2   Clustering Models for EDS Tasks

This section answers the research question RQ_1 regarding the role of clustering in EDS and its typical applications. We overview the main clustering models used in EDS and discuss several aspects of practical importance beyond the algorithm itself, like cluster model assumptions, complexity, and parameter selection.

2.2.1   Methodology of the Survey Process

We perform the following procedure to address RQ_1. First, we search the relevant works by a set of keywords in three well-known scientific databases (Google Scholar,[1] DBLP (Database Systems and Logic Programming),[2] and Scopus[3]) and select only quality studies on ranked venues; second, we categorize and summarize the educational tasks related to clustering in Sect. 2.2.2 and outline the different clustering models in Sect. 2.2.3.

In the first step, we identify the following relevant keywords: *bias*, *fairness*, *clustering*, *learning analytics*, *educational data mining*, and *educational datasets*. We define a set of synonyms and antonyms for each, as listed in Table 2.1. The search queries are performed on Google Scholar and DBLP with all possible combinations of the different synonyms (antonyms) of the keywords. After identifying the papers, we select only the published papers on ranked conferences (A*, A, B, C) or journals (Q1, Q2, Q3, Q4) that are indexed by Scopus. We refer to the updated databases, i.e., issued in 2021, of the Computing Research and Education Association of Australasia (CORE)[4] for conferences' ranking and Scimago Journal & Country Rank (SJR)[5] for journals' ranking. Because the related research of Dutt et al. [11] provided a systematic literature review on clustering algorithms and their

---

[1] https://scholar.google.com/

[2] https://dblp.org/

[3] https://www.scopus.com/

[4] http://portal.core.edu.au/conf-ranks/

[5] https://www.scimagojr.com/

Table 2.2 Venue-based distributions of publications

Type of venues	Q1/A/A*	Q2/B	Q3/C	Q4/Other
Journal	25	18	7	13
Conference	14	21	5	30

Table 2.3 The top ten venues

No.	Venues	#Publications	Ranking
1	Educational data mining (EDM)	11	B
2	*Journal of Physics: Conference Series*	7	Q4
3	International Conference on Artificial Intelligence in Education (AIED)	5	A
4	*IEEE Access*	4	Q1
5	Education and information technologies	3	Q1
6	Scientific programming	3	Q2
7	IEEE Frontiers in Education (FIE) Conference	3	C
8	Complexity	2	Q1
9	International Conference on Advanced Learning Technologies (ICALT)	2	B
10	International Conference on Intelligent Tutoring Systems (ITS)	2	B

applicability and usability in EDM from 1983 to 2016, we consider 133 publications from 2017 to June 2022.

In the next step, we analyze metadata on the selected publications. Table 2.2 provides a brief regarding the number of publications on venues w.r.t. ranking. As shown in Table 2.2, 29.3% (39 out of 133) of selected articles are published in leading conferences/journals, i.e., Q1/A*/A ranking. The papers are collected from 97 venues (47 conferences and 50 journals). The top ten venues are listed in Table 2.3; they account for 31.6% (42 out of 133) of the total selected papers. In addition, we count the number of publications by year and visualize it in Fig. 2.1. There has been a slight upward trend in the number of articles published in peer-reviewed journals over the years.

2.2.2 EDS Tasks Using Clustering Models

Clustering has been used in a variety of tasks in EDS, from analyzing students' behavior (Sect. 2.2.2.1) and performance (Sect. 2.2.2.2) to grade prediction (Sect. 2.2.2.3), recommendations (Sect. 2.2.2.4), supporting collaboration and teamwork (Sect. 2.2.2.5), and analyzing students' well-being (Sect. 2.2.2.6). Student data are collected from various sources, including traditional classrooms and learning management systems (LMSs), such as Moodle[6] or ITSs.

---

[6] https://moodle.org/

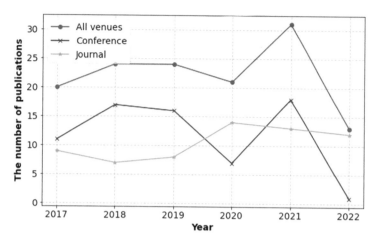

Fig. 2.1 Number of publications over the years (2022 is not yet complete)

2.2.2.1 Analyzing Students' Behavior, Interaction, Engagement, Motivation, and Emotion

Clustering models are used to discover the relationship between students' behaviors and their learning performance [31–35]. This allows teachers to get a good overview of their students and manage them accordingly [36]. For example, Waspada et al. [37] used k-means to cluster students' behavior and discovered that the activity of working on online quizzes affects the final scores. Students who participated more became more engaged in the course and achieved higher grades [38]. Jia et al. [39] clustered the behavioral records of reading tests in an online Chinese reading assessment system to identify students' reading abilities and testing strategies. Howlin and Dziuban [40] proposed a repeated fuzzy clustering algorithm for discovering student behaviors or outliers. Students' behavior might change over time, and therefore, it is necessary to investigate such changes and ensure an up-to-date clustering model. McBroom et al. [41] applied a hierarchical clustering model to detect the behavioral trends of students over time and produced clusters of students' behavior, which characterize exercises where many students give up. Shen and Chi [42] developed a temporal clustering framework to determine if a student's learning experience is "unprofitable" to offer a personal suggestion.

Since gamification is starting to play an increasingly important role in education activities, Ruipérez-Valiente et al. [43] analyzed the relationship between students' interactions and behaviors with the badge system, i.e., students earn badges through their interaction with engineering problems and the learning process. The results are considered as feedback or guidelines on how to provide appropriate game-based tasks to students, which can help them improve their learning. The interactions of university students in an e-learning system are clustered and analyzed by the hierarchical clustering algorithm and k-means algorithm in order to discover the relationship between the final grades and the use of the modules [44]. Besides, experiments were performed on several clustering methods to cluster students' interactions and investigate the associations between students' academic performance and social interactions [45]. The resulting clusters can help educators identify the special students and improve their learning design (LD) process and the teaching materials that need to be revised.

Using the expectation–maximization (EM) clustering algorithm, Orji and Vassileva [46] discovered that students' engagement is a good indicator of academic performance. Therefore, categorizing and identifying students with low levels of engagement are effective ways to help them improve their academic performance [47–51]. Besides, different levels of engagement are identified in students who use different learning strategies through Pearson correlation analysis on clusters resulting from the agglomerative hierarchical clustering method [52]. The k-means algorithm was used to cluster students based on 12 engagement metrics [53]. The authors discovered that the most representative of the students' engagement levels are the number of logins and the average duration to submit assignments.

Students' motivation is essential in the massive open online course (MOOC) environment [54]. k-Means was used to analyze the questionnaire results of the MOOC system [55] and discover the relationship between learning motivation and the learning behavior of the students. Besides, self-motivation in college students was investigated by using k-means clustering to reveal the effect of a hidden curriculum and character-building variables. The experiments were performed on the real dataset collected at the State University of Malang, Indonesia [56]. In addition, teachers should also pay attention to students' emotions [57] because there is a relationship between emotion and performance [58] as well as between the emotions and motivation of the students [59]. Students can obtain many emotional experiences through physical education activities, which improve learning interest and relationships among teachers and students [60].

2.2.2.2 Analyzing Students' Performance and Grading

Learning analytics (LA) is an important research area of EDS, which focuses on the collection and analysis of data about learners. In LA, students are often divided into groups based on their performance, which allows the educators to identify low-performing students and help them accordingly [61–63], whereas for high-performing students, enrichment tasks can be assigned to help them advance further. Moreover, analyzing student learning outcomes is one of the most effective methods for understanding the factors affecting students' learning ability [64, 65]. For example, procrastination is an important indicator of students' performance, with non-procrastinators tending to have higher performance [66, 67]. Using k-means, experimental results show that the academic performance of female students is affected by nutrition and health issues [68]. In this direction, the k-means clustering model was used to discover the factors contributing to students' performance [69], resulting in interpretable clusters of students based on their confidence entropy,[7] degree of over-/under-confidence, and other related variables.

A clustering model not only can help teachers analyze students' assignments, but it is also an effective method to support essay grading. For instance, interpretable clustering has been used to cluster and find the structure of students' solutions in a Python programming course [70]. With a similar purpose to automatically cluster programming assignments (C programming), k-means was applied in the study of Gao et al. [71]. Besides, Chang et al. [72] proposed a clustering method to cluster sentences to support grading the essays written in Finnish by bachelor students. In addition, Sobral and de Oliveira [73] used k-means to observe the change in self-assessment skills between two evaluation moments (after submitting their work/test and after knowing the correct answers) to investigate the role of self-evaluation in the student's final grade.

2.2.2.3   Predicting Students' Performance

Student performance prediction (including prediction on test scores, students at risk, dropout, etc.) is a common task in EDS [74]. Various EDS techniques, such as correlation, regression, and classification, have been applied to predict students' performance [10]. By using clustering models, studies have tried to achieve various objectives, which can be grouped broadly into three categories: (1) predicting the performance of students in terms of the actual grade as a regression problem, (2) considering student performance prediction as a classification problem, (3) predicting students at risk or student dropout.

To predict students' grades, clustering has been used as a preprocessing step to detect groups/clusters of similarly performing students before applying the predictive models. For example, k-means was used to reveal the unique types of students labeled as "proficient," "struggling," "learning," and "gaming" [75], which was helpful in predicting standardized test scores. Besides, Bayesian fuzzy clustering was used to cluster students into groups [76], and then the Kernel-based principal component analysis (KPCA) was applied to identify the best features from the data, which is used in the prediction phase with a Lion–Wolf deep belief network model. Recently, Hassan et al. [77] applied k-means to group students in terms of both static attributes like *sleep*, *stress*, and *study spaces* and spatiotemporal attributes like *latitude*, *longitude*, and *time* before transferring these clusters to several well-known regression models for grade prediction.

Casalino et al. [78] proposed an adaptive fuzzy clustering algorithm to process the educational data as data streams to predict student outcomes (pass, fail). They clustered nonoverlapping chunks of data and then investigated cluster prototypes to classify students into two classes (pass, fail). In addition, an Ensemble Meta-based Tree (EMT) model was introduced to classify students into four categories

---

[7] Students' confidence entropy is computed by the Shannon equation.

(excellent, very good, good, and satisfactory) [79]. In their architecture, k-means was used to find a set of homogeneous clusters before applying a classifier model for each cluster. Concerning predicting student enrollment in postgraduate studies, k-means was applied to reveal the presence of three coherent clusters of students [80]. In another approach, Francis and Babu [81] classified students into three groups (high, medium, low) by using four popular classifiers (support vector machine (SVM), naive Bayes, decision tree, and neural network). After that, k-means clustering and majority voting were used to predict the best accuracy of students. In the MOOC environment, Chu et al. [82] used clustering guided meta-learning to exploit clusters of frequent patterns in the students' clickstream sequences in order to predict students' in-video quiz performance.

Many studies predict student dropout and students at risk by applying clustering methods because student dropout is the primary concern of many educational institutes [74], especially in MOOCs. A soft subspace clustering algorithm has been applied to discover the interesting patterns and relationships with student dropout rates [83]. A link-based cluster ensemble was proposed to transform the data into a new form before applying some classification models [84] to improve the performance of student dropout prediction. k-Means was used to discover the students likely to drop out of school [85]. Regarding prediction on students at risk, a time-series clustering [86] was applied to capture the at-risk students. Besides, a clustering ensemble method on temporal educational data combining dynamic topic modeling and kernel k-means was introduced to predict early students in trouble [87].

2.2.2.4 Supporting Learning, Providing Feedback and Recommendation

Hierarchical clustering was applied to group students with similar learning styles in the classroom to support teachers in understanding the cognitive skills of students and selecting appropriate teaching methods for each group [88]. Besides, grouping students based on their knowledge patterns is a common strategy to allow lecturers to monitor students' performance easily and provide suitable learning materials to each group [89–91]. In addition, in the tutoring process, the appropriate assignment of students to teachers contributes to improving students' learning quality. Hence, Urbina Nájera et al. [92] proposed a clustering model based on k-means to cluster students and teachers according to their skills and affinities. Furthermore, an online annotation and clustering system has been developed to allow lecturers to generate students' online reading activity patterns and review their annotations [93].

In terms of generating automatic feedback for students answering a well-being survey, Kylvaja et al. [94] applied hierarchical clustering to identify the typical patterns in the well-being profiles. The automatic feedback was generated by searching the best matching cluster for an individual survey response. With the aim of providing feedback on psychological fitness, Li and Sun [95] applied the fuzzy c-means (FCM) algorithm to analyze the university students' psychological fitness data by extracting the characteristics of students' rebellious psychological phenomenon, e.g., talking during class and continually using profanity. Besides, in programming education, researchers suggested that feedback on students' assignments can be generated based on program repair techniques [96]. The authors proposed an automated program repair algorithm for a fundamental programming course. They built a clustering algorithm to divide similar correct solutions into a group and applied the existing correct solutions to repair the new incorrect assignments. In addition, concerning students' feedback for academic courses, Masala et al. [97] applied k-means on local contexts (where the keywords occurred) centered on specific keywords to find similar students' opinions.

As for the requirement of developing a course recommendation method to satisfy students' personal goals and interests in MOOC, Guo et al. [98] proposed a group-oriented course recommendation approach. They used a semantic k-means clustering algorithm to group students before applying a course recommendation algorithm based on similarity and expert knowledge to generate the final recommended courses. Following another approach, hierarchical clustering was used to divide students into different types, and a collaborative filtering artificial intelligence (AI) algorithm was utilized for lesson recommendations [99]. Furthermore, Liu et al. [100] introduced an incremental tensor-based adaptive clustering method based on correlative analysis and personalized recommendation to offer students appropriate learning resources in various contexts.

2.2.2.5 Supporting Collaboration Activities

Teamwork is a popular activity in educational settings and is an essential factor in improving students' engagement in the classroom [101]. Clustering methods have been applied to group learners in traditional classrooms and MOOC systems. In the traditional classroom, students are grouped into homogeneous and heterogeneous groups based on their knowledge levels to capture rich semantic information about the group [102]. They apply a spectral clustering algorithm for dividing students into groups with a group size of no more than eight. Besides, Pratiwi et al. [103] proposed their clustering method to automatically generate heterogeneous groups based on students' dissimilarity. The resulting clusters are comparable with the teachers' manual grouping solutions. In addition, k-means is applied to produce the training groups for American football student-athletes based on their roles and expected performance during the competition [104].

In MOOC systems, the problem of grouping students w.r.t. preferences and interests was introduced by Akbar et al. [105]. They introduced a grouping method based on hierarchical k-means clustering and a weighting formula (the priority of topic preferences) to satisfy students' interests and to ensure that students are grouped in a team with similar preferences. Furthermore, Wang and Wang [106] developed their clustering technique based on the enhanced particle swarm optimization algorithm to group students w.r.t. their knowledge state and interests.

2.2.2.6 Analyzing Physical and Mental Health

Health and well-being are determined by an individual's physical condition and the potential to contribute to society [107]. In their study, k-means was used to cluster and analyze the physical quality of college students based on their characteristics (body mass index (BMI), vital capacity, endurance quality, flexibility, strength quality, speed, and dexterity quality). From there, colleges and universities could optimize their education and teaching plans and perfect their talent training programs by using the study's results. In other aspects, a variety of clustering methods (k-means, hierarchical clustering, BIRCH (Balanced Iterative Reducing and Clustering using Hierarchies), etc.) were applied to identify several types of students based on fitness scores and BMI (fit, not fit, overweight and fit, normal weight and non-fit) [108]. Besides, since nutrition may affect both students' physical health and well-being, an exploratory analysis was performed to understand university students' foot habits by using k-means [109].

In terms of mental health analysis, psychological fitness was analyzed by using fuzzy c-means [95]. Besides, a fast clustering method was used to analyze college students' mental health data [110]. As a result, the findings from these studies may support school counselors and student managers in providing better mental health services for students. In addition, Li et al. [111] explained the relationship between mental health and career planning and examined how mental health education and career planning are integrated for college students using fuzzy feature clustering.

2.2.2.7 Miscellaneous Tasks

Differentiated institutions and diversity are key policy issues in higher education [112]. Hence, Wang and Zha [112] introduced three different methods to measure diversity in higher education: (1) using existing institutional classifications to measure systemic diversity, (2) measuring systemic diversity based on detecting optimal types of institutions (k-means and k-medians were applied to obtain the institutional types from the higher education institution data), (3) taking into consideration both within-group homogeneity and between-group heterogeneity when measuring systemic diversity. Experimental results based on data from Chinese universities in 1998 and 2011 showed that the operations and profiles of Chinese universities have become more diverse.

The common assumption of similarity-based clustering methods is that students should perform similarly on items that belong to the same knowledge component [113]. Therefore, the authors proposed a new item similarity measure, namely, Kappa Learning. As students progress through the items, their mastery of the skills underlying each item changes, and this measure indicates similarity between items. The experimental results on the K–6 math intelligent tutoring system showed that the new measure outperforms clustering based on popular similarity measures. In clustering, feature selection plays an essential role because it strongly affects the clustering quality [114]. They developed an efficient algorithm to select the most important features using an entropy-based feature selection method. Their proposed method was evaluated on an online course at Guangxi Radio and TV University and showed its effectiveness in choosing good-quality features and contribution to the interpretability of the resulting clusters.

Furthermore, concerning implementing an intelligent educational administration system, Liu [115] introduced a student achievement prediction model based on fuzzy c-means and collaborative filtering. The problem of grouping leadership variables in nonformal education has been investigated in the research of Rahmat [116] using hierarchical clustering. Toward the expansion of personalization in online learning environments, Ahmed et al. [117] used k-means to cluster students into similar groups based on their skills. In addition, grouping students can also be based on their learning style [118]. The DBSCAN (Density-Based Spatial Clustering of Applications with Noise) algorithm can be used to analyze the learning status of students, which influences teaching activities [119].

Summary of EDS Tasks Using Clustering Models Clustering is frequently used in EDS to analyze students' data related to students' behavior, interaction, motivation, physical and mental health, etc. Clustering has been used both as a preprocessing and as a stand-alone tool. As a preprocessing tool, clustering is useful to identify groups of interest for further investigation and analysis, as in cluster-based grade prediction. As a stand-alone tool, clustering allows the teachers to get a better overview of their students and adapt the learning activities to the needs and capabilities of each group. Moreover, clustering algorithms also assist in managing and administrating educational institutions. In terms of clustering methods, k-means seems to be the algorithm of choice when it comes to clustering, most probably because it is one of the easiest algorithms to implement and experiment with. In most of the approaches, clustering was applied upon static data; however, there exist approaches that also consider changes in students' data over the course of their studies.

2.2.3   Clustering Models

We identify the clustering models in the literature and record publications using the clustering algorithms in Table 2.8 (in "Appendix"). There are 23 clustering methods applied for EDS tasks. k-Means is the most prevalent approach, followed by hierarchical and fuzzy c-means clustering. The distribution of the clustering algorithms across observed publications is illustrated in Fig. 2.2. In this chapter, based on the popularity of the methods, we overview the algorithms used in at least three publications.

The goal of a clustering model is to group objects, e.g., students, into clusters where the objects in the same cluster are similar and the objects in different clusters

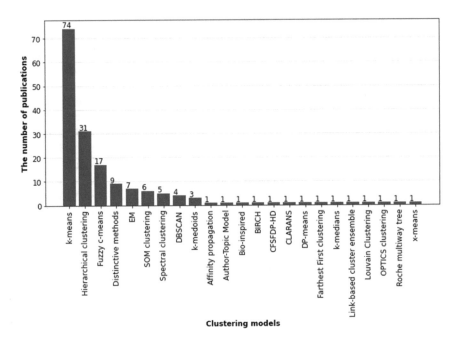

Fig. 2.2 Clustering models applied in EDS

are different. We denote $\mathcal{D} = \{x_1, x_2, \ldots, x_n\}$ to be a dataset of n objects[8] in d-dimensional space, and k is the number of clusters. We list the symbols used in our review in Table 2.4.

2.2.3.1 k-Means

k-Means [120] is the most popular partitioning-based clustering method that aims to partition n objects into k clusters. The number of clusters k is given in advance. A clustering C is a partition of dataset $\mathcal{D}$ into k disjoint subsets, $C = \{C_1, C_2, \ldots, C_k\}$, called clusters with $S = \{s_1, s_2, \ldots, s_k\}$ the corresponding cluster centers (centroids). The centroid of each cluster is computed by the mean of the cluster's members. Formally, the goal is to minimize the clustering cost:

$$L(C, \mathcal{D}) = \sum_{i=1}^{k} \sum_{x \in C_i} dist(x, s_i)^2 \tag{2.1}$$

where function $dist(x, s_i)$ returns the distance from any point $x \in C_i$ to the centroid s_i of cluster C_i.

---

[8] Because the objects or data points for clustering in the related work are mainly students, we use the terms "objects," "data points," and "students" interchangeably.

Table 2.4 Symbols and their descriptions

Symbol	Description
$\mathcal{D}$	Dataset $\mathcal{D} = \{x_1, x_2, \ldots, x_n\}$
C	A clustering $C = \{C_1, C_2, \ldots, C_k\}$
n	The number of data points
d	The number of attributes (dimensions)
k	The number of cluster centers (centroids, medoids)
S	A set of cluster centers $S = \{s_1, s_2, \ldots, s_k\}$
x or x_i	A data point
C_j	A cluster
s or s_j	A cluster center (centroid, medoid)
t	The number of iterations
$dist(,)$	A distance function
$\mathcal{L}(C, \mathcal{D})$	An objective function of clustering C on dataset $\mathcal{D}$

Algorithm Lloyd's version [121] is the most prevalent implementation of the k-means algorithm, described below:

1. Randomly choose k points as initial centroids $\{s_1, s_2, \ldots, s_k\}$.
2. Assign each point in $\mathcal{D}$ to its closest centroid.
3. Update the center of each cluster based on the new point assignments.
4. Repeat until convergence.

In this chapter, there are several variant versions of k-means used in the related work, including k-medians [122] (the median of points in a cluster is used to compute its centroid) and x-means (replicates partitions and separates the best results until a certain threshold is reached) [123].

Complexity The computational complexity of k-means is $O(tkn)$, where t is the number of iterations needed until convergence, and usually $k, t \gg n$.

Parameter Selection To choose the best value for the number of clusters k, researchers run the clustering with several values for k and select the one that does not result in a significant improvement in the compactness of the cluster at any further increase in k [109]. In the study of Yang [107], k was determined based on the experimental analysis of the real application of clustering results for students' development, optimizing education and teaching. In another approach, researchers determined the optimal k based on tenfold cross-validation implemented on the training set [75]. Besides, the silhouette coefficient is also used as an essential criterion to select the best number of clusters [53, 64, 77, 124]. The pre-implemented package, namely, *NbClust*,[9] is also used to find the optimal k [48, 49]. In addition, Zhang et al. [125] proposed a method to select the initial center points by partitioning the dataset into k domains and choosing the points with the

---

[9]https://cran.r-project.org/web/packages/NbClust/index.html

greatest density in each domain as the initial cluster centers. Aghababyan et al. [69], Nazaretsky et al. [113], and Ahmed et al. [117] applied the sum of squared error (SSE) between data points in clusters to select the number of clusters.

2.2.3.2 *k*-Medoids

k-Medoids clustering is introduced by Kaufman and Rousseeuw [126] with the partition around medoids (PAM) algorithm. Similar to *k*-means, the goal of *k*-medoids is to minimize the clustering cost (Eq. 2.1). However, *k*-medoids selects the actual data points as cluster centers (namely, medoids).

Algorithm The PAM algorithm uses a greedy search strategy, which is presented below:

1. Arbitrarily select *k* data points as the medoids.
2. Assign each data point to the closest medoid.
3. While the clustering cost decreases

    a. For each medoid *s* and for each non-medoid *x*

        i. Check whether *x* could replace *s*, and compute the cost change.
        ii. If the cost change is the current best, remember (*s*, *x*) combination.

    b. Perform the best $swap(s_{best}, x_{best})$ if it decreases the clustering cost; else, terminate the algorithm.

Complexity The computation complexity of the original PAM algorithm per iteration (step 3) is $O\left(k(n-k)^2\right)$. The complexity can be reduced to $O(n^2)$ with several improved algorithms (CLARA (Clustering in LARge Applications), CLARANS (Clustering Large Applications based on RANdomized Search)) [127], in which, the CLARANS algorithm was evaluated in the study of Vasuki and Revathy [128] to predict the degree of student achievement in placement.

Parameter Selection The Euclidean distance is usually used as the dissimilarity measure [129], and the number of medoids *k* is determined by using the average silhouette coefficient [129, 130].

2.2.3.3 Hierarchical Clustering

Hierarchical clustering is a set of nested clusters organized as a hierarchical tree that can be visualized as a dendrogram [131]. There are two main types: agglomerative (bottom-up approach) and divisive (top-down approach). Hierarchical agglomerative clustering is applied in the related work. BIRCH (T. [132]), a version of hierarchical agglomerative clustering that is suitable for very large databases, is used by Dovgan et al. [108] for their analysis. This section presents the basic hierarchical agglomerative clustering method [126]. There are many approaches for computing the proximity of two clusters (linkage algorithms): single linkage, complete linkage, average linkage, centroid distance, Ward's method [133], etc.

Algorithm
1. Compute the proximity matrix.
2. Each data point is considered as a cluster.
3. *Repeat*

    a. Merge two closest clusters.
    b. Update the proximity matrix.

    Until only a single cluster.

Complexity The time complexity of the basic agglomerative algorithm is $O(n^3)$. The complexity can be reduced to $O(n^2 \log n)$ by using the heap data structure.

Parameter Selection In the related work, Euclidean distance is a similar common dissimilarity measure [88, 94, 134, 135]. Yotaman et al. [88] used Manhattan distance and several linkage algorithms—average linkage, complete linkage, single linkage, Ward's linkage, and median linkage—in their experiments. However, Ward's linkage is the popular linkage applied for proximity matrix computation [42, 94, 134, 135]. The number of clusters was chosen using the local maximum average [135] or based on the meaning of the resulting clusters [94]. In other approaches, the gap statistics (comparing total intra-cluster variation against expected values under distributions exhibiting no obvious clustering patterns) were applied to evaluate the resulting clusters and determine the number of clusters [134]. The knee position, which is described by the local change of slope of the SSE significantly, was applied to identify the number of clusters [42, 88].

2.2.3.4   Fuzzy *c*-Means Clustering

Fuzzy *c*-means (FCM) clustering [136] is a soft clustering approach where each data point can belong to more than one cluster. Each data point is assigned a weight (membership grade) indicating the degree to which the data point belongs to the cluster. We denote a matrix $W = \{w_{ij}\} \in [0, 1]$, where w_{ij} refers to the degree to which data point x_i belongs to cluster C_j, $i = 1, \ldots, n; j = 1, \ldots, k$. The goal of the FCM algorithm is to minimize the following objective function:

$$L(C, \mathcal{D}) = \sum_{i=1}^{n} \sum_{j=1}^{k} w_{ij}^m dist(x_i, s_j)^2 \qquad (2.2)$$

where

$$w_{ij}^m = \cfrac{1}{\sum_{h=1}^{k}\left(\cfrac{dist(x_i, s_j)}{dist(x_i, s_h)}\right)^{\frac{2}{m-1}}} \tag{2.3}$$

m is the fuzzy parameter indicating the degree of fuzziness, $m \in (1, +\infty)$. The centroid s_j of cluster C_j is computed by

$$s_j = \frac{\sum_{i=1}^{n} w_{ij}^m x_i}{\sum_{i=1}^{n} w_{ij}^m} \tag{2.4}$$

Algorithm
1. Select the number of clusters k.
2. Assign weights randomly for data points.
3. *Repeat*

   a. For each cluster, compute its new centroid.
   b. For each data point, compute its new weight.

   Until convergence (the weights' change between two iterations is no more than a given ε).

Complexity The time complexity of fuzzy c-means is similar to k-means, i.e., $O(tkn)$, where t is the number of iterations [137].

Parameter Selection The parameters are set by a heuristic strategy, with a threshold $\varepsilon = 10e\text{-}5$ (Tang [138]). Usually, the fuzzy parameter is set at $m = 2$ [40, 138, 139]. Besides, the number of clusters is set in a range of values from 0 to 11, and the optimal k is determined by using the gap statistics method [50] or based on the variation trend of the error of the predictive model [115]. Howlin and Dziuban [40] used validity indices implemented in the *FClust*[10] R package to select the optimal k. In addition, Oladipupo and Olugbara [49] applied the *NbClust* package to find the number of clusters.

2.2.3.5 EM Clustering

EM is a soft clustering approach to find the maximum likelihood estimates of parameters in probabilistic models [140]. There are two important steps: *expectation* (*E*) and *maximization* (*M*). EM uses the finite Gaussian mixture model to estimate parameters until convergence. Each cluster corresponds to one of k probability distributions in the mixture. Similar to fuzzy c-means clustering, each data point is assigned a membership probability for each cluster [141]. The EM algorithm is summarized as follows.

---

[10] https://cran.r-project.org/web/packages/fclust/index.html

Algorithm
1. Initialize with two randomly placed Gaussians (mean and standard deviation).
2. Refine the parameters iteratively with two alternating steps until convergence:

   a. *E* step: Re-estimate the membership possibility for each instance based on the current model.
   b. *M* step: Re-estimate the model parameters based on the new membership possibilities.

3. Assign data points to the cluster with their membership probability.

In the *E* step, the membership possibility for each instance x_i w.r.t. cluster C_j [142] is computed by

$$P(C_j|x_i) = \frac{P(x_i|C_j)P(C_j)}{\sum_j^k P(x_i|C_j)P(C_j)} \tag{2.5}$$

where $P(x_i|C_j)$ is the probability density function of the Gaussian distribution.

In the *M* step, the probability of data points coming from the cluster C_j (cluster density) is determined by

$$P(C_j) = \frac{1}{N}\sum_{i=1}^n P(C_j|x_i) \tag{2.6}$$

Update centroid s_j for each cluster C_j

$$s_j = \frac{\sum_{i=1}^n x_i P(C_j|x_i)}{\sum_{i=1}^n P(C_j|x_i)} \tag{2.7}$$

and cluster covariances (distance computations) are determined as

$$\Sigma_j = \frac{\sum_{i=1}^n (x_i - s_j)(x_i - s_j)' P(C_j|x_i)}{\sum_{i=1}^n P(C_j|x_i)} \tag{2.8}$$

Complexity In the EM algorithm, for each iteration, distance computations require $O(nk)$. Hence, the distance computation takes $O(d^3)$ where d is the dimension. Therefore, in general, the time complexity of the EM algorithm for each iteration is $O(nkd^3)$.

Parameter Selection Zhang et al. [135] used the R package *mixsmsn*[11] to execute the EM clustering (with fitting finite mixtures of uni- and multivariate scale mixtures of skew-normal distributions) for the categorization of events. A probabilistic

---

[11] https://cran.r-project.org/web/packages/mixsmsn/index.html

mixture model with Poisson distribution within the EM algorithm was applied to cluster students' procrastination behavior [67]. They used the Bayesian approach and Gamma prior distributions for the rate parameters to fit the mixture model with two clusters ($k = 2$) on the students' dataset. In addition, the log-likelihood measure and Schwarz's Bayesian criterion are used for EM clustering in the study of Ruipérez–Valiente et al. [43], with the number of clusters set at $k = 3$ based on cluster quality regarding cohesion and separation. With an assumption that the resulting clusters of students should not exceed 20 members, Urbina Nájera et al. [92] applied the EM to divide students into $k = 14$ groups. Preetha [68] set the number of clusters at $k = 4$ to group students based on their performance. Moreover, the *NbClust* pre-implemented package is also used to determine the number of clusters [49].

2.2.3.6   SOM Clustering

Self-organizing map (SOM) [143] is a center-based clustering scheme. The idea of SOM clustering is for each data point $x_i \in \mathcal{D}; i = 1, \ldots, n$, we aim to represent x_i with d dimensions through an output space with two dimensions. SOM clustering creates a two-dimensional output space where each node in the output space is associated with a "weight" vector. The weight vector contains the position of the node in the input space (dataset), and the weight vectors will move toward the input data when the best matching unit (BMU) is found based on the Euclidean distance from the observed data point to all weight vectors. The summary of the SOM clustering algorithm is described as follows [144].

Algorithm   Apart from the symbols described in Table 2.4, let W be a grid of m nodes, w_{uv} (u and v are their coordinates of the grid); t be the iteration; $\alpha(t) \in [0, 1]$ be the learning rate factor; $r(t)$ be the radius of the neighborhood function $h(w_{uv}, w_{min}, t)$; $\alpha(t)$ and $r(t)$ decrease monotonically over time. The Gaussian is a commonly used neighborhood function [145]:

$$h(w_{min}, w_{uv}, t) = \alpha(t)e^{\left(-\frac{dist\left(w_{min}, w_{uv}\right)^2}{2r(t)^2}\right)} \tag{2.9}$$

1. Initialize the weight vectors of nodes in the output space randomly.
2. *Repeat*

   a. For each data point x_i

      i. For all $w_{uv} \in W$ calculate $d_{uv} = dist(x_i, w_{uv})$.
      ii. Select the node that minimizes d_{uv} as w_{min}.
      iii. Update node $w_{uv} \in W$: $w_{uv} = w_{uv} + \alpha h(w_{min}, w_{uv}, t)dist(x_i, w_{uv})$.

   b. Increase t.

   Until the number of iterations t is reached.

Complexity The computation complexity of SOM clustering is $O(nmdt)$, where n is the number of data points in the dataset $\mathcal{D}$, m is the number of nodes of the output space, d is the dimension of each data point, and t is the number of iterations [146].

Parameter Selection To determine the number of nodes in the output space, Delgado et al. [147] trained different SOM models with a different number of nodes and selected the one with the lowest value of the topographic function (the average number of leftover neighborhood connections per output node). Besides, the number of clusters is chosen by using the connectivity indices *Conn_Index* [148] and Davies–Bouldin index (DBI) methods. The quality of clustering is evaluated by silhouette coefficient, DBI, and Dunn index (DI) measures [149]. In addition, the number of nodes could be chosen as $m = \sqrt[3]{n}$ or equal to the number of clusters [150]. Ahmad et al. [151] set the number of nodes as $m = 5\sqrt{n}$ and the neighborhood radius $r = \frac{max(m)}{4}$ after training ten times.

2.2.3.7   Spectral Clustering

In spectral clustering [152, 153], the spectrum (eigenvalues) of the similarity matrix of the data is used to reduce dimensionality before clustering in fewer dimensions. A similarity matrix is provided as the input and represents quantitative measurements of the similarity between two points. We summarize a well-known version of spectral clustering proposed by Ng et al. [152] as follows.

Algorithm Given a dataset $\mathcal{D} = \{x_1, x_2, \ldots, x_n\}$ and k as the number of clusters

1. Create the similarity matrix $A \in \mathbb{R}^{n \times n}$ where $A_{ij} = e^{\left(-\frac{dist(x_i,x_j)^2}{2\sigma^2}\right)}$ if $i \neq j$; else, $A_{ij} = 0$. The scaling parameter σ^2 is used to decide how quickly A_{ij} falls off with the distance between x_i and x_j.
2. Create the diagonal matrix G with G_{ii} as the sum of column i of matrix A. Define the Laplacian $L = G^{-1/2}AG^{-1/2}$.
3. Find k largest eigenvectors of L, denoted as u_1, u_2, $\ldots$, u_n. They should be orthogonal in the case of repeated eigenvalues. Then, we create the matrix $U = [u_1 u_2 \ldots u_k] \in \mathbb{R}^{n \times k}$ by columnarizing the eigenvectors.
4. Renormalize each row of U to have unit length to create matrix V, where

$$V_{ij} = \frac{U_{ij}}{\sqrt{\sum_j U_{ij}^2}}$$

5. Consider each row of matrix V as a point in $\mathbb{R}^k$ and use k-means (or any other clustering methods) to cluster them into k clusters.
6. If and only if row i of the matrix V is assigned to cluster j, we assign the original point x_i to cluster j.

Complexity In comparison with other clustering methods, spectral clustering has a high computational complexity of $O(n^3)$ in general. Therefore, spectral clustering is not suitable for large datasets [154].

Parameter Selection In the related work, the cluster size is determined by $\sqrt{\frac{n}{2}}$ [102]. Besides, the elbow method is used to find the optimal number of clusters [66] by comparing the distortion score with k in a range of [2, 4].

2.2.3.8 DBSCAN Clustering

DBSCAN is a density-based clustering that discovers the clusters and the noise in a spatial database [155]. There are two parameters: *Eps* (the maximum radius of the neighborhood) and *MinPts* (the minimum number of points in an *Eps*-neighborhood of that point). In DBSCAN clustering, the points are categorized as *core points* (points inside of the cluster), *border points* (points on the border of the cluster), and *outliers*.

- A point p is a core point if at least *MinPts* points are within distance *Eps* of p (including p), i.e., $| N_{Eps}(p) = \{q \,|\, dist(p, q) \leq Eps\} | \geq MinPts$.
- A point p is a border point if it has fewer than *MinPts* points within *Eps* radius, but it is in the neighborhood of a core point.
- *Outliers* are any points that cannot be reached from any other point.
- A point p is directly density-reachable from point p w.r.t. *Eps*, *MinPts* if $p \in N_{Eps}(q)$ and p is a core point.
- A point p is density-reachable from a point p w.r.t. *Eps*, *MinPts* if there is a chain of points $p_1, \ldots, p_n$, where $p_1 = q$, $p_n = p$, such that p_{i+1} is directly density-reachable from p_i.
- A point p is density-connected to a point p w.r.t. *Eps*, *MinPts* if there is a point o such that both p and p are density-reachable from o w.r.t. *Eps*, *MinPts*.

A cluster is a maximal set of density-connected points. Based on the above notations, DBSCAN clustering can be summarized as follows.

Algorithm
1. Start at a random point p.
2. Find all points density-reachable from p w.r.t. *Eps*, *MinPts*.
3. If p is a core point, form a cluster starting with p and expand the cluster through neighbors of p.
4. If p is a border point, and no point is density-reachable from p, DBSCAN visits the next point of the dataset.
5. Repeat the process until all points are processed.

Complexity In general, $O(n \times$ time to find points in the Eps $-$ neighborhood) is the runtime complexity. In the worst case, it is $O(n^2)$. The average time complexity can be reduced to $O(n \log n)$ by using an efficient data structure (e.g., kd-trees).

Parameter Selection A heuristic method is used to find the optimal *MinPts* [124] by using the graphs w.r.t. different *MinPts*, where the graph is the visualization of a function mapping each sample in the dataset to the distance from its *MinPts*[th] nearest neighbor. Optimal *MinPts* is determined by selecting a minimum value whose *MinPts*-dist graph exhibits no significant difference from the rest. As a result, a maximal *Eps* can be obtained by finding the *MinPts*-dist value of the sample in the first "valley" in the graph of optimal *MinPts*. In the work of Kausar et al. [130], the *k*-nearest neighbor distances (the average distance of every data point to its *k*-nearest neighbors) in a matrix of data points were applied to determine the optimal values of the *Eps* and *MinPts* parameters. Besides, Du et al. [119] computed the *k*-distance of each data point and then sorted these distances in ascending order before visualizing them in a scatter chart. The *Eps* value is decided by the value of *k*-distance corresponding to the position that changes sharply. The parameter *MinPts* is chosen based on the minimum number of students with higher similarity in the class.

Summary of Clustering Models in EDS The goal of clustering models is to divide similar objects, i.e., students, into clusters. Center-based methods such as *k*-means, *k*-medoids, and fuzzy *c*-means are the commonly used algorithms due to easy implementation and low computational complexity. However, the disadvantage of these methods is that they are sensitive to outliers, and the user has to specify the desired number of clusters. Fortunately, there are preinstalled packages that efficiently find the optimal number of clusters. Besides, the hierarchical clustering method is an effective technique for discovering the structure of clusters. The result of this method can be represented as a dendrogram when the number of elements is not much. However, computational complexity is a significant obstacle when applying hierarchical clustering on a large dataset. In addition, studies also show that other clustering techniques such as EM, SOM, spectral clustering, or DBSCAN are also effective methods in EDS. The interesting point is that DBSCAN does not require the number of clusters as an input parameter. Therefore, choosing the appropriate clustering method will depend on many factors, such as the problem or analysis task to be performed, the algorithm's computational complexity, the data type, as well as the representation of the results.

2.3 Fair Clustering Models (for EDS)

In this section, we find the answer to the research question RQ_2 by summarizing the popular fairness notions for clustering and well-known fair clustering models and investigating the applicability of those models in EDS.

2.3.1 Fairness Notions

In general, the fairness notions depend on the application and specific context. There are 20 fairness notions used for fair clustering summarized in a survey of fairness in clustering [156]. In their review, the fairness notions are categorized into four types: *group-level*, *individual-level*, *algorithm-agnostic*, and *algorithm-specific*. Based on the popularity of the fairness notions[12] [156], we summarize the following four notions: *balance*, *bounded representation*, *social fairness*, and *individual fairness*.

In the next sections, we denote G as the protected attribute with two values $\{F, M\}$, e.g., *gender* = $\{female, male\}$, in which, F is the discriminated group (or protected group), e.g., "*female*," and M is the non-discriminated group (or non-protected group), e.g., "*male*." Besides, we continue using the symbols listed in Table 2.4.

2.3.1.1 Balance

Balance is the most popular group-level fairness notion used in studies in fair clustering, which was introduced by Chierichetti et al. [158]. Given a clustering $C = \{C_1, C_2, \ldots, C_k\}$ with k clusters, the *balance* of a cluster C_i is the minimum ratio between the number of objects in the protected group and the non-protected group. The *balance* of a clustering is the minimum *balance* score of all clusters. If each cluster has a balance of at least θ as defined by the balance requirement θ, then clustering is fair. The *balance* measure can be applied in any fair clustering model.

$$balance(C_i) = min\left(\frac{\#F \in C_i}{\#M \in C_i}, \frac{\#M \in C_i}{\#F \in C_i}\right) \qquad (2.10)$$

$$balance(C) = min_{i=1}^{k}(C_i) \qquad (2.11)$$

2.3.1.2 Bounded Representation

Bounded representation is a generalization of disparate impact and was introduced by Ahmadian et al. [159]. This group-level notion aims to reduce imbalances in cluster representations of protected attributes (e.g., gender). Let α be the overrepresentation parameter; a cluster C_i is fair if the fractional representation of each group (protected, non-protected group) in the cluster is at most α.

---

[12] We only take into account the fairness notions introduced in the published papers. Because the fairness notions may be turned into measures [157], therefore, in this review we use the term "fairness notion" and "fairness measure" interchangeably.

$$\frac{|\{x \in C_i | \mathcal{G} = g\}|}{|C_i|} \leq \alpha, \quad \text{where } g \in \{F, M\} \tag{2.12}$$

Then, a clustering C is fair if all clusters satisfy the representation constraint. Similar to the *balance* notion, all fair clustering models can apply the bounded representation measure. Besides, *bounded representation* is generalized with two parameters α and β in the study of Bera et al. [160], where for each value of the protected attribute, the fractional representation of it in the cluster must be between β and α.

2.3.1.3 Social Fairness

Social fairness was first introduced by Ghadiri et al. [161], which aims to provide equitable costs for different clusters. In the k-means algorithm, the target is to minimize the objective function (recall Eq. 2.1):

$$L(C, \mathcal{D}) = \sum_{i=1}^{k} \sum_{x \in C_i} dist(x, s_i)^2$$

We denote $\mathcal{D}_F$ and $\mathcal{D}_M$ as two subsets of dataset $\mathcal{D}$, which contain values F and M of the protected attribute, respectively; $\mathcal{D} = \mathcal{D}_F \cup \mathcal{D}_M$. Then, the fair k-means objective is the larger average cost (social fairness cost).

$$\Phi(C, \mathcal{D}) = max\left\{\frac{L(C, \mathcal{D}_F)}{|\mathcal{D}_F|}, \frac{L(C, \mathcal{D}_M)}{|\mathcal{D}_M|}\right\} \tag{2.13}$$

The goal of the fair k-means is to minimize the social fairness cost $\Phi(C, \mathcal{D})$. A disadvantage of this measure is that it can only be applied to center-based clustering because it is modeled on the objective function of a particular algorithm [156].

2.3.1.4 Individual Fairness

Chakrabarti et al. [162] introduced two individual-level fairness notions for k-center clustering, namely, *Per-Point Fairness* and *Aggregate Fairness*. Given a dataset $\mathcal{D}$, the goal is to choose a set $S \subseteq \mathcal{D}$ of at most k centers and then find an assignment $\lambda : \mathcal{D} \mapsto S$. We denote $\alpha \geq 1$ as a parameter. For all data points $x \in \mathcal{D}$, $I_x \subseteq \mathcal{D}$ is denoted as the group of data points that are similar to x. The fairness notions are described as follows:

Per-Point Fairness: For all data points $x \in \mathcal{D}$, *Per-Point Fairness* is satisfied if the distance from x to its center is at most α time the distance of the closest point in that cluster to the center $\lambda(x)$.

$$dist(x, \lambda(x)) \leq \alpha min_{y \in \mathcal{I}_x}\{dist(y, \lambda(y))\} \qquad (2.14)$$

Aggregate Fairness: For all data points $x \in \mathcal{D}$, *Aggregate Fairness* is satisfied if the distance from x to its center is at most α time the *average* distance of the points in that cluster to the center $\lambda(x)$.

$$dist(x, \lambda(x)) \leq \alpha \frac{\sum_{y \in \mathcal{I}_x} dist(y, \lambda(y))}{|\mathcal{I}_x|} \qquad (2.15)$$

2.3.2 Fair Clustering Models

In this section, we overview the fair clustering models based on the popularity of their corresponding traditional clustering models. We hypothesize that if a clustering model is widely used in the education setting, its corresponding fair model will also be more applicable. Therefore, we focus on center-based clustering (such as k-means, k-center, k-medoids), hierarchical clustering, and spectral clustering.

2.3.2.1 Fair Center-Based Clustering

Fair k-Center Clustering The first work on group-level fair clustering was introduced by Chierichetti et al. [158] to ensure an equal representation of each protected attribute in every cluster. They defined a new fairness notion, *balance*, described in Sect. 2.3.1.1. A two-phase approach was proposed: (1) fairlet decomposition, clustering all instances into *fairlets*, which are small clusters guarantying fairness constraint, and (2) applying vanilla clustering methods (k-center, k-median) on those *fairlets* to obtain the final resulting fair clusters. Figure 2.3 illustrates the fair clustering method using the *fairlets* concept.

Besides, Ahmadian et al. [159] introduced a new fairness measure (bounded representation) by providing an upper-bound constraint for fairness in resulting clusters, applied for k-center clustering. Jones et al. [163] proposed an algorithm with a linear time complexity and obtained a 3-approximation for the fair k-center summaries problem. Recently, Chakrabarti et al. [162] presented two individual fairness notions that guarantee each data point has a similar quality of service. They proposed approximation algorithms for the k-center objective.

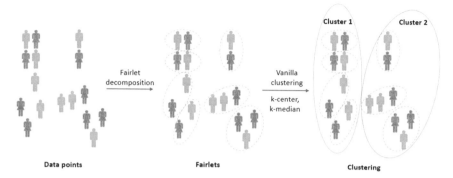

Fig. 2.3 Fair clustering through fairlets

Fair k-Means Clustering Schmidt et al. [164] proposed a fair k-means clustering by using the *coresets* concept. Essentially, a *coreset* is a summary of a point set that approximates any possible solution's cost function well. They extended the approach of Chierichetti et al. [158] to compute the fairlet for the k-means algorithm. Their experiments show that the new method can be applied to big data. Besides, Ghadiri et al. [161] introduced the social fairness notion, which focuses on minimizing the clustering cost across groups of the protected attribute. They proposed a fair clustering version of the well-known Lloyd k-means and reported results through clustering cost. Abraham et al. [165] introduced a fair k-means clustering model, namely, *FairKM*, which combines the optimization of the classical clustering objective and a novel fairness loss term. Their model aims to achieve a trade-off between clustering quality (on the non-protected attributes) and cluster fairness (on the multiple protected attributes).

Fair k-Medoids Clustering k-Medoids clustering and fairlet decomposition were applied in the study of Le Quy et al. [30]. They introduced the grouping problem in the educational setting, where each cluster has a limitation in terms of cardinality and the clustering must be fair w.r.t. the protected attribute. They applied the fairlet decomposition [158] to obtain the fairness w.r.t. the protected attribute, and k-medoids clustering was used to satisfy the cardinality constraint. To the best of our knowledge, this is the first work taking into account fairness and clustering in the educational environment.

Fair Fuzzy c-Means Clustering A fair fuzzy c-means clustering was proposed by Xia et al. [166]. Their goals are to minimize the objective function $L(C, D)$ (Eq. 2.2) and maximize the ratio of each group (e.g., *female*, *male*), i.e., as close as possible to the ratio in the original dataset. They defined a new *fair loss function* to measure the fair loss in clusters and optimized the assignment phase according to the objective function to obtain a fair clustering.

Parameter Selection In the fair k-means version, Schmidt et al. [164] evaluated their proposed approach with the number of clusters in a range of [20, 100], and

Ghadiri et al. [161] used the range of [4, 16] for k. Regarding k-medoids clustering, Le Quy et al. [30] experimented their fair-capacitated clustering based on the k-medoids model with the number of clusters in a range of [3, 20]. In addition, Chakrabarti et al. [162] performed experiments with the number of clusters k in {2, 4, 8, 16, 32, 64, 128} for their k-center fair clustering. A set of k in {3, 4, 6, 8, 10, 12, 14, 16, 18, 20} was selected for experiments in the work of Chierichetti et al. [158]. However, the optimal value of k was not discussed in their studies. In the fair fuzzy c-means clustering, the number of clusters k was chosen in {4, 6, 8} [166].

2.3.2.2 Fair Hierarchical Clustering

Ahmadian et al. [167] defined a fair hierarchical clustering for any fairness constraint, in which "a hierarchical clustering is fair if all of its clusters (besides the leaves) are fair." They extended the fairlet decomposition [158] for upper-bounded representation fairness and proved the results for the revenue, value, and cost objectives. In the educational domain, Le Quy et al. [30] proposed a fair-capacitated clustering and used hierarchical clustering to obtain the clusters with a given maximum capacity, while the fairness constraint was satisfied by using the fairlet decomposition approach.

Parameter Selection Ahmadian et al. [167] set the similarity to be $sim(x_i, x_j) = \frac{1}{1+dist(x_i, x_j)}$ where $dist(x_i, x_j)$ is the Euclidean distance. Le Quy et al. [30] varied the number of clusters between 3 and 20 to evaluate their fair-capacitated hierarchical clustering model.

2.3.2.3 Fair Spectral Clustering

Kleindessner et al. [168] incorporated the balance fairness notion in normalized and unnormalized constrained spectral clustering. They applied k-means clustering on a fair subspace generated by projecting the graph Laplacian. As mentioned in Sect. 2.2.3.7, spectral clustering is used widely in educational data analysis. Therefore, the fair version of this model is a potential candidate for EDS.

Parameter Selection Kleindessner et al. [168] performed their experiments with the number of clusters from 2 to 8 and from 1 to 15 for two datasets.

Summary of Fair Clustering Models Although fairness-aware clustering has only been studied in the last 5 years, quite a few fair clustering methods have been introduced. Center-based fair clustering methods still prevail over other approaches, such as hierarchical or spectral fair clustering. It is worth noting that there is little research on the development and use of fair clustering in EDS. To our knowledge, only k-medoids and hierarchical methods are used to group students in collaborative learning activities, which focus on ensuring fairness regarding protected attributes and capacity (fair-capacitated clustering). However, since the effectiveness of other fair clustering models such as k-center, k-means, and fair spectral clustering has been demonstrated in related work, we believe such approaches will also be effective in EDS. Moreover, a fair clustering would contribute to heterogeneity w.r.t. some protected attributes like gender and race because it can take into account the heterogeneous property of the resulting clusters. Therefore, researchers and educators should tackle the questions "what sort of groups are needed and what context or EDS task is considered" in order to choose the suitable fair clustering models.

2.4   Clustering Evaluation in EDS Toward Fairness-Aware Learning

In this section, we investigate the evaluation aspects of clustering models toward fairness, including evaluation measures and datasets to tackle RQ_3.

2.4.1   Evaluation Measures

This section presents the essential components to evaluate the results of (fair) clustering models, including clustering quality (Sect. 2.4.1.1) and fairness measures (Sect. 2.4.1.2).

2.4.1.1   Clustering Quality Measures

Table 2.5 outlines the measures used in related work to evaluate cluster validity. We summarize the most prevalent measures as follows:

Silhouette Coefficient This is a metric that measures cluster separation and compactness at the same time for both individual data points and clusters and clustering. The silhouette score of a data point x_i in cluster C_I is computed by

$$score_i = \frac{\beta_i - \alpha_i}{max(\alpha_i, \beta_i)} \tag{2.16}$$

and $score_i = 0$ if $|C_I| = 1$.

In Eq. 2.16, α_i is the average distance of the data point x_i to other points in its cluster C_I, i.e., $\alpha_i = \frac{\sum_{j \neq i, x_j \in C_I} dist(x_i, x_j)}{|C_I| - 1}$, and β_i is the minimum of the average distance of the point x_i to points in other clusters, $\beta_i = min_{J \neq I} \frac{1}{|C_J|} \sum_{x_j \in C_J} dist(x_i, x_j)$. Finally,

Table 2.5 Clustering quality measures

Measures	Used in
Silhouette coefficient	Battaglia et al. [169], Bharara et al. [64], Fang et al. [32], Khayi and Rus [89], Howlin and Dziuban [40], Abraham et al. [165], Maylawati et al. [170], Moubayed et al. [53], Šarić-Grgić et al. [171], Ahmed et al. [117], Li et al. [124], Palani et al. [50], Xia et al. [166], Zhang et al. [135], Talebinamvar and Zarrabi [172], Hassan et al. [77]
Sum of squared error (SSE)	Kurniawan et al. [173], Schmidt et al. [164], Rijati et al. [174], Abraham et al. [165], Chang et al. [31], Le Quy et al. [30], Xia et al. [166], Yin [65], Sobral and de Oliveira [73]
Davies–Bouldin index (DBI)	Chang et al. [31], Li et al. [124], Ahmed et al. [117], Palani et al. [50], Delgado et al. [147]
Calinski–Harabasz index (CHI)	Wang and Zha [112], Chang et al. [31], Li et al. [124], Palani et al. [50]
Analysis of variance (ANOVA) test	Shen and Chi [42], Aghababyan et al. [69]
Dunn index (DI)	Fang et al. [32], Ahmed et al. [117]
Normalized mutual information (NMI)	Vo and Nguyen [63], Tang et al. [138]
Adjusted Rand index (ARI)	Mishler and Nugent [175]
Chi-square	Wu et al. [102]
Clustering accuracy (ACC)	Tang et al. [138]
Fuzzy silhouette index	Howlin and Dziuban [40]
Heterogeneity score	Wu et al. [102]
Homogeneity score	Wu et al. [102]
Log-likelihood	Wu et al. [102]
Modified partition coefficient	Howlin and Dziuban [40]
Partition coefficient	Howlin and Dziuban [40]
Partition entropy	Howlin and Dziuban [40]
Simpson index	Wang and Zha [112]
t-Tests	Mojarad et al. [176]
Xie and Beni index	Howlin and Dziuban [40]

the silhouette coefficient of the entire dataset is the maximum value of the mean $score_i$ over all data points [126].

Sum of Squared Error (SSE) or Cluster Cohesion SSE measures how closely related objects are in a cluster, which is computed by the formula:

$$SSE = \sum_{i=1}^{k} \sum_{x \in C_i} dist(x, s_i)^2 \qquad (2.17)$$

where k is the number of clusters and s_i is the centroid of the cluster C_i.

The **Davies–Bouldin index (DBI)** is an internal evaluation scheme introduced by Davies and Bouldin [177]. DBI is calculated by the following formula:

$$DBI = \frac{1}{k}\sum_{i=1}^{k} max_{j \neq i}\left(\frac{\sigma_i + \sigma_j}{dist(s_i, s_j)}\right) \tag{2.18}$$

where k is the number of clusters, s_i is the centroid of cluster i^{th}, and σ_i is the average distance of all cluster members in cluster i to its centroid s_i.

The **Dunn index (DI)** was introduced by Dunn [178], which identifies clusters that are well-separated and dense. DI is computed by dividing the minimal inter-cluster distance by the maximal intra-cluster distance:

$$DI = \frac{min_{1 \leq i < j \leq k} dist(i,j)}{max_{1 \leq p \leq k} dist'(p)} \tag{2.19}$$

where $dist(i,j)$ is the distance between clusters i^{th} and j^{th} (e.g., the distance between two centroids) and $dist'(p)$ is the intra-cluster distance of cluster p (e.g., the maximal distance between any pair of data points in cluster p).

2.4.1.2 Fairness Measures for Fair Clustering

Balance is the most prevalent fairness measure used in the fair clustering models [30, 158, 164, 168]. Xia et al. [166] reported the clustering results in terms of fairness by the following measures: *balance* [160], *Euclidean distance of distribution vectors*, and *Wasserstein distance of distribution vectors*. Abraham et al. [165] used the following fairness measures to evaluate their model: *average Euclidean, average Wasserstein, max Euclidean, max Wasserstein*. Besides, Chakrabarti et al. [162] reported their experimental results on two fairness measures: *Per-Point Fairness* and *Aggregate Fairness*. In terms of clustering cost, Ghadiri et al. [161] presented their result on *social fairness*, which is the clustering cost of each group of the protected attribute. In addition, the *bounded representation* measure was used to report the resulting clustering for fair hierarchical clustering [167].

2.4.2 Datasets

In the related work, most datasets are non-public. The data are collected by the paper's authors or shared within their institute. Therefore, accessing educational datasets is a significant challenge. This section summarizes the educational datasets used in the references w.r.t. (fair) clustering. Due to the capacity of the book chapter, we only list the datasets used in the top journals and conferences, i.e., ranking Q1, A*/A. In Table 2.6, we outline 27 non-public educational datasets used in the related

Table 2.6 Non-public datasets used to evaluate (fair) clustering models

Dataset and description	Used in	Protected attributes	Data source	Data type
427,382 logs in the time period of 16 weeks and 546,966 cleaned logs in 18 courses on the Moodle online graduate program in the United States	Hung et al. [86]	Ethnicity, gender	Online learning platform	Tabular
Course "social aspects of information technology" in the iMooX platform with 459 matriculated undergraduates and 379 external students	Khalil and Ebner [48]	Gender	MOOC	Tabular
811 records covering the admission years of 2007–2009 at Mae Fah Luang University	Iam-On et al. [83]	Gender	Traditional classroom	Tabular
1199 university students and 35 teachers (277 students and 19 teachers are selected for experiments)	Urbina Nájera et al. [92]	Gender, marital status	Traditional classroom	Tabular
MITx introductory programming course with 12,973 correct and 4293 incorrect attempts	Gulwani et al. [96]	–	MOOC	Text
1093 students in two e-tutorials: SOWISO and MyStatLab (MSL) in 2016/2017	Tempelaar et al. [179]	Gender	Online learning platform	Tabular
Synthetic dataset of 600 students	Kausar et al. [130]	–	Synthetic	Tabular
438 and 617 Chinese public universities in 1998 and 2011, respectively	Wang et al. [112]	–	Traditional classroom	Tabular
64 students in projects in a graduate software engineering course	Akbar et al. [105]	–	MOOC	Tabular
10 million meals were consumed by about 82,871 students at the canteen of the University of Pisa	Natilli et al. [109]	Gender	Canteen of a university	Tabular
Real student dataset of various academic disciplines of higher educational institutions in Kerala, India	Francis et al. [81]	Gender	Traditional classroom	Tabular
Three groups of learners (sizes 30, 50, 80) in Hubei province of China (National Training Plan 2015)	Liu et al. [100]	–	Traditional classroom	Tabular
486 students attending undergraduate science courses at the University of Western Ontario in Canada	Moubayed et al. [53]	–	Online learning platform	Tabular
102 middle school students in the northeastern United States	Li et al. [134]	Race	Traditional classroom	Tabular
1062 students from Al-Balqa Applied University (BAU) in Jordan from 2006 to 2018	Almasri et al. [79]	Gender	Traditional classroom	Tabular
2000 students from 4 universities in China	Chang et al. [31]	Gender	Traditional classroom	Tabular

(continued)

Table 2.6 (continued)

Dataset and description	Used in	Protected attributes	Data source	Data type
Two online Python programming courses—intermediate (42,131 students) and beginners (7164 student)—in Australia	McBroom et al. [41]	–	Online learning platform	Tabular
9024 undergraduates at a university in Beijing during the spring of 2019	Li et al. [124]	–	Traditional classroom	Tabular
180 students	Yin [65]	Gender	Traditional classroom	Tabular
121 students in a programming didactic course at Stockholm University in 2020	Wu et al. [102]	–	Traditional classroom	Text
1,709,189 records of online students enrolled from 2015 to 2019 at Universidad Internacional de La Rioja (UNIR)	Delgado et al. [147]	–	Online learning platform	Tabular
8201 feedback responses for 168 distinct courses from the Politehnica University of Bucharest in 2019–2020	Masala et al. [97]	–	Survey	Text
Programming course in an online learning environment: training, 11 problems (2358 solutions); test, 11 new problems (2598 solutions)	Effenberger et al. [70]	–	Online learning platform	Text
Two datasets: 29 students from the international relations course and 50 students from the computer science and computer engineering courses	Sobral et al. [73]	Gender	Traditional classroom	Tabular
251 students' clickstream log data in an introductory physics course (fall 2020 semester)	Zhang et al. [135]	–	Online learning platform	Text
180 male and 458 female Iranian BA students involving 20 classes at Mazandaran University of Science and Technology	Talebinamvar et al. [172]	Gender	Online learning platform	Tabular
400 first-year students in a medium-sized metropolitan university in Dublin with two programming courses	Mai et al. [33]	–	Online learning platform	Tabular

work. Besides, seven public educational datasets are summarized in Table 2.7. The datasets are sorted in ascending order of the paper's publication year. The majority of the datasets are tabular and collected in MOOCs and e-learning platforms. Most of the datasets are small, and only seven datasets contain more than 10,000 instances.

Table 2.7 Public datasets used to evaluate (fair) clustering models

Dataset and description	Used in	Protected attributes	Data source	Data type
Dataset of shell commands with 18 cybersecurity training sessions, 8834 commands collected from 113 trainees[a]	Švábensky` et al. [180]	–	Online learning platform	Text
480 students collected by Kalboard 360 LMS (xAPI-Edu-data)[b]	Bharara et al. [64]	Gender	Online learning platform	Tabular
OULAD (Open University Learning Analytics) dataset with 32,593 students enrolled in 7 courses at an open university in England[c]	Casalino et al. [78], Le Quy et al. [30], Palani et al. [50]	Gender	Online learning platform	Tabular
Student performance dataset of two Portuguese secondary schools[d]	Jones et al. [163], Le Quy et al. [30]	Gender	Traditional classroom	Tabular
PISA test scores dataset with information on 5233 American students in 2009 from the Program for International Student Assessment (PISA)[e]	Le Quy et al. [30]	Gender	Traditional classroom	Tabular
MOOC dataset of 416,921 students in the MOOCs enrolled in the 16 edX courses offered by Harvard and MIT (Massachusetts Institute of Technology) during 2012–2013[f]	Le Quy et al. [30]	Gender	MOOC	Tabular
StudentLife dataset[g]	Hassan et al. [77]	–	Online learning platform	Tabular

[a] https://gitlab.ics.muni.cz/muni-kypo-trainings/datasets/commands
[b] https://www.kaggle.com/datasets/aljarah/xAPI-Edu-Data
[c] https://analyse.kmi.open.ac.uk/open_dataset
[d] https://archive.ics.uci.edu/ml/datasets/student+performance
[e] https://www.kaggle.com/econdata/pisa-test-scores
[f] https://github.com/kanika-narang/MOOC_Data_Analysis
[g] https://studentlife.cs.dartmouth.edu/dataset.html

2.5 Beyond Fairness Requirements for Clustering in EDS

In this section, we investigate other problems and needs apart from fairness for clustering in EDS to tackle the research question RQ_4.

Cardinality Constraints In collaborative learning, it is important to have comparable group sizes to distribute work fairly. However, many traditional and fair clustering models do not consider the cardinality of the resulting clusters. As a result, clustering models may produce groups of students of various sizes, which makes it difficult for teachers to assign work to groups. Therefore, clustering models need to ensure fairness and consider the resulting clusters' size to increase the usefulness and actionability of the clustering. The capacity constraint can be expressed as the minimum size, the maximum size of the clusters, or both

[181, 182]. Recently, the fair-capacitated clustering problem was introduced by Le Quy et al. [30], which ensures fairness and balanced cardinalities of the resulting clusters. They proposed two solutions to achieve the capacitated clustering on top of fairlet decomposition, namely, hierarchical-based and k-medoid-based approaches.

Explainability Explaining decisions is becoming increasingly important in education, especially in ML-based learning systems. ML algorithms, e.g., clustering models, are black boxes for decision-makers. However, because clustering algorithms rely on all data features, they produce cluster assignments that are difficult to explain. Hence, a fairly obvious requirement is that clustering models should provide explanations for the model's results, i.e., cluster assignments, in a way that humans can understand. Several studies are focusing on the explainability of well-known clustering models, such as k-means and k-medians [183], by using a small decision tree. Bandyapadhyay et al. [184] investigated the computational complexity of several variants of explainable clustering and introduced a new model of explainable clustering by a threshold tree after removing a subset of points. As mentioned above, fair clustering models aim to find the trade-off between the clustering quality and fairness in resulting clusters. Therefore, the goal of explainable models is to provide an explanation for these fair clustering models in terms of both cluster quality and fairness.

Homogeneous and Heterogeneous Clustering In educational settings, both homogeneous and heterogeneous groups play an essential role. Typically, the objective of clustering models is to divide objects into homogeneous clusters, for example, students with the same ability level. Hence, homogeneous groups are the base for providing adequate support in a learning situation and allowing ability-specific learning. Besides, in collaborative learning, the heterogeneous groups w.r.t. characteristics and level of ability of students are more important to prevent critical group building (winner and loser, the smart student and the other) or to allow learning in teams with different specific abilities [103, 185, 186]. In this way, collaborative learning can benefit from psychological factors in heterogeneous groups. Therefore, clustering algorithms need to consider the specific EDS task to determine whether the algorithm's output is homogeneous or heterogeneous clusters.

2.6 Conclusions and Outlook

Clustering techniques play a crucial role in analyzing students' performance, supporting learning, giving feedback and recommendation, and many EDS tasks. Studies on clustering algorithms and fairness in EDS can employ a variety of approaches and discussions.

First, in this chapter, we summarize the popular EDS tasks using clustering techniques and present the most prevalent clustering algorithms (traditional and fairness-aware models). We also discuss evaluation aspects w.r.t. clustering performance, datasets, and fairness measures. In this, improving clustering performance is an essential requirement of the clustering models because the clustering quality is the objective of any clustering algorithm. Hence, by increasing clustering quality, the clustering models are also more likely to find meaningful clusters, which are useful for cluster analysis. In addition, fair clustering models also need to take into account the trade-off between clustering quality and the fairness of the resulting clusters.

Second, because EDS has grown rapidly due to the emergence of both novel data and innovative methods [9], it is crucial to collect and develop a benchmark educational dataset for EDS. On the one hand, the vast majority of datasets are non-public and challenging to access due to their privacy issues, as demonstrated in our review. On the other hand, they are collected for different analytical purposes or problems. Therefore, the scope of use of these datasets is somewhat limited. Besides, generating a synthetic dataset for EDS is a potential solution to overcome the difficulties caused by the lack of benchmark educational datasets [187, 188].

Third, with the rapid development of MOOC systems, researchers can collect and process large volumes of data in structured and unstructured formats. Therefore, traditional clustering and corresponding fairness-aware models also need to be improved and upgraded to handle such big educational data. For example, (fair) clustering methods should be scalable with big data [189, 190] or be able to process high-dimensional data [191].

Fourth, bias and discrimination are common problems in education. In this chapter, we study the well-known fairness measures applied in clustering models in various domains. However, because fairness notions differ across disciplines, it isn't easy to evaluate the efficiency of fairness-aware clustering algorithms. Therefore, selecting or defining the appropriate fairness notions for the educational domain is crucial and necessary. There are several studies on fairness evaluation for classification algorithms in education [24, 192]. However, choosing the appropriate fairness measures for fair clustering models in EDS is still a significant challenge for researchers.

Fifth, because clustering models are widely applied in many EDS tasks, and each problem has different constraints or requirements, constrained clustering models are also applied. For example, the recommendation systems take into account students' preferences, or clustering models that support collaborative learning concern the groups' size. Moreover, the explainability and interpretability of (fair) clustering models are also interesting topics for future research.

To conclude, the construction and selection of clustering models (traditional or fair) in EDS is an important issue in improving the efficiency of educational data analysis and contributing to improving current educational systems. We overview the role of (fair) clustering models in EDS and investigate the use of existing clustering models in EDS. We believe our review can help both educational and computer scientists have a broad picture of clustering models and fairness in education to determine the appropriate clustering methods for their studies.

Acknowledgments The work of the first author is supported by the Ministry of Science and Culture of Lower Saxony, Germany, within the PhD program "LernMINT: Data-assisted teaching in the MINT subjects."

Appendix

Table 2.8 Clustering models used in EDS

Clustering models	#Papers	References
k-Means	73	Adjei et al. [75], Battaglia et al. [169], Ding et al. [36], Iam-On and Boongoen [83], Jia et al. [39], Khalil and Ebner [48], Hung et al. [86], López et al. [44], Salwana et al. [61], Rihák and Pelánek [193], Roy et al. [51], Urbina Nájera et al. [92], Zhang et al. [125], Aghababyan et al. [69], Akbar et al. [105], Bharara et al. [64], Esnashari et al. [38], Fang et al. [32], Kurniawan et al. [173], Kausar et al. [130], Gunawan et al. [56], Mengoni et al. [45], Mishler and Nugent [175], Mojarad et al. [176], Ninrutsirikun et al. [194], Natilli et al. [109], Nen-Fu et al. [55], Purba et al. [85], Tempelaar et al. [179], Rijati et al. [174], Vo and Nguyen [63], Wang and Zha [112], Dovgan et al. [108], Francis and Babu [81], Huang et al. [114], Nazaretsky et al. [113], Oladipupo and Olugbara [49], Phanniphong et al. [195], Wang et al. [196], Waspada et al. [37], Almasri et al. [79], Chang et al. [31], Chaves et al. [197], Kosztyán et al. [198], Maylawati et al. [170], Moubayed et al. [53], Pradana et al. [199], Qoiriah et al. [90], Shelly et al. [104], Tang et al. [200], Rijati et al. [201], Šarić-Grgić et al. [171], Ahmed et al. [117], Chi [202], Chu et al. [82], Iatrellis et al. [80], Li et al. [203], Li et al. [124], Masala et al. [97], Preetha [68], Putra et al. [204], Sobral and de Oliveira [73], Susanto et al. [205], Rauthan et al. [206], Wang [207], Yang [107], Yin [65], Zhang et al. [35], Guo et al. [98], Hassan et al. [77], Talebinamvar and Zarrabi [172], Vasuki and Revathy [128]
Hierarchical clustering	31	López et al. [44], Salwana et al. [61], Shen and Chi [42], Rahmat [116], Rihák and Pelánek [193], Akbar et al. [105], Fang et al. [32], Mengoni et al. [45], Mishler and Nugent [175], Kausar et al. [130], Wang and Zha [112], Cheng and Shwe [208], Dovgan et al. [108], Kylvaja et al.

(continued)

Table 2.8 (continued)

Clustering models	#Papers	References
		[94], Nazaretsky et al. [113], Oladipupo and Olugbara [49], Li et al. [134], Pradana et al. [199], Silva and Silla [91], Singelmann et al. [209], Popov et al. [210], Yotaman et al. [88], Chang et al. [72], Chu and Yin [110], Huang et al. [52], Pamungkas et al. [118], Preetha [68], Mai et al. [33], Wang and Lv [99], Zhang et al. [135], Khayi and Rus [89]
Fuzzy c-means	17	Güvenç and Çetin [47], Amalia et al. [139], Howlin and Dziuban [40], Casalino et al. [78], Oladipupo and Olugbara [49], Ramanathan et al. [76], Varela et al. [34], Supianto et al. [211], Thilagaraj and Sengottaiyan [62], Yadav [212], Li et al. [111], Liu [115], Palani et al. [50], Parvathavarthini et al. [213], Tang et al. [138], Li and Sun [95], Premalatha and Sujatha [214]
Distinctive clustering models	9	Pratiwi et al. [103], Gulwani et al. [96], Fasanya and Fathizadeh [101], Nguyen and Vo [87], Waluyo et al. [215], McBroom et al. [41], Effenberger and Pelánek [70], Wang and Wang [57, 106]
EM	7	Ruipérez-Valiente et al. [43], Urbina Nájera et al. [92], Mengoni et al. [45], Park et al. [67], Oladipupo and Olugbara [49], Preetha [68], Zhang et al. [135]
SOM clustering	6	Alias et al. [149], Salwana et al. [61], Bara et al. [150], Ahmad et al. [151], Purbasari et al. [216], Delgado et al. [147]
Spectral clustering	5	Mengoni et al. [45], Dovgan et al. [108], Gao et al. [71], Hooshyar et al. [66], Wu et al. [102]
DBSCAN	4	Kausar et al. [130], Oladipupo and Olugbara [49], Du et al. [119], Li et al. [124]
k-Medoids	3	Kausar et al. [130], Furr [129], Vasuki and Revathy [128]
Affinity propagation	1	Liu et al. [100]
Author topic model	1	Rakhmawati et al. [217]
Bio-inspired clustering	1	Chang et al. [93]
BIRCH	1	Dovgan et al. [108]

(continued)

Table 2.8 (continued)

Clustering models	#Papers	References
CFSFDP-HD (Clustering by Fast Search and Finding of Density Peaks via Heat Diffusion)	1	Kausar et al. [130]
CLARANS	1	Vasuki and Revathy [128]
DP-means	1	Khayi and Rus [89]
Farthest first clustering	1	Urbina Nájera et al. [92]
k-Medians	1	Wang and Zha [112]
Link-based cluster ensemble	1	Iam-On and Boongoen [84]
Louvain clustering	1	Pradana et al. [199]
OPTICS (Ordering Points To Identify the Clustering Structure) clustering	1	Švábensky` et al. [180]
Roche multiway tree	1	Yan and Su [218]
x-Means	1	Phanniphong et al. [195]

References

1. Dorans, N.J., Cook, L.L.: Fairness in Educational Assessment and Measurement. Routledge, New York (2016)
2. Zlatkin-Troitschanskaia, O., Schlax, J., Jitomirski, J., Happ, R., Kühling-Thees, C., Brückner, S., Pant, H.: Ethics and fairness in assessing learning outcomes in higher education. High Educ. Pol. **32**(4), 537–556 (2019). https://doi.org/10.1057/s41307-019-00149-x
3. Ford, M., Morice, J.: How fair are group assignments? A survey of students and faculty and a modest proposal. J. Inform. Technol. Educ. Res. **2**(1), 367–378 (2003)
4. Miles, J.A., Klein, H.J.: The fairness of assigning group members to tasks. Group Org. Manag. **23**(1), 71–96 (1998). https://doi.org/10.1177/1059601198231005
5. Rezaeinia, N., Góez, J.C., Guajardo, M.: Efficiency and fairness criteria in the assignment of students to projects. Ann. Oper. Res., 1–19 (2021). https://doi.org/10.1007/s10479-021-04001-7
6. Song, X.: The fairness of a graduate school admission test in China: voices from administrators, teachers, and test-takers. Asia Pac. Educ. Res. **27**(2), 79–89 (2018). https://doi.org/10.1007/s40299-018-0367-4
7. Xiao, W., Ji, P., Hu, J.: A survey on educational data mining methods used for predicting students' performance. Eng. Rep. (2021). https://doi.org/10.1002/eng2.12482
8. Meyer, K.: Education, Justice and the Human Good: Fairness and Equality in the Education System. Routledge, London (2014)
9. McFarland, D.A., Khanna, S., Domingue, B.W., Pardos, Z.A.: Education data science: past, present, future. AERA Open. **7** (2021). https://doi.org/10.1177/23328584211052055
10. Romero, C., Ventura, S.: Educational data science in massive open online courses. Wiley Interdisc. Rev. Data Min. Know. Discov. **7**(1), e1187 (2017). https://doi.org/10.1002/widm.1187
11. Dutt, A., Ismail, M.A., Herawan, T.: A systematic review on educational data mining. IEEE Access. **5**, 15991–16005 (2017). https://doi.org/10.1109/ACCESS.2017.2654247
12. Peña-Ayala, A.: Educational data mining: a survey and a data mining-based analysis of recent works. Expert Syst. Appl. **41**(4), 1432–1462 (2014). https://doi.org/10.1016/j.eswa.2013.08.042

13. Romero, C., Ventura, S.: Educational data mining and learning analytics: an updated survey. Wiley Interdisc. Rev. Data Min. Know. Discov. **10**(3), e1355 (2020). https://doi.org/10.1002/widm.1355
14. Del Bonifro, F., Gabbrielli, M., Lisanti, G., Zingaro, S.P.: Student dropout prediction. In: Proceedings of the International Conference on Artificial Intelligence in Education (AIED), pp. 129–140 (2020). https://doi.org/10.1007/978-3-030-52237-7_11
15. Kemper, L., Vorhoff, G., Wigger, B.U.: Predicting student dropout: a machine learning approach. Eur. J. High. Educ. **10**(1), 28–47 (2020). https://doi.org/10.1080/21568235.2020.1718520
16. Hutt, S., Gardner, M., Duckworth, A.L., D'Mello, S.K.: Evaluating fairness and generalizability in models predicting on-time graduation from college applications. In: Proceedings of the 12th International Conference on Educational Data Mining (EDM), pp. 79–88 (2019)
17. Livieris, I.E., Tampakas, V., Karacapilidis, N., Pintelas, P.: A semi-supervised self-trained two-level algorithm for forecasting students' graduation time. Intel. Decis. Technol. **13**(3), 367–378 (2019). https://doi.org/10.3233/IDT-180136
18. Fenu, G., Galici, R., Marras, M.: Experts' view on challenges and needs for fairness in artificial intelligence for education. In: International Conference on Artificial Intelligence in Education, pp. 243–255. Springer, Cham (2022). https://doi.org/10.1007/978-3-031-11644-5_20
19. Vasquez Verdugo, J., Gitiaux, X., Ortega, C., Rangwala, H.: FairEd: a systematic fairness analysis approach applied in a higher educational context. In: LAK22: 12th International Learning Analytics and Knowledge Conference, pp. 271–281 (Mar 2022). https://doi.org/10.1145/3506860.3506902
20. Ntoutsi, E., et al.: Bias in data-driven artificial intelligence systems—an introductory survey. Wiley Interdisc. Rev. Data Mining Know. Discov. **10**(3), e1356 (2020). https://doi.org/10.1002/widm.1356
21. Le Quy, T., Roy, A., Iosifidis, V., Zhang, W., Ntoutsi, E.: A survey on datasets for fairness-aware machine learning. Wiley Interdiscip. Rev. Data Min. Knowl. Disc., e1452 (2022). https://doi.org/10.1002/widm.1452
22. Mehrabi, N., Morstatter, F., Saxena, N., Lerman, K., Galstyan, A.: A survey on bias and fairness in machine learning. ACM Comput. Surv. (CSUR). **54**(6), 1–35 (2021). https://doi.org/10.1145/3457607
23. Bayer, V., Hlosta, M., Fernandez, M.: Learning analytics and fairness: do existing algorithms serve everyone equally? In: Proceedings of the International Conference on Artificial Intelligence in Education (AIED), pp. 71–75 (2021). https://doi.org/10.1007/978-3-030-78270-2_12
24. Gardner, J., Brooks, C., Baker, R.: Evaluating the fairness of predictive student models through slicing analysis. In: Proceedings of the 9th International Conference on Learning Analytics & Knowledge, pp. 225–234 (2019). https://doi.org/10.1145/3303772.3303791
25. Riazy, S., Simbeck, K., Schreck, V.: Systematic literature review of fairness in learning analytics and application of insights in a case study. In: Proceedings of the International Conference on Computer Supported Education, pp. 430–449 (2020). https://doi.org/10.1007/978-3-030-86439-2_22
26. Baker, R.S., Hawn, A.: Algorithmic bias in education. Int. J. Artif. Intell. Educ., 1–41 (2021). https://doi.org/10.1007/s40593-021-00285-9
27. Kizilcec, R.F., Lee, H.: Algorithmic fairness in education. In: Ethics in Artificial Intelligence in Education (2022)
28. Liu, S., d'Aquin, M.: Unsupervised learning for understanding student achievement in a distance learning setting. In: Proceedings of the IEEE Global Engineering Education Conference (EDUCON), pp. 1373–1377 (2017). https://doi.org/10.1109/EDUCON.2017.7943026
29. Zhang, N., Biswas, G., Dong, Y.: Characterizing students' learning behaviors using unsupervised learning methods. In: Proceedings of the International Conference on Artificial Intelligence in Education (AIED), pp. 430–441 (2017). https://doi.org/10.1007/978-3-319-61425-0_36

30. Le Quy, T., Roy, A., Friege, G., Ntoutsi, E.: Fair-capacitated clustering. In: Proceedings of the 14th International Conference on Educational Data Mining (EDM21), pp. 407–414 (2021)
31. Chang, W., Ji, X., Liu, Y., Xiao, Y., Chen, B., Liu, H., Zhou, S.: Analysis of university students' behavior based on a fusion k-means clustering algorithm. Appl. Sci. **10**(18), 6566 (2020). https://doi.org/10.3390/app10186566
32. Fang, Y., et al.: Clustering the learning patterns of adults with low literacy skills interacting with an intelligent tutoring system. In: Proceedings of the 11th International Conference on Educational Data Mining (EDM), pp. 348–354. ERIC (2018)
33. Mai, T.T., Bezbradica, M., Crane, M.: Learning behaviours data in programming education: community analysis and outcome prediction with cleaned data. Futur. Gener. Comput. Syst. **127**, 42–55 (2022). https://doi.org/10.1016/j.future.2021.08.026
34. Varela, N., et al.: Student performance assessment using clustering techniques. In: Proceedings of the International Conference on Data Mining and Big Data, pp. 179–188 (2019). https://doi.org/10.1007/978-981-32-9563-6_19
35. Zhang, S., Shen, M., Yu, Y.: Research on student big data portrait method based on improved k-means algorithm. In Proceedings of the 3rd International Academic Exchange Conference on Science and Technology Innovation (IAECST), pp. 146–150 (2021). https://doi.org/10.1109/IAECST54258.2021.9695501
36. Ding, D., Li, J., Wang, H., Liang, Z.: Student behavior clustering method based on campus big data. In: Proceedings of the 13th International Conference on Computational Intelligence and Security (CIS), pp. 500–503 (2017). https://doi.org/10.1109/CIS.2017.00116
37. Waspada, I., Bahtiar, N., Wibowo, A.: Clustering student behavior based on quiz activities on moodle lms to discover the relation with a final exam score. J. Phys. Conf. Ser. **1217**, 012118 (2019). https://doi.org/10.1088/1742-6596/1217/1/012118
38. Esnashari, S., Gardner, L., Watters, P.: Clustering student participation: implications for education. In: Proceedings of the 32nd International Conference on Advanced Information Networking and Applications Workshops (WAINA), pp. 313–318 (2018). https://doi.org/10.1109/WAINA.2018.00104
39. Jia, L., Cheng, H.N., Liu, S., Chang, W.C., Chen, Y., Sun, J.: Integrating clustering and sequential analysis to explore students' behaviors in an online Chinese reading assessment system. In: Proceedings of the 6th IIAI International Congress on Advanced Applied Informatics (IIAI-AAI), pp. 719–724 (2017). https://doi.org/10.1109/IIAI-AAI.2017.55
40. Howlin, C.P., Dziuban, C.D.: Detecting outlier behaviors in student progress trajectories using a repeated fuzzy clustering approach. In: Proceedings of the 12th International Conference on Educational Data Mining (EDM), pp. 742–747 (2019)
41. McBroom, J., Yacef, K., Koprinska, I.: DETECT: a hierarchical clustering algorithm for behavioural trends in temporal educational data. In: Proceedings of the International Conference on Artificial Intelligence in Education (AIED), pp. 374–385 (2020). https://doi.org/10.1007/978-3-030-52237-7_30
42. Shen, S., Chi, M.: Clustering student sequential trajectories using dynamic time warping. In: Proceedings of the 10th International Conference on Educational Data Mining (EDM), pp. 266–271 (2017)
43. Ruipérez-Valiente, J.A., Muñoz-Merino, P.J., Delgado Kloos, C., et al.: Detecting and clustering students by their gamification behavior with badges: a case study in engineering education. Int. J. Eng. Educ. **33**(2-B), 816–830 (2017)
44. López, S.L.S., Redondo, R.P.D., Vilas, A.F.: Discovering knowledge from student interactions: clustering vs classification. In: Proceedings of the 5th International Conference on Technological Ecosystems for Enhancing Multiculturality, pp. 1–8 (2017). https://doi.org/10.1145/3144826.3145390
45. Mengoni, P., Milani, A., Li, Y.: Clustering students interactions in e-learning systems for group elicitation. In: Proceedings of the International Conference on Computational Science and Its Applications, pp. 398–413. Springer (2018). https://doi.org/10.1007/978-3-319-95168-3_27

46. Orji, F., Vassileva, J.: Using machine learning to explore the relation between student engagement and student performance. In: Proceedings of the 24th International Conference Information Visualisation (IV), pp. 480–485. IEEE (2020). https://doi.org/10.1109/IV51561.2020.00083

47. Güvenç, E., Çetin, G.: Clustering of participation degrees of distance learning students to course activity by using fuzzy c-means algorithm. In: Proceedings of the 26th Signal Processing and Communications Applications Conference (SIU), pp. 1–4 (2018). https://doi.org/10.1109/SIU.2018.8404292

48. Khalil, M., Ebner, M.: Clustering patterns of engagement in massive open online courses (MOOCs): the use of learning analytics to reveal student categories. J. Comput. High. Educ. 29(1), 114–132 (2017). https://doi.org/10.1007/s12528-016-9126-9

49. Oladipupo, O.O., Olugbara, O.O.: Evaluation of data analytics based clustering algorithms for knowledge mining in a student engagement data. Intell. Data Anal. 23(5), 1055–1071 (2019). https://doi.org/10.3233/IDA-184254

50. Palani, K., Stynes, P., Pathak, P.: Clustering techniques to identify low-engagement student levels. In: Proceedings of the 13th International Conference on Computer Supported Education (CSEDU), pp. 248–257 (2021). https://doi.org/10.5220/0010456802480257

51. Roy, D., Bermel, P., Douglas, K.A., Diefes-Dux, H.A., Richey, M., Madhavan, K., Shah, S.: Synthesis of clustering techniques in educational data mining. In: Proceedings of the ASEE Annual Conference & Exposition (2017)

52. Huang, J.B., Huang, A.Y., Lu, O.H., Yang, S.J.: Exploring learning strategies by sequence clustering and analysing their correlation with student's engagement and learning outcome. In: Proceedings of the International Conference on Advanced Learning Technologies (ICALT), pp. 360–362. IEEE (2021). https://doi.org/10.1109/ICALT52272.2021.00115

53. Moubayed, A., Injadat, M., Shami, A., Lutfiyya, H.: Student engagement level in an e-learning environment: clustering using k-means. Am. J. Dist. Educ. 34(2), 137–156 (2020). https://doi.org/10.1080/08923647.2020.1696140

54. Hartnett, M.: The importance of motivation in online learning. In: Motivation in Online Education, pp. 5–32. Springer (2016). https://doi.org/10.1007/978-981-10-0700-2_2

55. Nen-Fu, H., et al.: The clustering analysis system based on students' motivation and learning behavior. In: Proceedings of the Learning with MOOCS (LWMOOCS), pp. 117–119 (2018). https://doi.org/10.1109/LWMOOCS.2018.8534611

56. Gunawan, I., et al.: Hidden curriculum and character building on self-motivation based on k-means clustering. In: Proceedings of the 4th International Conference on Education and Technology (ICET), pp. 32–35 (2018). https://doi.org/10.1109/ICEAT.2018.8693931

57. Wang, Z., Wang, J.: Analysis of emotional education infiltration in college physical education based on emotional feature clustering. Wirel. Commun. Mob. Comput. 2022 (2022). https://doi.org/10.1155/2022/7857522

58. Ashkanasy, N.M.: Emotion and performance. Human Perform. 17(2), 137–144 (2004). https://doi.org/10.1207/s15327043hup1702_1

59. Muñoz-Merino, P.J., Molina, M.F., Muñoz-Organero, M., Kloos, C.D.: Motivation and emotions in competition systems for education: an empirical study. IEEE Trans. Educ. 57(3), 182–187 (2014). https://doi.org/10.1109/TE.2013.2297318

60. Guo, H., Wang, M.: Analysis on the penetration of emotional education in college physical education based on emotional feature clustering. Sci. Program. 2022 (2022). https://doi.org/10.1155/2022/2389453

61. Salwana, E., Hamid, S., Yasin, N.M.: Student academic streaming using clustering technique. Malays. J. Comput. Sci. 30(4), 286–299 (2017). https://doi.org/10.22452/mjcs.vol30no4.2

62. Thilagaraj, T., Sengottaiyan, N.: Implementation of fuzzy clustering algorithms to analyze students performance using R-tool. In: Intelligent Computing and Innovation on Data Science, pp. 287–294. Springer, Berlin (2020). https://doi.org/10.1007/978-981-15-3284-9_31

63. Vo, C.T.N., Nguyen, P.H.: A weighted object-cluster association-based ensemble method for clustering undergraduate students. In: Proceedings of the Asian Conference on Intelligent

Information and Database Systems (ACIIDS), pp. 587–598 (2018). https://doi.org/10.1007/978-3-319-75417-8_55

64. Bharara, S., Sabitha, S., Bansal, A.: Application of learning analytics using clustering data mining for students' disposition analysis. Educ. Inf. Technol. **23**(2), 957–984 (2018). https://doi.org/10.1007/s10639-017-9645-7

65. Yin, X.: Construction of student information management system based on data mining and clustering algorithm. Complexity. **2021** (2021). https://doi.org/10.1155/2021/4447045

66. Hooshyar, D., Pedaste, M., Yang, Y.: Mining educational data to predict students' performance through procrastination behavior. Entropy. **22**(1), 12 (2019). https://doi.org/10.3390/e22010012

67. Park, J., Yu, R., Rodriguez, F., Baker, R., Smyth, P., Warschauer, M.: Understanding student procrastination via mixture models. In: Proceedings of the 11th International Conference on Educational Data Mining (EDM), pp 187–197 (2018)

68. Preetha, V.: Data analysis on student's performance based on health status using genetic algorithm and clustering algorithms. In: Proceedings of the 5th International Conference on Computing Methodologies and Communication (ICCMC), pp. 836–842 (2021). https://doi.org/10.1109/ICCMC51019.2021.9418235

69. Aghababyan, A., Lewkow, N., Baker, R.S.: Enhancing the clustering of student performance using the variation in confidence. In: Proceedings of the International Conference on Intelligent Tutoring Systems, pp. 274–279 (2018). https://doi.org/10.1007/978-3-319-91464-0_27

70. Effenberger, T., Pelánek, R.: Interpretable clustering of students' solutions in introductory programming. In: Proceedings of the International Conference on Artificial Intelligence in Education (AIED), pp. 101–112 (2021). https://doi.org/10.1007/978-3-030-78292-4_9

71. Gao, L., Wan, B., Fang, C., Li, Y., Chen, C.: Automatic clustering of different solutions to programming assignments in computing education. In: Proceedings of the ACM Conference on Global Computing Education, pp. 164–170 (2019). https://doi.org/10.1145/3300115.3309515

72. Chang, L.H., Rastas, I., Pyysalo, S., Ginter, F.: Deep learning for sentence clustering in essay grading support. In: The 14th International Conference on Educational Data Mining (EDM) (2021)

73. Sobral, S.R., de Oliveira, C.F.: Clustering algorithm to measure student assessment accuracy: a double study. Big Data Cognit. Comput. **5**(4), 81 (2021). https://doi.org/10.3390/bdcc5040081

74. Khan, A., Ghosh, S.K.: Student performance analysis and prediction in classroom learning: a review of educational data mining studies. Educ. Inf. Technol. **26**(1), 205–240 (2021). https://doi.org/10.1007/s10639-020-10230-3

75. Adjei, S., Ostrow, K., Erickson, E., Heffernan, N.T.: Clustering students in assistments: exploring system-and school-level traits to advance personalization. In: Proceedings of the 10th International Conference on Educational Data Mining (EDM), pp. 340–341 (2017)

76. Ramanathan, L., Parthasarathy, G., Vijayakumar, K., Lakshmanan, L., Ramani, S.: Cluster-based distributed architecture for prediction of student's performance in higher education. Clust. Comput. **22**(1), 1329–1344 (2019). https://doi.org/10.1007/s10586-017-1624-7

77. Hassan, Y.M., Elkorany, A., Wassif, K.: Utilizing social clustering-based regression model for predicting student's GPA. IEEE Access. **10**, 48948–48963 (2022). https://doi.org/10.1109/ACCESS.2022.3172438

78. Casalino, G., Castellano, G., Mencar, C.: Incremental and adaptive fuzzy clustering for virtual learning environments data analysis. In: Proceedings of the 23rd International Conference Information Visualisation (IV), pp. 382–387 (2019). https://doi.org/10.1109/IV.2019.00071

79. Almasri, A., Alkhawaldeh, R.S., Çelebi, E.: Clustering-based EMT model for predicting student performance. Arab. J. Sci. Eng. **45**(12), 10067–10078 (2020). https://doi.org/10.1007/s13369-020-04578-4

80. Iatrellis, O., Savvas, I.K., Fitsilis, P., Gerogiannis, V.C.: A two-phase machine learning approach for predicting student outcomes. Educ. Inf. Technol. **26**(1), 69–88 (2021). https://doi.org/10.1007/s10639-020-10260-x

81. Francis, B.K., Babu, S.S.: Predicting academic performance of students using a hybrid data mining approach. J. Med. Syst. **43**(6), 1–15 (2019). https://doi.org/10.1007/s10916-019-1295-4

82. Chu, Y.W., Tenorio, E., Cruz, L., Douglas, K., Lan, A.S., Brinton, C.G.: Click-based student performance prediction: a clustering guided meta-learning approach. In: Proceedings of the IEEE International Conference on Big Data (BigData), pp. 1389–1398 (2021). https://doi.org/10.1109/BigData52589.2021.9671729

83. Iam-On, N., Boongoen, T.: Generating descriptive model for student dropout: a review of clustering approach. HCIS. **7**(1), 1–24 (2017). https://doi.org/10.1186/s13673-016-0083-0

84. Iam-On, N., Boongoen, T.: Improved student dropout prediction in Thai university using ensemble of mixed-type data clusterings. Int. J. Mach. Learn. Cybern. **8**(2), 497–510 (2017). https://doi.org/10.1007/s13042-015-0341-x

85. Purba, W., Tamba, S., Saragih, J.: The effect of mining data k-means clustering toward students profile model drop out potential. J. Phys. Conf. Ser. **1007**, 012049 (2018). https://doi.org/10.1088/1742-6596/1007/1/012049

86. Hung, J.-L., Wang, M.C., Wang, S., Abdelrasoul, M., Li, Y., He, W.: Identifying at-risk students for early interventions—a time-series clustering approach. IEEE Trans. Emerg. Top. Comput. **5**(1), 45–55 (2017). https://doi.org/10.1109/TETC.2015.2504239

87. Nguyen, P., Vo, C.: Early in-trouble student identification based on temporal educational data clustering. In: Proceedings of the International Conference on Information Technology (ICIT), pp. 313–318 (2019). https://doi.org/10.1109/ICIT48102.2019.00062

88. Yotaman, N., Osathanunkul, K., Khoenkaw, P., Pramokchon, P.: Teaching support system by clustering students according to learning styles. In: Proceedings of the Joint International Conference on Digital Arts, Media and Technology with ECTI Northern Section Conference on Electrical, Electronics, Computer and Telecommunications Engineering (ECTI DAMT & NCON), pp. 137–140 (2020). https://doi.org/10.1109/ECTIDAMTNCON48261.2020.9090729

89. Khayi, N.A., Rus, V.: Clustering students based on their prior knowledge. In: Proceedings of the 12th International Conference on Educational Data Mining (EDM), pp. 246–251 (2019)

90. Qoiriah, A., et al.: Application of k-means algorithm for clustering student's computer programming performance in automatic programming assessment tool. In: Proceedings of the International Joint Conference on Science and Engineering (IJCSE 2020), pp. 421–425 (2020). https://doi.org/10.2991/aer.k.201124.075

91. Silva, D.B., Silla, C.N.: Evaluation of students programming skills on a computer programming course with a hierarchical clustering algorithm. In: Proceedings of the IEEE Frontiers in Education Conference (FIE), pp. 1–9 (2020). https://doi.org/10.1109/FIE44824.2020.9274130

92. Urbina Nájera, A.B., De La Calleja, J., Medina, M.A.: Associating students and teachers for tutoring in higher education using clustering and data mining. Comput. Appl. Eng. Educ. **25**(5), 823–832 (2017). https://doi.org/10.1002/cae.21839

93. Chang, M.H., Kuo, R., Essalmi, F., Chang, M., Kumar, V., Kung, H.Y.: Usability evaluation plan for online annotation and student clustering system—a tunisian university case. In: Proceedings of the International Conference on Digital Human Modeling and Applications

in Health, Safety, Ergonomics and Risk Management, pp. 241–254 (2017). https://doi.org/10.1007/978-3-319-58463-8_21

94. Kylvaja, M., Kumpulainen, P., Konu, A.: Application of data clustering for automated feedback generation about student Well-being. In: Proceedings of the 1st ACM SIGSOFT International Workshop on Education Through Advanced Software Engineering and Artificial Intelligence, pp. 21–26 (2019. https://doi.org/10.1145/3340435.3342720

95. Li, Y., Sun, X.: Data analysis and feedback system construction of university students' psychological fitness based on fuzzy clustering. Wirel. Commun. Mob. Comput. **2022** (2022). https://doi.org/10.1155/2022/6019803

96. Gulwani, S., Radiček, I., Zuleger, F.: Automated clustering and program repair for introductory programming assignments. ACM SIGPLAN Not. **53**(4), 465–480 (2018). https://doi.org/10.1145/3296979.3192387

97. Masala, M., Ruseti, S., Dascalu, M., Dobre, C.: Extracting and clustering main ideas from student feedback using language models. In: Proceedings of the International Conference on Artificial Intelligence in Education (AIED), pp. 282–292 (2021). https://doi.org/10.1007/978-3-030-78292-4_23

98. Guo, Y., Chen, Y., Xie, Y., Ban, X.: An effective student grouping and course recommendation strategy based on big data in education. Information. **13**(4), 197 (2022). https://doi.org/10.3390/info13040197

99. Wang, M., Lv, Z.: Construction of personalized learning and knowledge system of chemistry specialty via the internet of things and clustering algorithm. J. Supercomput. **78**(8), 10997–11014 (2022). https://doi.org/10.1007/s11227-022-04315-8

100. Liu, H., Ding, J., Yang, L.T., Guo, Y., Wang, X., Deng, A.: Multi-dimensional correlative recommendation and adaptive clustering via incremental tensor decomposition for sustainable smart education. IEEE Trans. Sustainable Comput. **5**(3), 389–402 (2019). https://doi.org/10.1109/TSUSC.2019.2954456

101. Fasanya, B. K., & Fathizadeh, M.: Clustering from grouping: a key to enhance students' classroom active engagement. In: 2019 ASEE Annual Conference & Exposition (2019). https://doi.org/10.18260/1-2-32511

102. Wu, Y., Nouri, J., Li, X., Weegar, R., Afzaal, M., Zia, A.: A word embeddings based clustering approach for collaborative learning group formation. In: Proceedings of the International Conference on Artificial Intelligence in Education (AIED), pp. 395–400 (2021). https://doi.org/10.1007/978-3-030-78270-2_70

103. Pratiwi, O.N., Rahardjo, B., Supangkat, S.H.: Clustering multiple mix data type for automatic grouping of student system. In: Proceedings of the International Conference on Information Technology Systems and Innovation (ICITSI), pp. 172–176 (2017). https://doi.org/10.1109/ICITSI.2017.8267938

104. Shelly, Z., Burch, R.F., Tian, W., Strawderman, L., Piroli, A., Bichey, C.: Using k-means clustering to create training groups for elite American football student-athletes based on game demands. Int. J. Kinesiol. Sports Sci. **8**(2), 47–63 (2020). https://doi.org/10.7575//aiac.ijkss.v.8n.2p.47

105. Akbar, S., Gehringer, E., Hu, Z.: Poster: improving formation of student teams: a clustering approach. In: Proceedings of the IEEE/ACM 40th International Conference on Software Engineering: Companion (ICSE-Companion), pp. 147–148 (2018)

106. Wang, Y., Wang, Q.: A student grouping method for massive online collaborative learning. Int. J. Emerg. Technol. Learn. **17**(3), 18–33 (2022). https://doi.org/10.3991/ijet.v17i03.29429

107. Yang, Y.: Evaluation model and application of college students' physical fitness based on clustering extraction algorithm. In: Proceedings of the 4th International Conference on Information Systems and Computer Aided Education, pp. 547–552 (2021). https://doi.org/10.1145/3482632.3482748

108. Dovgan, E., Leskošek, B., Jurak, G., Starc, G., Sorić, M., Luštrek, M.: Enhancing BMI-based student clustering by considering fitness as key attribute. In: Proceedings of the International

Conference on Discovery Science, pp. 155–165 (2019). https://doi.org/10.1007/978-3-030-33778-0_13

109. Natilli, M., Monreale, A., Guidotti, R., Pappalardo, L.: Exploring students eating habits through individual profiling and clustering analysis. In: Proceedings of the MIDAS/PAP@PKDD/ECML 2018, pp. 156–171 (2018). https://doi.org/10.1007/978-3-030-13463-1_12

110. Chu, Y., Yin, X.: Data analysis of college students' mental health based on clustering analysis algorithm. Complexity. **2021** (2021). https://doi.org/10.1155/2021/9996146

111. Li, Y., Liu, C., Zhao, X.: Research on the integration of college students' mental health education and career planning based on feature fuzzy clustering. In: Proceedings of the 4th International Conference on Information Systems and Computer Aided Education, pp. 56–59 (2021). https://doi.org/10.1145/3482632.3482644

112. Wang, C., Zha, Q.: Measuring systemic diversity of Chinese universities: a clustering-method approach. Qual. Quant. **52**(3), 1331–1347 (2018). https://doi.org/10.1007/s11135-017-0524-5

113. Nazaretsky, T., Hershkovitz, S., Alexandron, G.: Kappa learning: a new item-similarity method for clustering educational items from response data. In: Proceedings of the 12th International Conference on Educational Data Mining (EDM), pp 129–138 (2019)

114. Huang, L., Wang, X., Wu, Z., Wang, F.: Feature selection for clustering online learners. In: Proceedings of the 8th International Conference on Educational Innovation Through Technology (EITT), pp. 1–6 (2019). https://doi.org/10.1109/EITT.2019.00009

115. Liu, F.: Design and implementation of intelligent educational administration system using fuzzy clustering algorithm. Sci. Program. **2021** (2021). https://doi.org/10.1155/2021/9485654

116. Rahmat, A.: Clustering in education. Eur. Res. Stud. J. **20**(3) (2017)

117. Ahmed, A., Zualkernan, I., Elghazaly, H.: Unsupervised clustering of skills for an online learning platform. In: Proceedings of the International Conference on Advanced Learning Technologies (ICALT), pp. 200–202 (2021). https://doi.org/10.1109/ICALT52272.2021.00066

118. Pamungkas, A.A.P., Maryono, D., Budiyanto, C.W.: Cluster analysis for student grouping based on index of learning styles. J. Phys. Conf. Ser. **1808**, 012023 (2021). https://doi.org/10.1088/1742-6596/1808/1/012023

119. Du, H., Chen, S., Niu, H., Li, Y.: Application of dbscan clustering algorithm in evaluating students' learning status. In: Proceedings of the 17th International Conference on Computational Intelligence and Security (CIS), pp. 372–376 (2021). https://doi.org/10.1109/CIS54983.2021.00084

120. MacQueen, J.: Some methods for classification and analysis of multivariate observations. In: Proceedings of the 5th Berkeley Symposium on Math., Stat., and Prob, p. 281 (1965). http://projecteuclid.org/euclid.bsmsp/1200512992

121. Lloyd, S.: Least squares quantization in PCM. IEEE Trans. Inf. Theory. **28**(2), 129–137 (1982). https://doi.org/10.1109/TIT.1982.1056489

122. Jain, A.K., Dubes, R.C.: Algorithms for Clustering Data. Prentice-Hall, Inc., Hoboken (1988). https://doi.org/10.1080/00401706.1990.10484648

123. Pelleg, D., Moore, A.W., et al.: X-means: extending k-means with efficient estimation of the number of clusters. In: Proceedings of the International Conference on Machine Learning (ICML), vol. 1, pp. 727–734 (2000)

124. Li, X., Zhang, Y., Cheng, H., Zhou, F., Yin, B.: An unsupervised ensemble clustering approach for the analysis of student behavioral patterns. IEEE Access. **9**, 7076–7091 (2021). https://doi.org/10.1109/ACCESS.2021.3049157

125. Zhang, T., Yin, C., Pan, L.: Improved clustering and association rules mining for university student course scores. In: Proceedings of the 12th International Conference on Intelligent Systems and Knowledge Engineering (ISKE), pp. 1–6 (2017). https://doi.org/10.1109/ISKE.2017.8258808

126. Kaufman, L., Rousseeuw, P.J.: Finding Groups in Data: An Introduction to Cluster Analysis. John Wiley & Sons, New York (1990). https://doi.org/10.1002/9780470316801

127. Schubert, E., Rousseeuw, P.J.: Fast and eager k-medoids clustering: O(k) runtime improvement of the PAM, CLARA, and CLARANS algorithms. Inf. Syst. **101**, 101804 (2021). https://doi.org/10.1016/j.is.2021.101804

128. Vasuki, M., Revathy, S.: Analyzing performance of placement students record using different clustering algorithm. Indian J. Comput. Sci. Eng. **13**(2), 410–419 (2022). https://doi.org/10.21817/indjcse/2022/v13i2/221302083

129. Furr, D.: Visualization and clustering of learner pathways in an interactive online learning environment. In: Proceedings of the 12th International Conference on Educational Data Mining (EDM) (2019)

130. Kausar, S., Huahu, X., Hussain, I., Wenhao, Z., Zahid, M.: Integration of data mining clustering approach in the personalized e-learning system. IEEE Access. **6**, 72724–72734 (2018). https://doi.org/10.1109/ACCESS.2018.2882240

131. Patel, S., Sihmar, S., Jatain, A.: A study of hierarchical clustering algorithms. In: Proceedings of the 2nd International Conference on Computing for Sustainable Global Development (INDIACom), pp. 537–541 (2015)

132. Zhang, T., Ramakrishnan, R., Livny, M.: BIRCH: an efficient data clustering method for very large databases. ACM SIGMOD Rec. **25**(2), 103–114 (1996). https://doi.org/10.1145/235968.233324

133. Ward Jr., J.H.: Hierarchical grouping to optimize an objective function. J. Am. Stat. Assoc. **58**(301), 236–244 (1963). https://doi.org/10.1080/01621459.1963.10500845

134. Li, S., Chen, G., Xing, W., Zheng, J., Xie, C.: Longitudinal clustering of students' self-regulated learning behaviors in engineering design. Comput. Educ. **153**, 103899 (2020). https://doi.org/10.1016/j.compedu.2020.103899

135. Zhang, T., Taub, M., Chen, Z.: A multi-level trace clustering analysis scheme for measuring students' self-regulated learning behavior in a mastery-based online learning environment. In: Proceedings of the 12th International Learning Analytics and Knowledge Conference (LAK), pp. 197–207 (2022). https://doi.org/10.1145/3506860.3506887

136. Dunn, J.C.: A fuzzy relative of the ISODATA process and its use in detecting compact well-separated clusters. J. Cybernet. **3**(3), 32–57 (1973). https://doi.org/10.1080/01969727308546046

137. Zhang, P., Shen, Q.: Fuzzy c-means based coincidental link filtering in support of inferring social networks from spatiotemporal data streams. Soft. Comput. **22**(21), 7015–7025 (2018). https://doi.org/10.1007/s00500-018-3363-y

138. Tang, Q., Zhao, Y., Wei, Y., Jiang, L.: Research on the mental health of college students based on fuzzy clustering algorithm. Secur. Commun. Net. **2021** (2021). https://doi.org/10.1155/2021/3960559

139. Amalia, N., et al.: Determination system of single tuition group using a combination of fuzzy c-means clustering and simple additive weighting methods. In: IOP Conference Series: Materials Science and Engineering, vol. 536, p. 012148 (2019). https://doi.org/10.1088/1757-899X/536/1/012148

140. Dempster, A.P., Laird, N.M., Rubin, D.B.: Maximum likelihood from incomplete data via the em algorithm. J. R. Stat. Soc. Ser. B. **39**(1), 1–22 (1977). https://doi.org/10.1111/j.2517-6161.1977.tb01600.x

141. Jin, X., Han, J.: In: Sammut, C., Webb, G.I. (eds.) Expectation Maximization Clustering. Springer US, Boston, MA (2010). https://doi.org/10.1007/978-0-387-30164-8_289

142. Tan, P.N., Steinbach, M., Kumar, V.: Introduction to Data Mining. Pearson Education India (2016)

143. Kohonen, T.: Self-organized formation of topologically correct feature maps. Biol. Cybern. **43**(1), 59–69 (1982). https://doi.org/10.1007/BF00337288

144. Bação, F., Lobo, V., Painho, M.: Self-organizing maps as substitutes for k-means clustering. In: Proceedings of the International Conference on Computational Science, pp. 476–483 (2005). https://doi.org/10.1007/11428862_65

145. Natita, W., Wiboonsak, W., Dusadee, S.: Appropriate learning rate and neighborhood function of self-organizing map (SOM) for specific humidity pattern classification over southern Thailand. Int. J. Model. Optimiz. **6**(1), 61 (2016). https://doi.org/10.7763/IJMO.2016.V6.504

146. Melka, J., Mariage, J.J.: Efficient implementation of self-organizing map for sparse input data. In: International Joint Conference on Computational Intelligence (IJCCI), pp. 54–63 (2017). https://doi.org/10.5220/0006499500540063

147. Delgado, S., Morán, F., San José, J.C., Burgos, D.: Analysis of students' behavior through user clustering in online learning settings, based on self organizing maps neural networks. IEEE Access. **9**, 132592–132608 (2021). https://doi.org/10.1109/ACCESS.2021.3115024

148. Tasdemir, K., Merényi, E.: A validity index for prototype-based clustering of data sets with complex cluster structures. IEEE Trans. Syst. Man Cybern. B Cybern. **41**(4), 1039–1053 (2011). https://doi.org/10.1109/TSMCB.2010.2104319

149. Alias, U.F., Ahmad, N.B., Hasan, S.: Mining of e-learning behavior using SOM clustering. In: Proceedings of the 6th ICT International Student Project Conference (ICT-ISPC), pp. 1–4 (2017). https://doi.org/10.1109/ICT-ISPC.2017.8075350

150. Bara, M.W., Ahmad, N.B., Modu, M.M., Ali, H.A.: Self-organizing map clustering method for the analysis of e-learning activities. In: Majan International Conference (MIC), pp. 1–5 (2018). https://doi.org/10.1109/MINTC.2018.8363155

151. Ahmad, N.B., Alias, U.F., Mohamad, N., Yusof, N.: Principal component analysis and self-organizing map clustering for student browsing behaviour analysis. Procedia Comput. Sci. **163**, 550–559 (2019). https://doi.org/10.1016/j.procs.2019.12.137

152. Ng, A., Jordan, M., Weiss, Y.: On spectral clustering: analysis and an algorithm. Adv. Neural Inf. Proces. Syst. **14** (2001)

153. Shi, J., Malik, J.: Normalized cuts and image segmentation. IEEE Trans. Pattern Anal. Mach. Intell. **22**(8), 888–905 (2000). https://doi.org/10.1109/34.868688

154. Yan, D., Huang, L., Jordan, M.I.: Fast approximate spectral clustering. In: Proceedings of the 15th ACM SIGKDD International Conference on Knowledge Discovery and Data Mining (KDD), pp. 907–916 (2009)

155. Ester, M., Kriegel, H.P., Sander, J., Xu, X., et al.: A density-based algorithm for discovering clusters in large spatial databases with noise. In: Proceedings of the ACM SIGKDD International Conference on Knowledge Discovery and Data Mining (KDD), vol. 96, pp. 226–231 (1996)

156. Chhabra, A., Masalkovaite, K., Mohapatra, P.: An overview of fairness in clustering. IEEE Access. (2021). https://doi.org/10.1109/ACCESS.2021.3114099

157. Žliobaitė, I.: Measuring discrimination in algorithmic decision making. Data Min. Knowl. Disc. **31**(4), 1060–1089 (2017). https://doi.org/10.1007/s10618-017-0506-1

158. Chierichetti, F., Kumar, R., Lattanzi, S., Vassilvitskii, S.: Fair clustering through fairlets. In: Neural Information Processing Systems, pp. 5036–5044 (2017)

159. Ahmadian, S., Epasto, A., Kumar, R., Mahdian, M.: Clustering without over-representation. In: Proceedings of the 25th ACM SIGKDD International Conference on Knowledge Discovery and Data Mining (KDD), pp. 267–275 (2019). https://doi.org/10.1145/3292500.3330987

160. Bera, S., Chakrabarty, D., Flores, N., Negahbani, M.: Fair algorithms for clustering. In: Proceedings of the Neural Information Processing Systems Conference (NIPS 2019), p. 32 (2019)

161. Ghadiri, M., Samadi, S., Vempala, S.: Socially fair k-means clustering. In: Proceedings of the 2021 ACM Conference on Fairness, Accountability, and Transparency (ACM FAccT), pp. 438–448 (2021). https://doi.org/10.1145/3442188.3445906

162. Chakrabarti, D., Dickerson, J.P., Esmaeili, S.A., Srinivasan, A., Tsepenekas, L.: A new notion of individually fair clustering: α-equitable k-center. In: Proceedings of the International Conference on Artificial Intelligence and Statistics, pp. 6387–6408 (2022)

163. Jones, M., Nguyen, H., Nguyen, T.: Fair k-centers via maximum matching. In: Proceedings of the International Conference on Machine Learning (ICML), pp. 4940–4949 (2020)

164. Schmidt, M., Schwiegelshohn, C., Sohler, C.: Fair corescts and streaming algorithms for fair k-means. In: Proceedings of the International Workshop on Approximation and Online Algorithms, pp. 232–251 (2019). https://doi.org/10.1007/978-3-030-39479-0_16
165. Abraham, S.S., Padmanabhan, D., Sundaram, S.S.: Fairness in clustering with multiple sensitive attributes. In: EDBT/ICDT 2020 Joint Conference, pp. 287–298 (2020). https://doi.org/10.5441/002/edbt.2020.26
166. Xia, X., Hui, Z., Chunming, Y., Xujian, Z., Bo, L.: Fairness constraint of fuzzy c-means clustering improves clustering fairness. In: Proceedings of the Asian Conference on Machine Learning (ACML), pp. 113–128 (2021)
167. Ahmadian, S., et al.: Fair hierarchical clustering. Adv. Neural Inf. Proces. Syst. **33**, 21050–21060 (2020)
168. Kleindessner, M., Samadi, S., Awasthi, P., Morgenstern, J.: Guarantees for spectral clustering with fairness constraints. In: Proceedings of the International Conference on Machine Learning (ICML), pp. 3458–3467 (2019)
169. Battaglia, O.R., Di Paola, B., Fazio, C.: K-means clustering to study how student reasoning lines can be modified by a learning activity based on feynman's unifying approach. Eur. J. Math. Sci. Technol. Educ. **13**(6), 2005–2038 (2017). https://doi.org/10.12973/eurasia.2017.01211a
170. Maylawati, D.S., Priatna, T., Sugilar, H., Ramdhani, M.A.: Data science for digital culture improvement in higher education using k-means clustering and text analytics. Int. J. Electr. Comput. Eng. **10**(5), 2088–8708 (2020). https://doi.org/10.11591/ijece.v10i5.pp4569-4580
171. Šarić-Grgić, I., Grubišić, A., Šerić, L., Robinson, T.J.: Student clustering based on learning behavior data in the intelligent tutoring system. Int. J. Dist. Educ. Technol. **18**(2), 73–89 (2020). https://doi.org/10.4018/IJDET.2020040105
172. Talebinamvar, M., Zarrabi, F.: Clustering students' writing behaviors using keystroke logging: a learning analytic approach in efl writing. Lang. Test. Asia. **12**(1), 1–20 (2022). https://doi.org/10.1186/s40468-021-00150-5
173. Kurniawan, C., Setyosari, P., Kamdi, W., Ulfa, S.: Electrical engineering student learning preferences modelled using k-means clustering. Global J. Eng. Educ. **20**(2), 140–145 (2018)
174. Rijati, N., Sumpeno, S., Purnomo, M.H.: Multi-attribute clustering of student's entrepreneurial potential mapping based on its characteristics and the affecting factors: preliminary study on Indonesian higher education database. In: Proceedings of the 10th International Conference on Computer and Automation Engineering, pp. 11–16 (2018). https://doi.org/10.1145/3192975.3193014
175. Mishler, A., Nugent, R.: Clustering students and inferring skill set profiles with skill hierarchies. In: Proceedings of the 11th International Conference on Educational Data Mining (EDM) (2018)
176. Mojarad, S., Essa, A., Mojarad, S., Baker, R.S.: Data-driven learner profiling based on clustering student behaviors: learning consistency, pace and effort. In: Proceedings of the International Conference on Intelligent Tutoring Systems, pp. 130–139 (2018). https://doi.org/10.1007/978-3-319-91464-0_13
177. Davies, D.L., Bouldin, D.W.: A cluster separation measure. IEEE Trans. Pattern Anal. Mach. Intell. **2**, 224–227 (1979). https://doi.org/10.1109/TPAMI.1979.4766909
178. Dunn, J.C.: Well-separated clusters and optimal fuzzy partitions. J. Cybernet. **4**(1), 95–104 (1974). https://doi.org/10.1080/01969727408546059
179. Tempelaar, D., Rienties, B., Mittelmeier, J., Nguyen, Q.: Student profiling in a dispositional learning analytics application using formative assessment. Comput. Hum. Behav. **78**, 408–420 (2018). https://doi.org/10.1016/j.chb.2017.08.010
180. Švábensky`, V., Vykopal, J., Čeleda, P., Tkáčik, K., Popovič, D.: Student assessment in cybersecurity training automated by pattern mining and clustering. Educ. Inf. Technol. **1–32** (2022). https://doi.org/10.1007/s10639-022-10954-4
181. Bradley, P.S., Bennett, K.P., Demiriz, A.: Constrained K-Means Clustering, p. 20. Microsoft Research, Redmond (2000)

182. Mulvey, J.M., Beck, M.P.: Solving capacitated clustering problems. Eur. J. Oper. Res. **18**(3), 339–348 (1984). https://doi.org/10.1016/0377-2217(84)90155-3

183. Moshkovitz, M., Dasgupta, S., Rashtchian, C., Frost, N.: Explainable k-means and k-medians clustering. In: Proceedings of the International Conference on Machine Learning (ICML), pp. 7055–7065 (2020)

184. Bandyapadhyay, S., Fomin, F., Golovach, P.A., Lochet, W., Purohit, N., Simonov, K.: How to find a good explanation for clustering? In: Proceedings of the AAAI Conference on Artificial Intelligence, vol. 36, pp. 3904–3912 (2022). https://doi.org/10.1609/aaai.v36i4.20306

185. Wang, D.-Y., Lin, S.S., Sun, C.-T.: DIANA: a computer-supported heterogeneous grouping system for teachers to conduct successful small learning groups. Comput. Hum. Behav. **23**(4), 1997–2010 (2007). https://doi.org/10.1016/j.chb.2006.02.008

186. Watson, S.B., Marshall, J.E.: Heterogeneous grouping as an element of cooperative learning in an elementary education science course. Sch. Sci. Math. **95**(8), 401–405 (1995). https://doi.org/10.1111/j.1949-8594.1995.tb10192.x

187. Flanagan, B., Majumdar, R., Ogata, H.: Fine grain synthetic educational data: challenges and limitations of collaborative learning analytics. IEEE Access. **10**, 26230–26241 (2022). https://doi.org/10.1109/ACCESS.2022.3156073

188. Vie, J.-J., Rigaux, T., Minn, S.: Privacy-preserving synthetic educational data generation. In: Proceedings of the EC-TEL 2022 (2022)

189. Backurs, A., Indyk, P., Onak, K., Schieber, B., Vakilian, A., Wagner, T.: Scalable fair clustering. In: Proceedings of the International Conference on Machine Learning (ICML), pp. 405–413 (2019)

190. Fahad, A., et al.: A survey of clustering algorithms for big data: taxonomy and empirical analysis. IEEE Trans. Emerg. Top. Comput. **2**(3), 267–279 (2014). https://doi.org/10.1109/TETC.2014.2330519

191. Assent, I.: Clustering high dimensional data. Wires Data Mining Know. Discov. **2**(4), 340–350 (2012). https://doi.org/10.1002/widm.1062

192. Le Quy, T., Nguyen, T.H., Friege, G., Ntoutsi, E.: Evaluation of group fairness measures in student performance prediction problems. In: Machine Learning and Principles and Practice of Knowledge Discovery in Databases: International Workshops of ECML PKDD 2022, pp. 119–136 (2022). https://doi.org/10.1007/978-3-031-23618-1_8

193. Rihák, J., Pelánek, R.: Measuring similarity of educational items using data on learners' performance. In: Proceedings of the 10th International Conference on Educational Data Mining (EDM), pp. 16–23 (2017)

194. Ninrutsirikun, U., Watanapa, B., Arpnikanondt, C., Watananukoon, V.: A unified framework for student cluster grouping with learning preference associative detection for enhancing students' learning outcomes in computer programming courses. In: Proceedings of 2018 Global Wireless Summit (GWS), pp. 266–271 (2018). https://doi.org/10.1109/GWS.2018.8686665

195. Phanniphong, K., Nuankaew, P., Teeraputon, D., Nuankaew, W., Boontonglek, M., Bussaman, S.: Clustering of learners performance based on learning outcomes for finding significant courses. In: Proceedings of the Joint International Conference on Digital Arts, Media and Technology with ECTI Northern Section Conference on Electrical, Electronics, Computer and Telecommunications Engineering (ECTI DAMT-NCON), pp. 192–196 (2019). https://doi.org/10.1109/ECTI-NCON.2019.8692263

196. Wang, X., Zhang, Y., Yang, Y., Liu, K., Gao, B.: Research on relevance analysis and clustering algorithms in college students' academic performance. In: Proceedings of the 10th International Conference on Information Technology in Medicine and Education (ITME), pp. 730–733 (2019). https://doi.org/10.1109/ITME.2019.00167

197. Chaves, V.E.J., García-Torres, M., Alonso, D.B., Gómez-Vela, F., Divina, F., Vázquez-Noguera, J.L.: Analysis of student achievement scores via cluster analysis. In: Proceedings of the International Conference on European Transnational Education, pp. 399–408 (2020). https://doi.org/10.1007/978-3-030-57799-5_41

198. Kosztyán, Z.T., Orbán-Mihálykó, É., Mihálykó, C., Csányi, V.V., Telcs, A.: Analyzing and clustering students' application preferences in higher education. J. Appl. Stat. **47**(16), 2961–2983 (2020). https://doi.org/10.1080/02664763.2019.1709052

199. Pradana, C., Kusumawardani, S., Permanasari, A.: Comparison clustering performance based on moodle log mining. IOP Conf. Ser. Mater. Sci. Eng. **722**, 012012 (2020). https://doi.org/10.1088/1757-899X/722/1/012012

200. Tang, P., Wang, Y., Shen, N.: Prediction of college students' physical fitness based on k-means clustering and SVR. Comput. Syst. Sci. Eng. **35**(4), 237–246 (2020). https://doi.org/10.32604/csse.2020.35.237

201. Rijati, N., Purwitasari, D., Sumpeno, S., Purnomo, M.: A decision making and clustering method integration based on the theory of planned behavior for student entrepreneurial potential mapping in Indonesia. Int. J. Intell. Eng. Syst. **13**(4), 129–144 (2020). https://doi.org/10.22266/ijies2020.0831.12

202. Chi, D.: Research on the application of k-means clustering algorithm in student achievement. In: Proceedings of the IEEE International Conference on Consumer Electronics and Computer Engineering (ICCECE), pp. 435–438 (2021). https://doi.org/10.1109/ICCECE51280.2021.9342164

203. Li, G., Alfred, R., Wang, X.: Student behavior analysis and research model based on clustering technology. Mob. Inf. Syst. **2021** (2021). https://doi.org/10.1155/2021/9163517

204. Putra, A.A.N.K., Nasucha, M., Hermawan, H.: K-means clustering algorithm in web-based applications for grouping data on scholarship selection results. In: Proceedings of the International Symposium on Electronics and Smart Devices (ISESD), pp. 1–6 (2021). https://doi.org/10.1109/ISESD53023.2021.9501716

205. Susanto, R., Husen, M.N., Lajis, A., Lestari, W., Hasanah, H.: Clustering of student perceptions on developing a physics laboratory based on information technology and local wisdom. In: Proceedings of the 8th International Conference on Information Technology, Computer and Electrical Engineering (ICITACEE), pp. 68–73 (2021). https://doi.org/10.1109/ICITACEE53184.2021.9617483

206. Rauthan, A., et al.: Impact on higher education in pandemic: analysis k-means clustering using urban & rural areas. In: Proceedings of the 3rd International Conference on Advances in Computing, Communication Control and Networking (ICAC3N), pp. 1974–1980 (2021). https://doi.org/10.1109/ICAC3N53548.2021.9725709

207. Wang, Q.: Application of the intra cluster, characteristic of k-means clustering method in English score analysis in colleges. J. Phys. Conf. Ser. **1941**, 012001 (2021). https://doi.org/10.1088/1742-6596/1941/1/012001

208. Cheng, W., Shwe, T.: Clustering analysis of student learning outcomes based on education data. In: 2019 IEEE Frontiers in Education Conference (FIE), pp. 1–7 (2019). https://doi.org/10.1109/FIE43999.2019.9028400

209. Singelmann, L., Alvarez, E., Swartz, E., Pearson, M., Striker, R., Ewert, D.: Innovators, learners, and surveyors: clustering students in an innovation-based learning course. In: IEEE Frontiers in Education Conference (FIE), pp. 1–9 (2020). https://doi.org/10.1109/FIE44824.2020.9274235

210. Popov, A., Ovsyankin, A., Emomaliev, M., Satsuk, M.: Application of the clustering algorithm in an automated training system. J. Phys. Conf. Ser. **1691**, 012120 (2020). https://doi.org/10.1088/1742-6596/1691/1/012120

211. Supianto, A.A., et al.: Improvements of fuzzy c-means clustering performance using particle swarm optimization on student grouping based on learning activity in a digital learning media. In: Proceedings of the 5th International Conference on Sustainable Information Engineering and Technology, pp. 239–243 (2020). https://doi.org/10.1145/3427423.3427449

212. Yadav, R.S.: Application of hybrid clustering methods for student performance evaluation. Int. J. Inf. Technol. **12**(3), 749–756 (2020). https://doi.org/10.1007/s41870-018-0192-2

213. Parvathavarthini, S., Sharvanthika, K., Jagadeesh, M., Kishore, B.: Analysis of student performance in e-learning environment using crow search based fuzzy clustering. In:

Proceedings of the 2nd International Conference on Smart Electronics and Communication (ICOSEC), pp. 1784–1787 (2021). https://doi.org/10.1109/ICOSEC51865.2021.9591920

214. Premalatha, N., Sujatha, S.: Prediction of students' employability using clustering algorithm: a hybrid approach. Int. J. Model. Simul. Sci. Comput. **2250049** (2022). https://doi.org/10.1142/S1793962322500490

215. Waluyo, E., Djeni, D., Pratama, L., Anggraini, V.: Clustering based on sociometry in Pythagoras theorem. J. Phys. Conf. Ser. **1211**, 012058 (2019). https://doi.org/10.1088/1742-6596/1211/1/012058

216. Purbasari, I., Puspaningrum, E., Putra, A.: Using self-organizing map (SOM) for clustering and visualization of new students based on grades. J. Phys. Conf. Ser. **1569**, 022037 (2020). https://doi.org/10.1088/1742-6596/1569/2/022037

217. Rakhmawati, N.A., Faiz, N., Hafidz, I., Raditya, I., Dinatha, P., Suwignyo, A.: Clustering student Instagram accounts using author-topic model. Int. J. Bus. Intell. Data Min. **19**(1), 70–79 (2021). https://doi.org/10.1504/IJBIDM.2021.115954

218. Yan, Q., Su, Z.: Evaluation of college students' English performance considering Roche multiway tree clustering. Int. J. Electric. Eng. Educ. (2021). https://doi.org/10.1177/00207209211004207

Chapter 3
Educational Data Science: An "Umbrella Term" or an Emergent Domain?

Alejandro Peña-Ayala

Abstract Among the recent domains specialized in the arena of tracking, examining, and interpreting educational big data (EBD), *educational data science* (EDS) emerges as a domain that fosters teaching and learning settings, particularly those that use computers and mobile devices linked to the Internet with the aim of adapting and personalizing educational practice according to learners' profiles. However, so far there is no a clear concept of what really is EDS. Hence, in order to give an answer to the question, this chapter shapes a *landscape of EDS* that covers from the background and baseline to the trends. In this pathway, a profile of some EDS-related works is outlined, and a *taxonomy of EDS* is proposed to organize EDS labor and shed light on the nature of the novel domain. As a result, one of the findings reveals EDS is coined from a dual view, where the first covers related fields and the second pursues the definition of its own identity to gain a distinctive place in the arena. Hence, one conclusion acknowledges the convenience of both concepts, *incl*usive and *exclusive*, for practical and technical purposes, respectively.

Keywords Educational data science · Data science · Big data · Teaching · Learning

Abbreviations

EBD Educational big data
EDM Educational data mining
EDS Educational data science

The original version of the chapter has been revised. A correction to this chapter can be found at https://doi.org/10.1007/978-981-99-0026-8_8

A. Peña-Ayala (✉)

Artificial Intelligence in Education Lab, WOLNM & Sección de Estudios de Posgrado e Investigación, ESIME-Z, Instituto Politécnico Nacional, CDMX, Mexico
e-mail: apenaa@ipn.mx

95
A. Peña-Ayala (ed.), *Educational Data Science: Essentials, Approaches, and Tendencies*, Big Data Management, https://doi.org/10.1007/978-981-99-0026-8_3

ITS   Intelligent tutoring systems
KDDB   Knowledge discovery in databases
LA   Learning analytics
ML   Machine learning
MOOC   Massive open online courses

3.1 Introduction

The growth of the *digital revolution* in human daily life [1] is featured by fresh facilities and functionalities, such as wireless Internet access across the world [2], intensive use of computers and smart mobile devices for all ages [3, 4], concurrent operation of distributed information systems and apps for any kind of purposes [5, 6], and the accelerated explosion and spread of multisource heterogeneous data and social media about all sorts of topics [7, 8].

Such a revolution recreates a *connected society* [9], thanks to a complex network of resources and the eruption of technologies that upgrade traditional ones, as those oriented to overcrowd data processing [10] through *cloud computing* [11], management of massive data [12] based on *big data* [13], and generation and visualization of relevant patterns and trends stemmed from databases [14] by means of *knowledge discovery in databases* (KDDB) [15], *data mining* [16], and *analytics* [17].

Deep in the outlined vortex, *data science* arises as a transdisciplinary field [18] to study the data DNA and scrutinize the phenomena and rules of the data themselves [19]. Data science strives for the systematic labor of the organization, characterization, and analysis of data, to produce inference [20]. In short, data science is the application of quantitative and qualitative methods to solve key problems and predict outcomes [21], whose goal is to improve decision-making based on insights extracted from huge data [22]. As data science is a term in "vogue," its application targets diverse fields, for example, science [23], healthcare [24], environmental sciences [25], management [26], business [27], politics [28], cybersecurity [29], biomedicine [30], finance [31], and many more fields including education [32].

In this context, *EDS* is a domain of research, development, and application in classroom and distance learning settings that benefits from the stated technologies. For example, data science offers frames and principles for extraction of information and knowledge from complex educational datasets [33]. Concerning big data, it is able to manage a vast volume of heterogeneous educational data that is generated in several formats at an alarming velocity from diverse sources and gathered through diverse channels as a digitized representation [34]. As for KDDB, it extracts main, implicit, and unknown knowledge from wide educational datasets by means of data collection, preparation, generation, interpretation, and exploitation of outcomes [35]. Briefly, EDS is a nascent, transdisciplinary field, relying on both data science practices and existing knowledge from the learning sciences [36]. Even though EDS is a novel logistic field, it has been used in diverse targets, such as learning design [37], virtual labs [38], social sciences [39], prediction of graduation [40], online education [41], and many more.

Nowadays, despite the wide diffusion of the buzzword *EDS* and the triggered expectations, its meaning is unclear and faces similar issues to those dealt by its root, data science. Provost and Fawcett [42] claim, "There is a confusion to say what exactly data science is, which yields to misconception as the term diffuses into meaningless buzz," whereas Irizarry [43] declares, "Data science has been defined as an *umbrella* term to depict the entire complex and multistep processes used to extract value from data" and Van der Laan and Rose [44] assert, "Data science is a rapidly evolving field with fuzzy boundaries to narrow the scope." Therefore, in the same way, the notion for EDS is seen from a twofold abstraction, *inclusive* to embrace affined domains and *exclusive* in pursuit of acquiring a proper identity that distinguishes it from other allied domains.

In this concern, the present study focuses on such a conceptual issue for EDS, which is an incipient field whose features are gradually emerging to tailor its own personality and little by little occupying a spot in the arena oriented to discover knowledge from educational big data (EBD). Hence, the *core problem* addressed in this research, whose goal is to facilitate the dissemination of EDS, is defined as follows: what really is EDS? Additionally, some questions are also made to deepen the study: (1) What about the research being accomplished in EDS? (2) What is the baseline that grounds EDS? (3) What are the main work lines that distinguish EDS endeavor? (4) What are the key features that profile EDS labor? (5) What are the expectations and shortcomings to be taken into account for boosting the progress of EDS?

With the purpose of contributing to outline the nature of EDS and answer the research inquiries, a solution is given: *to draw a conceptual landscape of the EDS field*. Hence, the deployment of the solution pursues the accomplishment of five goals: (1) to review a sample of relevant EDS-related works, (2) to figure out a way to organize EDS labor, (3) to feature the main properties of EDS approaches, (4) to draw a state of the art for EDS, (5) to clearly define the EDS nature. As a result of the study, four main outcomes are generated: (1) a sample of valuable EDS-related works, (2) a taxonomy for classifying EDS labor, (3) an analysis of the pros and cons of EDS, (4) a stand to approach the issue of the EDS nature.

The remainder of this chapter presents in Sect. 3.2 the settlement of the study, where the method, materials, and the taxonomy of EDS are described, while in Sect. 3.3 a profile of representative related works is edited according to the taxonomy of EDS. In Sect. 3.4 the discussion of the results is presented through an evaluation of the outcomes, an analysis of the EDS field, a comparison of the achievements to similar works, and responses to the research questions. Finally, in Sect. 3.5 the conclusions are uncovered to manifest the key achievements and shortages, as well as identify future work.

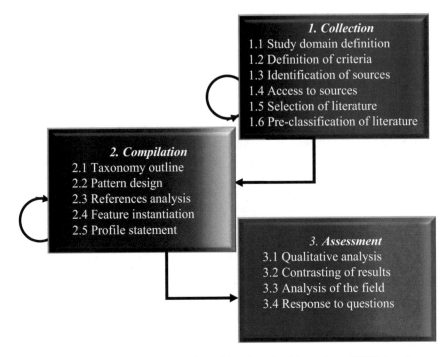

Fig. 3.1 Method used to carry out the review of the educational data science (EDS) domain

3.2   Review Settlement

With the idea of coping with the research problem and its questions, the conceived solution relies on a baseline that supports the formulation of a representative landscape for EDS. Such a settlement embraces three items: the first corresponds to the method that guides the study of the EDS field, the second identifies the sample of related works used to illustrate EDS endeavor, and the last shows the proposed taxonomy of EDS oriented to organize the accomplished duty.

3.2.1   Method

The development of the review is inspired in a method composed of three stages that define the task to be fulfilled to reach specific milestones. The guide is illustrated in Fig. 3.1, where stage 1 is oriented to gather, organize, and provide the references that represent EDS labor, stage 2 examines and sketches the EDS landscape according to the proposed taxonomy of EDS, and stage 3 points out the analysis and interpretation of the results, as well as responds to the research problem and specific questions.

3.2.2   Resources

Concerning the statement of the domain of research, it corresponds to EDS as the arena whose landscape is the object of definition by means of the analysis of related literature. With this in mind, the criteria followed for seeking and choosing publications consist of the following: (1) Only get works whose topic includes EDS. (2) Most of the articles should be published by journals indexed by Clarivate Analytics®, particularly in the Journal Citation Reports™ database. (3) Gather books and chapters concerned with data science and EDS. (4) Obtain selected papers from proceedings.

Based on those guidelines, the main sources of literature considered are the bibliographic database Scopus® and Google Scholar®. Although the EDS term and similar ones (e.g., education data science) were sought as part of the publications' title, keywords, abstract, and body text, the search for EDS-related works spuriously generated quite a few publications, the first of which appeared in 2013! In this regard, the "pioneer" work corresponds to an analysis of the nascent job named *educational data scientist*, where insights of the authors are revealed [45], and by contrast only two mentions are made of EDS, one as an author's academic evolution and the other as a trend to develop animated, multidimensional techniques for mining and visualizing.

As a consequence of the cited finding, the selection of literature triggers an alarm to take care of the "facade" EDS works that only make a scarce allusion to EDS. Therefore, particular interest demands the references that devote special attention to define the background, nature, or baseline of EDS, as well as describe an approach to cope a given target. Based on these factors, the articles are pre-classified to configure out a potential structure that classifies EDS labor.

As a result of the collected sample of *153* EDS-related works, Table 3.1 shows the annual frequency of papers published since 2013 until 2022 of the literature instances corresponding to the subcategories "1.3 Incipient" up to "3.2 Application" that appear in Table 3.2, whose details are explained in Sect. 3.2.3.

It is seen that the total number of publications reported in Table 3.1 is 153 in contrast with the accumulation informed in Table 3.2, 168, and the reason is: there are several articles that are considered for more than one *instance* of the *taxonomy*, but for the estimation of the annual summation of works, such papers are counted just one time, eliminating their repetitions.

Table 3.1 Annual frequency of related works

Sum	2013	2014	2015	2016	2017	2018	2019	2020	2021	2022
153	2	3	3	9	11	11	17	36	35	26

Table 3.2 Taxonomy of educational data science

$\sum$	Category	$\sum$	Subcategory	$\sum$	Instance
168					
51	1 Context	–	1.1 Baseline	–	1.1.1 Big data
				–	1.1.2 Probability and statistics
				–	1.1.3 Data analysis
				–	1.1.4 Machine learning
				–	1.1.5 Data mining
				–	1.1.6 Knowledge discovery in databases
				–	1.1.7 Analytics
				–	1.1.8 Data science
		–	1.2 Related field	–	1.2.1 Educational big data
				–	1.2.2 Educational data mining
				–	1.2.3 Learning analytics
		51	1.3 Incipient	20	1.3.1 Facade
				31	1.3.2 Quotation
76	2 Profile	13	2.1 Introductory	8	2.1.1 Pioneer
				5	2.1.2 Concepts
		25	2.2 Scope	14	2.2.1 Umbrella
				11	2.2.2 Concrete
		31	2.3 Expression	26	2.3.1 Mention
				5	2.3.2 Depth
		7	2.4 Overview	2	2.4.1 Editorial
				5	2.4.2 Review
41	3 Approach	16	3.1 Logistic	2	3.1.1 Frames
				6	3.1.2 Frameworks
				8	3.1.3 Conceptual
		25	3.2 Application	6	3.2.1 Instruction
				6	3.2.2 Apprenticeship
				7	3.2.3 Evaluation
				6	3.2.4 Resources

3.2.3   Taxonomy of Educational Data Science: A Proposal

With regard to the sample of gathered EDS-related works, the analysis of their content gives birth to a classification of literature that facilitates the organization of the labor being achieved. The view considers the main subject addressed in the *body text* of the article. Thus, the explicit mention of EDS as part of the title, abstract, keywords, references, and affiliation is not enough for an article to be a real valuable work.

This is why the design of a *taxonomy of EDS* considers the core of the alluded topic in the body text of the publication, continuing to label its essence by means of a representative term. Once all the works have been examined, and the main topics concerning EDS identified, the frequency of their appearance in the references is estimated to reveal the subjects most mentioned in the EDS sources. Later, the analysis of logical relationships among the subjects is made to look for hierarchies that cluster sibling topics under a representative *term*, giving as a result the proposal of the *taxonomy of EDS* outlined in Table 3.2, whose columns have the syymbol "∑" to unveil the number of works classified for the subcategory "1.3 Incipient" up to the subcategory "3.2 Application" with their respective instances. In contrast, the counts for subcategories 1.1 and 1.2 are ignored because they do not correspond to EDS being that their content is oriented to other domains.

The taxonomy of EDS contains three tiers, where the first names the *category* that is split into *subcategories* that encompass *instances* as follows. The first layer contains three categories, in which the first defines the *context* that establishes the boundaries of the EDS domain through the statement of three subcategories: *baseline* (i.e., it puts eight bricks: *big data, probability and statistics, data analysis, machine learning (ML), data mining, KDDB, analytics*, and *data science), related field* (i.e., it represents fields that strive for harnessing datasets to use an approach to gain knowledge: *EBD, educational data mining (EDM)*, and *learning analytics (LA))*, and *incipient* (i.e., it reveals two sorts of works: *facade*, whose content just labels the EDS term in a supplementary way, and *quotation*, where the article only shows statements said by another author, but does not contribute to extend EDS).

With respect to the second category of the first layer, it concerns the *profile* that represents a particular or general feature of EDS by means of four subcategories: *introductory* (i.e., it presents an elementary glimpse of EDS by identifying its *pioneer* works and giving some basic *concepts), scope* (i.e., it tackles the disjunctive view given for EDS, inclusive and exclusive, that corresponds to those works that consider EDS as an *umbrella* term that embraces related fields and as a *concrete* term to gather works that distinguish it as a specific domain), *expression* (i.e., it reveals the literature that superficially or broadly alludes to the EDS topic in their body text, which, respectively, corresponds to a light *mention* of the domain or in-*depth* talk about it), and *overview* (i.e., it sketches a general perspective of EDS according to two types of publication, *editorial* or *review*, where the former intro-duces a journal issue or book volume about EDS and the latter is a wide research that compiles a series of related works to draw the background, state of the art, and trends about EDS).

Lastly, the third category corresponds to the *approach* that organizes diverse contributions to extend EDS labor into two subcategories: *logistic* (i.e., it offers three theoretical constructs of EDS: *frames* and *frameworks* to lead the development of approaches and *conceptual* to express a given viewpoint to be taken into account in the EDS domain) and *application* (i.e., it depicts the nature and outcomes of specific developments oriented to cope with particular issues of three targets, *instruction, apprenticeship*, and *evaluation*, as well as those that report the experience of using some *resources* that are necessary for achieving EDS labor).

3.3 Educational Data Science: A Glimpse

With the idea of tailoring a landscape of EDS labor, as well as responding to the
research problem and specific questions, particularly the answer for what really is
EDS, in this subsection an account of the collected works is given. Based on the
proposed *taxonomy of EDS*, this section is organized according to the three-tier
structure, where the first layer corresponds to the three categories.

3.3.1 Context

As a first reference to understand what EDS means, in this subsection three subcategories are described. The first reveals the *baseline* to support EDS through eight
constructs, the second identifies some *related fields*, and the third shows *incipient*
works that mention EDS as a *facade* or *quote* a statement said by others.

3.3.1.1 Baseline

Concerning the roots that settle EDS, eight related fields provide essential views,
concepts, theoretical constructs, and approaches that nurture the basis of EDS. With
this in mind, Fig. 3.2 sketches a four-tier hierarchy to reveal how the fields that make
up a lower layer ground the ones that compose upper levels. With this in mind, both
big data and *probability and statistics* form the essential baseline to support the
remaining fields, while both *data analysis* and *ML* boost *data mining*, *KDDB*, and
analytics. Thereafter, the seven mentioned instances frame *data science*, where one
of its application domains is EDS.

Big Data

Data, information, and knowledge are the building blocks for *information science*,
whose concept is interrelated, but their meaning often is overlapped. With this in
mind, Zins [46] briefs the definitions given to those terms by 45 experts, where his

Fig. 3.2 Underlying disciplines that compose the educational data science baseline

own is stated from the subjective perspective (i.e., knowledge exists in the person's mind) as follows: *data* "are the sensory stimuli perceived through senses"; *information* "is the meaning of the sensory stimuli, considering empirical perception"; *knowledge* "is a thought based on the belief it is true." Moreover, Jifa and Lingling [47] give their definitions as follows: *data* "is raw, exists in any form, usable or not, and lacks of meaning of itself"; *information* "is data that has been given significance by way of relational ties"; *knowledge* "is the suitable group of information, whose intent is to be useful." What is more, Zhu and Xiong [19] point out, "*Data* is the formal statement of nature in computer systems; *information* is the phenomena of nature, society, and thinking; and *knowledge* is experience gained through practice; thus data are featured as symbols and representations of information and knowledge; but never being equivalent to information and knowledge."

In this context, *information explosion* has been the object of study since 1944, when Rider [48] warned, "The increase in publications of all kinds and the breeding of new fields of specialization. . .are feeders to reservoirs near the bursting point." He also estimated that "American university libraries double every sixteen years." Later as a consequence of the advent of streams of *digital data*, Lesk [49] wondered, "How much information is there in the world?" and estimated ". . .of disk and tape sales, and size of all human memory, there may be a few thousand petabytes of information all told." According to Selwyn [50], digital data "is produced from a diversity of sources and take several forms, where some are generated deliberately, others automatically, and more volunteered by people when they are using digital resources." In this concern, *datafication* represents the rendering of natural and social worlds in machine-readable digitable format, stored as datasets of commercial transactions, social media, etc. [51]. For instance, Google manages 100 petabytes of data, and Facebook monthly generates log data of over 10 petabytes[52].

As a result of such an explosion, in 1977 the term *big data* was coined by Cox and Ellsworth [53] to highlight a couple of issues: collection (i.e., aggregates of many datasets stemmed from multiple sources, many databases, and several disciplines and distributed among diverse physical platforms) and objects (i.e., single datasets, which are too large to be processed by typical software and algorithms on the available hardware). Later, Mashey [54] asserts "storage growing bigger faster. . .net continues raising user expectations more difficult data (e.g., image, audio, graphics, video. . .)." Simultaneously, Bryson et al. [55] address the issue that surges when datasets used for scientific visualization surpass the available size of computers' memory by both sparse traversal and compression techniques. Naeem et al. [13] assert, "Big data is defined in terms of '5 v_s', *volume, variety, veracity, value*, and *velocity*." According to Gökalp et al. [56], big data pursues effectively and efficiently to collect large, complex, (un)structured, and continuous data from a large number of disparate data sources, as well as store and analyze such data on distributed architectures. Inclusively, Saggi and Jain [57] characterize big data into "seven v_s": *volume, velocity, variety, valence, veracity, variability*, and *value*. Concerning the challenges that face big data, Rahman and Slepian [58] identify the following issue: the difficulty to connect structured datasets to text-based content and unstructured data, quite a few skilled experts are available for working with big data toolsets, the uncertainty to estimate the cost of big data applications, the lack of big data culture, privacy issues and security concerns.

Probability and Statistics

Both *probability* and *statistics* represent essential underlying items of data science. According to Von Mises [59], "*mathematical probability theory* asserts that given infinite sequences $\{x_j\}$ of labels in which each distinct label a_i has a limiting frequency p_i called the *chance of a_i within the given sequence* $\{x_j\}$." Probability theory is the ground of statistical inference, which is necessary for data analysis biased by chance, to estimate with actual data the expected value of random variables of interest. One reason probability comes in data science is that many computational problems are hard when algorithms need to be efficient on all possible datasets, so that domain knowledge enables the statement of stochastic models of data [60].

Concerning *statistics*, in the words of Johnson and Kuby [61], "it is more than just numbers, it is data, what is done to data, what is learned from data, and the resulting conclusions; it is said, the science of gathering, featuring, and interpreting data." With respect to the ties between statistics and data science, in 1977 Jeff Wu proposed for statistics to be renamed "data science" and therefore statisticians as "data scientists," while the International Association for Statistical Computing tried to link statistical methodology, computer technology, and the knowledge of domain experts to transform data into information and knowledge [62]. Even more, Olhede and Wolfe [63] envision as the future of statistics and data science that "affordability of computing, low cost of data storage, and ubiquity of sensing devices produce a volume and variety of heterogeneous huge data sets that generate four effects for statistics: (1) data sets are related in some sense to human behavior; (2) data regulation should concern of what we are allowed to do with the data; (3) the increasing complexity of algorithms corresponds to a growing variety and complexity of data; (4) the types of data sets are far from meeting the ordinary statistical premises of identically distributed and independent observations."

Data Analysis

Although *data analysis* is recognized as being a part of data science's baseline, where probability and statistics are clearly essential building blocks, *data analysis* is really the antecedent for *data science*. Similarly, as data science would do it later, whose essence smells of statistics with the addition of new flavors to deal with data, in 1961 John W. Tukey shifted his aim from inductive inference to data analysis backed by statistics to grasp the traits of science rather than those of mathematics and in this way to focus on both discovery and confirmation [64].

In this sense, Tukey [65] fostered data analysis with procedures for analyzing data, techniques for interpreting results, data collection planning, and the resources and outcomes of the statistics that analyze data to make it easier and more accurate, all of this to highlight data analysis as an empirical science, whose mission is "learning from data." Such statement inspired Donoho [66] to say, "More than 50 years ago, Tukey prophesied that something like today's *data science* moment would be coming." In addition, he also textually mentioned the comment made by Peter Huber: "Tukey stands data analysis as a term for what applied statisticians do, differentiating this name from statistical inference; but he stretched the term to such an extent that includes all of statistics." This seminal work was later extended when Tukey [67] claimed for the need of using data to formulate a hypothesis to test and that both exploratory data analysis and confirmatory data analysis must proceed side by side. It is said *exploratory data analysis* describes the act of looking at data to see what it seems to say [68], while confirmatory data analysis is more about confirming or rejecting hypotheses derived from theory [69]. Both exploratory and confirmatory data analyses are regularly applied in sequence: firstly, exploratory data analysis helps to shape a hypothesis to be tested; afterward, confirmatory data analysis offers the tools to uphold if that hypothesis holds true [70].

Machine Learning

Another brick to settle data science is ML, which was coined by Samuel [71] with the intention of programming computers to automatically learn and improve from experience. With this goal he proposed two methods: the first is a neural net for inducing learned behavior into a connected switching net as a result of reward-punishment policy, and the second is a network that learns only specific things. This layout was tested to demonstrate that given only the rules of the game, direction sense, and a list of parameters, a computer is able to learn to play a better game of checkers than the author of such a program.

Later, Hormann [72, 73] took the ML incipient torch to lead the design of learning systems pulling apart both problem-solving techniques and the learning process from specific problem content, whose behavior claims response patterns, and three mechanisms for community unit, induction, and planning. In this approach, the community unit mechanism offers higher-level programs enabled to achieve requested tasks, the induction mechanism manages classes of problems and leverages acquired experiences, and the planning mechanism splits the task into subtasks to sketch a course of action to lead the community unit.

Conceptually speaking, ML is a part of the artificial intelligence's kernel, lying at the intersection of computer science and statistics, where the essentials of statistical, computational, and information theoretical constructs ground the development of learning systems [74]. ML is a requirement for artificial intelligence; specifically, when a system is an evolving setting, it should detect and adapt to changes by itself [75]. An intelligent system that offers artificial intelligence skills relies on ML, which backs its capacity to learn from problem-specific training data to automate the process of analytical model building and solve related tasks [76]. Essentially, there are four types of ML algorithms: supervised (i.e., learns a function that maps an input to an output based on sample input-output pairs, e.g., classification, regression), unsupervised (i.e., analyzes unlabeled datasets without human assistance, e.g., clustering, association), semi-supervised (i.e., operates on both labeled and unlabeled data (e.g., classification, clustering), and reinforcement learning (i.e., enables agents to automatically evaluate the optimal behavior in a given context to enhance its efficiency (e.g., positive reward and negative penalty) [77].

Data Mining

During the 1960s, "data dredging," "data fishing," and other terms were used by statisticians to label the analysis of data without a given hypothesis [78]. It is said the main aim was to scan huge data and find out patterns on the premise of there was no clear a assumption prior to starting the process [79]. With the goal to back that aim, some constructs had been joined to arrange a baseline, as for instance: classic statistics for measuring data, artificial intelligence to apply the human thought process to statistical endeavors, and ML to merge experience-based techniques with statistical analysis [80]. In this context the term "data mining" first appeared in the 1980s, particularly in the economy, where data issues and statistical calculus were faced [81–86].

Data analysis has been the object of study and development in the arena of probability and statistics, whilst computer science cares of data mining, where some issues correspond to usability, scalability, and deployment [87]. Data mining is a stage in KDDB that seeks for a series of hidden patterns in databases, often involving a repeated iterative application of descriptive or predictive methods [88]. The data mining term depicts the analogy to the mining field, where a valuable resource (e.g., knowledge) is extracted from mining deposits (databases) [89]. In the words of Han et al. [16], "Data mining is the process of discovering interesting patterns, models, and other sorts of knowledge in huge data sets." Data mining tailors a couple of models, *descriptive* and *predictive*, which, respectively, use unsupervised learning to find out patterns that characterize data to be interpreted by users, and apply supervised learning to estimate unknown values of some variables of interest [90].

Knowledge Discovery in Databases

In 1989 the compound term *KDDB* was coined by Piatetsky-Shapiro [78] as an alternative name for *data mining* to *highlight* the essence of the endeavor, *discovery*, and articulate its object, *knowledge*. With this purpose, he and his colleagues (e.g., Jaime Carbonell, William Frawley, Michael Siegel, etc.) organized the Knowledge Discovery in Databases workshop at IJCAI-89 in the USA. In this context, KDDB refers to the overall process of discovering valuable knowledge from massive data, while data mining is the mathematical kernel of the KDDB process, where a given algorithm is applied for exploring data, developing models, and extracting implicit and explicit patterns of data that are the core of useful knowledge [91, 92].

Formally speaking, Frawley et al. [93] define *knowledge discovery* as "Given a set of facts F, a language L, and a measure of certainty C, a *pattern* is known as a statement S in L that depicts relationships among a subset F_S of F with a certainty c, such that S is simpler than the enumeration of all facts in F_S". Hence, *knowledge* is a pattern that is both interesting and certain enough according to a user-imposed interest measure. In this way, *discovered knowledge* is the result of a program that scans the set of facts in a database and generates patterns in this form. Thus, KDDB is the nontrivial process of identifying novel, useful, and understandable knowledge in data [94].

KDDB is a subfield of ML oriented to discovery from huge amounts of possible uncertain data, whose techniques range from statistics, ML, pattern recognition, and databases to domain knowledge to find out a logical description that uncovers a complex set of patterns in data [95, 96]. KDDB aims at developing strategies and methods for stemming meaningful data, where the challenge is mapping low-level data into another more abstract, compact, and useful form [97]. For instance, when an application is facing complex data that makes difficult the capture of multiple traits and the underlying structure of data, a framework based on ensemble learning integrates data modeling, data mining, and data fusion to merge the informative knowledge to achieve predictive performance [98].

Analytics

Although *business analytics* was born in the mid-1950s, Davenport [17, 99] traces the evolution of *analytics* through four eras, where the first depicts the use of business intelligence in data warehouses, the second exploits big data stemmed from Internet-based and social network sources to amass and analyze new types of information, the third is the era of data-enriched offerings that analyze datasets to back customers and markets as well as embed analytics and optimization into every business decision made at the front lines of the organization operations, and the fourth is the era of artificial intelligence to enable greater use of autonomy in the execution of methods. From the view of Piatetsky-Shapiro [78], a breakpoint surged in December 2005 when the term *analytics* reached great popularity as a result of the introduction of Google Analytics and by the publication of the book *Competing on Analytics* [100, 101].

Essentially, "Analytics is a data-driven decision making," as Cooper [102] reports from EDUCAUSE, who also provides his own concept: "Analytics is the process of developing actionable insights through problem statement and the use of statistical models and analysis against existing or simulated future data." As an example of progressive analytics processes, Turkay et al. [103] tailor the following series: consider human time shortcomings, apply progressive algorithms, deploy interaction mechanisms, and design features for visualizations. Given the usefulness of analytics, diverse additional terms are put before *analytics* to highlight a specific feature, for example: (1) to qualify, descriptive [104], predictive [105], prescriptive [106], immersive [107]; (2) to associate with other domains, data [108], big data [109]; (3) to identify the application target, business [110], social media [111], game [112], supply chain [113]; (4) to label the provider, Google [114].

Data Science

Regarding the background of data science, Naur [115] addressed a letter to the editor, where he defined *datalogy* as "the science of the nature and use of data." Afterward, Naur [116] published a book about data processing methods, where he made a couple of definitions, one for data as "A representation of facts or ideas in a formalized way capable of being communicated or manipulated by some process" and another for data science as "The science of dealing with data, once they are established, meanwhile the relation of the data to what they represent corresponds to other sciences." Later, the International Federation of Classification Societies enti-tled its 1996 conference "Data Science, Classification, and Related Methods" [117]. Later, Jeff Wu proposed to rename statistics as "data science," as well as its practitioners, statisticians, as "data scientists" [43]. Lastly, Van Dyk et al. [118] reported on the statement on the role of statistics in data science: (1) there is a lack of a consensus on what data science means; (2) three communities pertaining to computer science or statistics are appearing to ground data science (e.g., database management, statistics, and ML to transform data to knowledge and distributed and parallel systems).

Before the accelerated growth and vast amount of *datanature* (i.e., all the multi-source and complex data produced in cyberspace [19]; (e.g., every day 2.5 quintil-lion bytes of data are generated [119]), data DNA features key knowledge, insights, and potential to become an essential integrant of all data-based organisms, which demand suitable comprehension by data science and its cornerstone, *analytics* [120]. Data science combines three essential functions of *data analysis*: (1) to understand that real-world problems are the object of data analysis, (2) to encode real problems through mathematical computing, (3) to leverage the performance gained by mathematical computing to reach a value in the real world, where these functions define the spectrum of *analytics* that is the core of data science [121]. In the words of Cao [120], "data science originally corresponded to *data analysis*, but actually is the next generation of *statistics* that goes beyond of *ML* and *data mining*, so that is a trans-disciplinary field to study data following data science thinking."

As regards the foundations of data science, Blum et al. [122] published an essential book, where eight underlying topics are outlined to ground data science—high-dimensional space, subspaces and singular value decomposition, random walks and Markov chains, ML, massive data algorithms, clustering, random graphs, and graphical models—as well as other topics. Moreover, Fan et al. [123] also published a book about the statistical essentials for data science, for instance, multiple and nonparametric regression, penalized least squares, generalized linear models and penalized likelihood, penalized m-estimators, high-dimensional inference, feature screening, covariance regularization and graphical models, covariance learning and factor models, (un)supervised learning, and deep learning. Additionally, Igual and Santi [124] include descriptive and inferential statistics, network analysis, recommender systems, natural language processing, and parallel computing.

3.3.1.2   Related Field

Once the eight underlying fields that ground data science have already been stated, now three of their derivate "subfields" are illustrated in Fig. 3.3 as related ones for EDS. Thus in this section the identity of the alluded subfields, also known as "application fields," is recognized with the aim of clarifying their nature and avoiding misconceptions and assumptions, as for example: they are synonymous to, equivalents to, or parts of EDS.

Educational Big Data

Big data in education, also labeled *educational big data* (EBD), is an incipient trend with the availability of managing a huge amount of educational data stored in institutional databases (e.g., data obtained from learning management systems, massive open online courses (MOOC), social media, etc.) [33]. EBD expresses *administrative* data and *learning process* data, where the former depict demographic, behavioral, and achievement data gathered through governmental agencies and schools, and the latter are (near) continuous, fine-grained records stemmed from digital interactions of student behaviors during learning endeavor [125]. The EBD setting is seen as a *mind map* [126] composed of four themes (i.e., learner's behavior and performance, modeling and data warehouses, big data and curriculum, educational system), which contain subthemes (e.g., learner strategies, cluster analysis, statistical tools). In this scenery, EBD contributes to personalized learning, avoiding plagiarism, educational content authoring, decision-making, preventing academic failure, promoting quality of education, and improving learners' attention needs, academic outcomes, employability, school organization, and selection of teachers [127].

Data generated by digital traces of learning processes can be gathered, organized, and analyzed according to the following three granularity tiers proposed by Fischer

Fig. 3.3 Related fields for educational data science

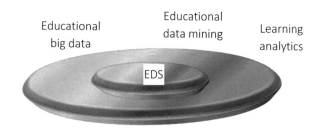

Educational big data   Educational data mining   Learning analytics   EDS

et al. [128]: (1) micro-level (i.e., data produced every second between actions, e.g., clickstream data) is suitable for identifying knowledge components, metacognitive and self-regulated learning skills, affective states, evaluating learners' knowledge, actionable knowledge, and clustering learners; (2) meso-level (i.e., data stemmed from learners' writing tasks and online discussions with varying durations—from minutes up to hours—frequencies, and regularities (e.g., text data) is used for cognitive functioning, examining social processes, behavioral engagement, and affective constructs); (3) macro-level characterizes institutional data generated over multiyear time spans that correspond to learners' demographics, course enrollment, course schedules and descriptions, grade records, admission and campus living data, degree and major requirement information, which is ad hoc for early-alert systems, course information, guidance systems, and administration-facing analytics.

Educational Data Mining

Some of the earlier works that applied data mining to educational affairs pursued to discover enrollment knowledge [129] and web access patterns [130], in addition to access log systems to evaluate computer-based education [131–133]. Later, in the birth of the twenty-first century, educational data mining (EDM) emerged as a "discipline oriented to develop methods for exploring the unique and increasingly large-scale data that come from educational settings and using those methods to better understand students, and the settings which they learn in," according to the words of the International Educational Data Mining Society.[1] Such a society was founded in 2011 to promote, foster, and disseminate EDM labor by means of the annual International Conference on Educational Data Mining[2] since 2008, whose antecedent corresponds to some workshops organized since 2005 and the publication of the *Journal of Educational Data Mining* since 2009.[3]

As a taxonomy to depict EDM approaches, Peña-Ayala [134] defines the following series of hierarchical features: (1) discipline—statistics, ML, etc.; (2) model—descriptive and predictive; (3) tasks—to implement the model—classification, clustering, etc.; (4) methods and techniques—to deploy the task—Bayes' theorem, decision trees, logistic and linear regressions, etc.; (5) algorithms, equations, and frames—J48, k-means, prescriptive and descriptive statistics, Bayesian networks, etc. For instance, Tsiakmaki et al. [135] depict diverse approaches for predicting learner outcomes oriented to performance (e.g., task—classification and so on; method—naïve Bayes, decision trees, etc.), dropout (task—classification and so on; method—J48, support vector machine, etc.), and grade (task—regression and so on; method—k-nearest neighbor, random forest, etc.). Concerning the most

---

[1] Visited on September 28, 2022, https://educationaldatamining.org/
[2] Visited on September 28, 2022, https://educationaldatamining.org/conferences/
[3] Visited on September 28, https://jedm.educationaldatamining.org/index.php/JEDM/index

targeted EDM topics, Du et al. [136] brief, in descending order, prediction performance, decision support for teachers, behavior patterns, algorithm evaluation, dropout analysis, and text and video classification. Basically, optimization of predictive models for learners' academic performance and detection of behaviors of learners for timely intervention is the most considered by EDM approaches [137]. As some of the main EDM purposes, one is necessary to improve the quality of education by predicting learners' academic performance and another demands fostering those learners who are at risk of dropout [138].

Learning Analytics

A pioneer work in the *learning analytics* (LA) arena is the one achieved by Tinto [139] to analyze causes and cures of college students' attrition. Afterward, the LA term is mentioned in a briefing about the e-learning industry infancy given by Mitchel and Costello [140], while Berk [141] tailors the state of the art for LA related to the effectiveness of corporate training, Whilst Moore [142] defines how LA assesses the effectiveness of learning according to its pedagogical design, Retails et al. [143] strive on networked LA toward knowledge constructions, and Bach [144] proposes a framework for the development of LA considering ethics.

LA is seen as "the measurement, collection, analysis, and reporting of data about learners and their contexts, for purposes of understanding and optimizing learning and the environments in which it occurs"[4] as the Society for Learning Analytics Research,[5] founded in 2014, declares. In 2014, such a Society took over the organization of the annual International Conference on Learning Analytics & Knowledge,[6] which started in 2011, as well as the publication of the *Journal of Learning Analytics*,[7] launched in 2014. As for the baseline that grounds LA, Peña-Ayala et al. [145] define three fields: computers in education, knowledge-based systems, and computer-based systems. Meanwhile, the alluded Society asserts, "LA sits at the convergence of learning (e.g. educational research, learning and assessment sciences, educational technology), analytics (e.g. statistics, visualization, computer/data sciences, artificial intelligence), and human-centered design (e.g. usability, participatory design, sociotechnical systems thinking)."[8]

In relation to the landscape of LA, Peña-Ayala [146] sketches a taxonomy for LA composed of three categories (e.g., profile, underlying factors, applications) that are split into subcategories (e.g., reviews, contextual surrounding, nature; legal and theoretical learning paradigms and settings; resources, learner analysis, functionalities), which, respectively, embrace instances (e.g., specialized items and others;

---

[4] Visited on September 28, 2022, https://www.solaresearch.org/about/what-is-learning-analytics/

[5] Visited on September 28, 2022, https://www.solaresearch.org/

[6] Visited on September 28, 2022, https://www.solaresearch.org/events/lak/

[7] Visited on September 28, 2022, https://www.solaresearch.org/publications/journal/

[8] Visited on September 28, 2022, https://www.solaresearch.org/

evolution and others; ethics, privacy, etc.; frameworks and methods and others; self-regulated learning, tools, etc.; behavior and others; feedback and others). Meanwhile, Lang et al. [147] publish a handbook for LA that organizes 30 chapters into four sections: foundational concepts, techniques and approaches, applications, institutional strategies and system perspectives. Moreover, Kew and Tasir [148] offer a review that classifies LA duty through three criteria: learning settings (e.g., learning management system and others), focus of LA (e.g., monitoring, prediction, etc.), types of student data (e.g., behavioral, emotional, etc.). On the other hand, another review is edited by Du et al. [149] to report the most used LA methods: descriptive (e.g., statistics, visualization, clustering, rule mining, social networking analysis, text miming) and predictive (e.g., regression, decision tree, naïve Bayes, k-nearest neighbor, etc.).

3.3.1.3 Incipient

A sample of initial EDS-related works has been found, where some are a kind of *facade* because they only present the term as a part of the supplementary sections (e.g., title, keywords, etc.) but it is absent in the main area of the publication, *body text*. Other papers just *quote* the statement made by third-party authors about any aspect of EDS, in consequence such papers do not really contribute to extend EDS. Both *facade* and quotation works are identified in the following subsections.

Facade

It is found that some articles include the term *EDS* just for *label* aims, because it only appears in a supplementary part of the paper (e.g., title, affiliations, abstract, keywords, conclusions, or references). However, in the body text (i.e., from the introduction to the discussion or similar sections), there is no additional statement of EDS or insights of how it is applied. The reason of this pattern is: the approach relies on another related domain (e.g., EDM, LA, etc.). As an example of these cases, Table 3.3 shows the citation and the part of the paper where EDS is edited.

Quotation

Other EDS works just quote a statement said by other authors with its respective citation and reference. These works only consider EDS without contributing to advance the domain. A sample of such works is shown in Table 3.4, where only a part of the quotation is edited without showing the alluded citation, so that interested readers should seek the respective reference in the mentioned work due to the aim is to identify this sort of "shy" work and not the former source.

Table 3.3 Sample of facade works that do not present the *EDS* term in the body text

Id	Authors	Year	Section
1	Paiva et al. [150]	2018	Keywords, references
2	Kiss et al. [151]	2019	Abstract
3	Alonso and Casalino [152]	2019	Keywords, references
4	Qu [153]	2019	Keywords, references
5	Yücel and Erol [154]	2019	Keywords
6	Albó and Hernández-Leo [155]	2019	Abstract
7	Zahedi et al. [156]	2020	Affiliations
8	Nguyen et al. [157]	2020	Affiliations
9	Naseer et al. [158]	2020	References
10	Marrhich et al. [159]	2020	References
11	Michos and Hernández-Leo [160]	2020	References
12	Tsai et al. [161]	2020	Keywords
13	Baranyi and Molontay [162]	2021	Keywords, affiliations
14	Sreenivasulu et al. [163]	2022	Affiliations, references
15	Yang et al. [164]	2022	Affiliations
16	Perez-Alvarez et al. [165]	2022	Affiliations
17	Maraza-Quispe et al. [166]	2022	References
18	Göktepe Körpeoğlu and Göktepe Yıldız [167]	2022	References
19	Ince [168]	2022	References
20	Buchanan [169]	2022	References

3.3.2 Profile

Now that the context for EDS has been stated, EDS research is shown from this subsection up to the next one according to four subcategories: *introductory* to identify its origin and definition, *scope* to reveal its dual view, *expression* to denounce papers that express quite a few mentions or a lot of mentions about EDS, and *overview* to recognize editorial and review articles that depict a small or broad sample of related works.

3.3.2.1 Introductory

As for the origin and nature of EDS, a pair of instances are edited here, where one briefs a collection of papers published from 2013 up to 2015 to acknowledge the *pioneer* works, while the other offers a sample of *concepts* given for EDS.

Table 3.4 Sample of quotation works that repeat statements expressed in other EDS papers

Id	Authors	Year	Quotation
1	Daud et al. [170]	2017	"EDS clarifies how different disciplines and researchers with... research interests and backgrounds... work..."
2	Guo and Zhang [171]	2017	"EDS is... needed to analyze learners' social interactions, predict... dropout, provide feedback and make recommendations."
3	Selwyn [172]	2018	"Links... to the rise of EDS and... to derive insights from the multitude of perceived indicators... generated... technology."
4	Deborah et al. [173]	2018	"EDS is a field of power and its exponents access these forms of capital as part of its ascendancy in educational domains..."
5	Wise and Cui [174]	2018	"...what EDS and LA methods have to offer our understanding of learning and ability to support it..."
6	Knox [175]	2018	"...the increasing prevalence of EDS is now embedding algorithmic processing in the everyday, mundane practices..."
7	Jarboui et al. [176]	2019	"EDS research, this problem was studied either testing correlations given conjectures or identify communities of look-a-likes."
8	Johanes and Thille [177]	2019	"We did not, however, find a uniform zeal for data collection at all costs... major concerns about the future of EDS..."
9	Kör et al. [178]	2019	"...focused on EDS, EDM, LA, academic analytics, institutional analytics, learner analytics ...are the topics examined in EDS."
10	Khalid et al. [179]	2019	"Moreover, Moodle log analysis is very useful and insightful in researches relating to EDS..."
11	Wasson and Kirschner [180]	2020	"Highlight (Stanford's) eager to develop a data-driven science of learning' or EDS that married educational research..."
12	Khatri et al. [181]	2020	"LA has been described as a part of... EDS which is rapidly emerging and comprises business intelligence...EDM... ML."
13	Cappello and Rizzuto [182]	2020	"...the role of LA and EDS in schools and higher education, the cognitive classrooms..."
14	Baranyi et al. [183]	2020	"Also in EDS some papers already successfully applied deep neural networks for student performance and dropout prediction."
15	Mor et al. [184]	2020	"Recent years have witnessed a growing interest in the promise of EDS, a term coalescing LA..."
16	Deshmukh et al. [185]	2020	"EDS grew due to MOOCs which generated huge amounts of data on participant's interaction with educational content..."
17	Alrmah and Lokman [186]	2020	"...data science in education and learning is recognized as EDS, which functions along with data coming from academic settings."

(continued)

Table 3.4 (continued)

Id	Authors	Year	Quotation
18	Jandrić	2021	"EDS has migrated from the academic lab to the commerce, ownership of the means to produce educational data analyses."
19	Lingard et al. [187]	2021	"Uses science and technology... to provide a... genealogy of developments in EDS ... the emergence of... learning engineer."
20	de Andrade et al. [188]	2021	"...educational data has been a field of study generating terms associated with this educational data exploration... EDS."
21	Parreira do Amaral and Hartong [189]	2021	"Datafication is the installment of data practices and interoperable data infrastructures, informed by values, skill of EDS."
22	Kolber and Heggart [190]	2021	"Linked to this is the growth of EDS. This has led to personalization for all students, opportunities for disadvantaged groups."
23	Wyatt-Smith et al. [191]	2021	"Concerns remain about the adequacy of student privacy and data protection policies & frameworks in relation to the rise of EDS."
24	Castellanos-Reyes [192]	2021	"The takeaway from the EDS conference was the collaboration between educational researchers and computer scientist..."
25	Fincham et al. [193]	2021	"EDS holds promise, and the existing literature offers an array of methods for predicting key outcomes such as student success."
26	Mangina and Psyrra [194]	2021	"In the bibliography the terminology is referred within a wide range of related terms such as academic analytics, institutional analytics, teaching analytics, data-driven education, data-driven decision-making in education, EBD, and EDS..."
27	Nawang et al. [195]	2022	
28	Okoye et al. [196]	2022	
29	Uttamchandani and Quick [197]	2022	"The need to involve practitioners in LA and EDS in the design of analytic systems through reflection on ethical issues..."
30	Gupta and Mishra [198]	2022	"A new term... EDS is the field which explores the educational data... to perform academic, predictive and learning analytics."
31	Gourlay [199]	2022	"...EDS has become a 'trans-sector enterprise', with ownership and power moving over to commercial vendors..."

Pioneer

As an early EDS conceptual work, Buckingham-Shum et al. [45] shape a view for EDS and its specialists, named *educational data scientists*. As for EDS, they assert, "It is a great deal: the foundational roles of developments in *statistical* computing for data *analytics*; the centrality of...interactive data visualization for exploratory *data analysis*; the vital roles of *ML*; more powerful and distributed computing architectures that enable sense-making and prediction with *big data*. . .and effective *LA* and *EDM* methodologies to suite the questions and data. . .," while the profile for an educational data scientist is "has an understanding of the cognitive, contextual, and design aspects of the transactions that generate educational data and the interdisciplinary sciences, which contribute to an understanding. . .that advance the sciences and practices of education and learning." On the other hand, Finzer [200] sees *data science education* as a domain knowledge and claims statistics educators should reach out to educators in other science and math disciplines to understand how data science is integrated into K–12 curricula.

Another conceptual work represents the broader view of EDS stated by Piety et al. [201], who consider EDM, LA, and other analytics that "seem to develop independently around diverse educational tiers (e.g., K–12, Higher Education. . .) and scales of educational contexts (e.g., individuals, cohorts. . .), but all encountering alike issues, which together are considered as the initial cells of EDS that conform a combined community." As a result of their convergence, they identify several features: rapid evolution, boundary issues, inquiries about visualization and others, ethics and privacy concerns, and disruption in practices. With this in mind, authors claim for a unified view for EDS through the consideration of some areas, value the unique EDS character, adopt interdisciplinary viewpoints, acknowledge social and temporal tiers, identify digital fluidity, and comprehend values embedded in information.

As a primary EDS approach, Gibson and Webb [202] discuss how EDS backs four assessments: dealing with change over time, interaction of the performance space's tiers with the learner's actions, generation of tiers of interpretation from translations of atomistic data, and representation of the interaction between learners.

In addition to the prior profound EDS articles, four more just make a superficial statement for EDS. Firstly, Mitchell [203] claims "data sharing and learner privacy are a pair of challenges for EDS," while Williamson [204] asserts, "EDS methods. . .to make future–tense predictions of their outcomes," Selwyn [50] affirms "interest is growing in. . .the possibilities of EDS," and Hood et al. [205] involve EDS as a MOOC to study how context influences learning.

Concepts

In the literature there are a plethora of terms to name EDS, such as *data science education* [200], *education data sciences* [201], *educational data science* and *education data science* both edited by Williamson [36], *data science applications in education* [161], *data science in education* [32], *education and data science* [206], and others. Thus, in this chapter all of them are seen as a single term, *educational data science* (EDS), to avoid misconceptions. Now that the clarification has been made about the diversity of terms, a concept for EDS is defined by means of representative works that are chronologically introduced as follows.

First of all, Williamson [36] affirms, "EDS is a hybrid of data scientific practices drawing from statistics, computer science, information science, and ML, combined with expertise in psychology and neuroscience from the existing field of the learning sciences." Later, Williamson [207] signals, "EDS is an emerging transdisciplinary field formed from an amalgamation of data science and elements of biological, psychological, and neuroscientific knowledge about learning, or learning science." In contrast, Kitto et al. [208] affirm, "EDS is at a point of transition, with focus turning from analytics for data collected by individual tools and towards analysis of student behaviour across learning ecosystems, consisting of a wide range of tools and educational technology products," while Rosenberg et al. [37] think EDS as "A sort of capabilities related to quantitative methods in educational research, computer science, programming capabilities, teaching, learning, and educational systems." In this sense, Nuankaew et al. [209] define EDS as "Brings educational data (e.g., learners' behavior, motivation, styles, interests, and achievement) to be analyzed through ML used for multidimensional analysis (e.g., teacher–student analysis, forecasting tools, feature analysis)."

3.3.2.2 Scope

With regard to the stated core problem, and as a sample of EDS-related works, some efforts have been made to precise the boundary for EDS that reveals its nature. In this concern a pair of instances is identified, *umbrella* and *concrete*. The former is an *inclusive* view of EDS as regards other related domains, whereas the latter sees EDS as an *exclusive* field in comparison with other ones.

Umbrella

As for the *inclusive* outlook that alludes to related fields and sees EDS as a whole, EDS is thought as an "*umbrella*" to implicitly or explicitly reference other allied fields, for example, *generic job* as an education data scientist to embed skills, specific mindset, and career path to deal with leaners' data that includes being an educational data miner or LA engineer [45]; *meta-domain* to say, "EDS embraces several fields (e.g., EDM, data-driven decision making, and four types of analytics)" [201]; *viewpoint* to declare, "LA community is different from other EDS sights because of its own view of learning as a whole complex activity" [210]; *domain* to say, "There are works on EDS focused on domain knowledge, grade prediction, student retention..." [211]; *flag* to claim, "EDS is a broad banner...that refers to areas of data-intensive research (e.g. LA, EDM, and intelligent tutoring systems) aimed at providing analysis and feedback based on learners' interaction logs and other sources of evidence" [212]; *multi-form* field to recognize "LA, EDM...as other forms of EDS" [213]; and *cover* to join some fields (e.g., learning, academic, and teaching analytics, EDM, EBD) to back a construct (e.g., a teaching outcome model) [214, 215].

In addition, Inan and Ebner [216] simply define, "LA is an educational data implementation science"; Aljawarneh and Lara [217] affirm, "There are a series of

disciplines related to EDS, such as EDM and LA, and all of them are of importance";
McFarland et al. [218] believe "EDS is an umbrella that approaches computational
techniques (e.g., stemmed from ML, natural language processing, network analysis)
being used to identify new forms of data, measures, descriptions, predictions, and
trials in education"; Srinivasa and Kurni [219] point out in their EDM and LA
chapter, "The first big challenge is to turn raw and inchoate data streams into usable
variables in data mining and other data science fields." Moreover, Gupta et al. [220]
assert, "The two broad fields that deal with this educational data and help...in
data-driven decision making are LA and EDM. Actually, the umbrella term being
used for these is EDS." In this regard, Susnjak et al. [221] affirm, "LA, EDM, and
academic analytics intersect at various points and share underlying data...that could
all be grouped under the same umbrella as EDS, although they differ in the
stakeholders that they target."

Concrete

By contrast, the *exclusive* outlook focuses on EDS's essence to distinguish it from
other similar fields, as for instance, "methodological field of power" that is creating
its own distinction fostered by a sociotechnical image of a data scientific future for
educational research and knowledge production [207]; "support," where LA
employs EDS methods to detect underlying patterns and relationships among variables and cases [222]; and "cooperation" to assert EDS provides a framework to
compare and assess sources of error stemming from survey methods and from
ML [223].

In this scenery, EDS is also seen as "primary" to work with data gathered from
educational settings to solve educational problems [224] and "associate" as a
discipline and area of practice closely related to LA and EDM for educational
analysts [37], whose evolution of EDS and LA has made advancements by means
of applying algorithms and developing applications for interpreting online interactions and so on [179], where LA is closely related to other concepts, such as EDM,
EBD, EDS, etc. [225].

However, as a contradictory dual view for EDS considered in the same article,
Jaakonmäki et al. [226] declare, "EDS cuts through almost all educational technology disciplines" and after say, "Data science extracts knowledge and insights from
data, and data science in the educational context is termed LA." So the following
questions are made: Is EDS a kind of *umbrella* term? Or is EDS LA? According to
the present review, the response for the first question is, *yes*, it is one of the views
assumed by diverse EDS works, whereas the answer for the second is *no*, due to both
EDS and LA owning their proper identities!

In another vein, McFarland et al. [169] believe EDS identifies the use of tools and
perspectives from statistics and computer science for educational phenomena, but
Williamson [206] acknowledges the EDS essence to inspire the idea of being an
objective science of learning that becomes authoritative in the production of

educational knowledge. Meanwhile, Fancsali et al. [227] review some EDS/EDM studies that close the loop between data-driven modeling and learning outcomes.

3.3.2.3 Expression

Once that the related works that correspond to the introductory and scope topics have been outlined, which, respectively, explain the origin and definition of EDS and its two conceptual boundaries, now diverse articles that contain just one *mention* or offer a *depth* perspective of EDS and therefore extend its domain are presented as follows.

Mention

In the *body text* of some works, an authors' simple clause about how EDS is involved in their research is found, without additional insights in the remainder of the paper. Although this can be a kind of a timid expression, Table 3.5 shows articles that approach EDS in this way, editing a part of their mention.

Depth

In contrast with the prior section, where quite a few real and full EDS works were introduced, in this one, five main EDS articles are chronologically briefed. These papers focus on a given view of EDS or tailor a glimpse of the field. Thus, the publication has plenty of statements about EDS and contributes to extend the field.

Firstly, Williamson [36] sees EDS as a biopolitical strategy enacted by the progress of neuro-pedagogies of brain empowerment, psycho-pedagogies of emotional maximization, and bio-pedagogies of body optimization. With this in mind, he proposes the idea of *transcoding* to consider the involved processes in the evolution of the bio-digital child. This approach relies on digital devices, data practices, pedagogies, and scientific knowledge that together drive logical constructs about children that are consequential to their evolution as bio-digital people, whose qualities and capacities are stated by means of practices of emotion analytics, neurocomputation, and bio-sensing combined with scientific knowledge. In another vein, Williamson [207] asserts EDS leverages algorithm-driven processes of personalization to produce findings and knowledge from EBD under the belief that EDS migrates from academy to the business sector. Moreover, he wonders how education is to be conceived in regard to epistemology and algorithms and how the political economy of education is moving toward data-driven commercial organizations. As a conclusion, EDS is a field of power in educational research, knowledge production, and theory development, which represents the role of a data scientist or an entrepreneurial algorithmist in the educational scenery.

Table 3.5 Sample of papers that lightly mention the *EDS* term

Id	Authors	Year	Mention
1	Williamson [228]	2016	"EDS is methodological and powerful: It is a site of innovation in conducting educational research… seriously consequential…"
2	Gibson and Ifenthaler [229]	2017	"…calling to data mining, model-based methods, ML, and data science for education researchers and the inclusion of topics…"
3	Klašnja-Milićević et al. [32]	2017	"…full impact and expand of data sciences in education is expected to happen in near future…"
4	Chopade et al. [230]	2019	"…advances in the fields of EDS… have allowed researchers to analyze and score a wider range of response data…"
5	Martins et al. [231]	2019	"…technologies and methods such as …and EDS will be adopted in the project to develop a form of learning intervention…"
6	Rosé et al. [232]	2019	"Explanatory learner models: Why machine learning (alone) is not the answer…"
7	Davis et al. [233]	2020	"This paper follows other work in EDS aimed at determining predictors of success in college coursework…"
8	Williamson and Eynon [234]	2020	"Artificial intelligence is located in a history of knowledge production, theory generation, and the development of epistemic expertise in the learning sciences and EDS…"
9	Arthurs and Alvero [235]	2020	"A more equitable EDS using text should therefore consider linguistic variation at the forefront of analysis…"
10	Munoz-Najar Galvez et al. [223]	2020	"EDS provides a framework to compare and assess sources of error stemming from our survey methods and from ML…"
11	Jasim et al. [236]	2021	"Related terms have been applied including data-driven education institutional, academic, and teaching analytics, EBD, and EDS"
12	Fancsali et al. [237]	2021	"We illuminate avenues for research at the intersection of EDS and learning engineering at scale in adaptive learning platform."
13	Madeira et al. [238]	2021	"In the midst of this educational revolution many terms have come about data-driven decision-making in education, EDS."
14	Dowell et al. [239]	2021	"…we provide an assessment of student-generated values affirmation essays using EDS and LA techniques…"
15	Lee et al. [240]	2021	"Researchers have analyzed MOOCs data by incorporating big data and … gained prominence in ML and data science…"
16	Zhang et al. [241]	2021	"…data science needs to consider a wider range of data types in education research…"
17	Fancsali et al. [242]	2021	"Literature in EDS considers data driven methods for improving existing instructional content in adaptive instructional systems."

(continued)

Table 3.5 (continued)

Id	Authors	Year	Mention
18	Panwong et al. [243]	2021	"For higher education the issue of student retention and personalized treatment grasped attention of data science researchers."
19	Aljawfi et al. [244]	2021	"Data science and engineering audiences were targeted for Developing work in the field of educational data science…"
20	Maldonado et al. [245]	2021	"This has stimulated studies where data science methods have been proposed to improve early prevention of student dropout."
21	Chaturapruek et al. [246]	2021	"With a data set of this nature and magnitude, the study of course consideration becomes a worthy subject of EDS…"
22	Hou et al. [247]	2022	"Provide invaluable data to researchers, about learning paths, common patterns of mistakes, which drive research in EDS."
23	Yang et al. [248]	2022	"In addition to addressing topics and threads, EDS researchers focused on investigating posts and comments in MOOC forums."
24	Hidalgo et al. [249]	2022	"EDS has been relevant and profuse due to the success of online teaching in academic and professional environments…"
25	Childs and Taylor [250]	2022	"…strands in the K12 related to EDS and twenty-first century methods of school measurement to understand how K-12 utilize internet."
26	Barbosa-Manhães et al. [251]	2022	"EDS to benefit the educational institutions' academic staff and academic managers to improve the quality of higher education."

A dual view of EDS is approached by Rosenberg et al. [37], both as a research methodology for education and as a teaching and learning content. In addition, they define three work lines: (1) a synergy for the pair of views including tools for professionals and learners, (2) inclusivity access for the community, (3) the use of data science to study teaching and learning about data science. Also, authors envision the evolution of EDS toward learning, design, and technology. Another contribution made by Williamson [206] represents the claim for data-intensive ways of KDDB that are promoted by educational data scientists. With this aim, he uses a genealogical method to feature the exploitation of EBD and to question how educational data scientists mine and analyze EBD. As such, three topics are considered: (1) big data imagery to foster education research, practice, and policy; (2) infrastructural arrangements of organizations, technologies, and methods framed by specific contexts and political, social, and economic concerns; (3) particular practices of educational data scientists.

In another vein, Rus et al. [252] report the conceptualization phase of the Learner Data Institute in pursuit of two objectives: the design of a framework for data-intensive science and engineering and its application in learning science. As a hypothesis, they suppose that "Emerging learning ecologies based on *adaptive instructional systems* are able to provide affordable, effective, efficient…individualized assistance for both learners and instructors." With the goal of backing the proposition, authors gamble for applying both data science and such systems to transform learning ecosystems according to five references: (1) individualized instruction being more effective than the traditional one; (2) the capability to gather, manage, and exploit big learner data; (3) secure and privacy-preserving ways to access and process educational data; (4) awesome new achievements in data science; (5) affordable, powerful, and scalable cloud-based resources for processing EBD.

3.3.2.4 Overview

Another collection of EDS publications offers a glimpse of the arena by means of a dedicated *editorial*, whose preface introduces a series of related works that focus on diverse topics, and a *review*, which summarizes the labor reported by a sample of papers and tailors a state of the art that characterizes the achieved labor. Both kinds of instances are worthy of being shared as follows.

Editorial

Quite a few special issues of a given journal consider EDS as one of their topics to be presented, where the *editorial* paper introduces the central theme, shapes a profile of each work, and tailors a view of the arena. In this concern, two prefaces are uncovered to show the editorial duty that makes some comments about EDS.

In the special issue "Learning in the Age of Algorithmic Cultures," published in the *E-Learning and Digital Media* journal, Jandrić et al. [253] expose five selected papers, where just one is related to EDS [207] and their preface edits scarce mentions for EDS (e.g., "...correspond to the corporate takeover of EDS," "researchers...increasingly rely on EDS to leverage the ability for large-scale research," and "EDS seems to be at the brink of a more mature phase, where compromise, balance, and appropriate solutions are sought"). Moreover, Aljawarneh and Lara [217] edited the special issue "Data Science for Analyzing and Improving Educational Processes" for the *Journal of Computing in Higher Education*, which presents ten papers. However, only two articles make a simple mention of data science, Tsiakmaki et al. [254] twice and Bekmanova et al. [255] once. Moreover, in the body text of the preface, just two mentions of EDS appear, and the last one represents a questionable assertion: "...10 selected papers have been included that present important advancements in the area of EDS."

Review

Another sort of overview is the reviews made for a specific feature or general sight of EDS or in conjunction with another domain. In this sense, a sample of eight publications is summarized to sketch the particular perspective of their authors as follows. The first focuses on the potential of ML and data science in primary and secondary education, where Donaldson et al. [256] back the statement that both EDM and LA pertain to the broader field of EDS. With this in mind, 50 papers are chosen to draw some judgments, such as "to study curricular designs. . .in a sequence of learning; lack of understanding about EDS and its benefits; privacy and ethical considerations for the collection, use and ownership of learners' data." In regard to EDS in MOOC, Romero and Ventura [224] report four communities to conform EDS (i.e., EDM, and learning, learner, academic, and institutional analytics). Thus, considering those domains, they gather 61 works to identify problems to be solved (e.g., analyzing students' interactions, predicting dropout, grading, assessing feedback, adapting learning, and recommendation making).

As regards big data and data science in education, Daniel [33] analyzes 68 papers to identify critical issues: ontological concerns, ethics and privacy, lack of expertise and academic development chances to prepare educational researchers, diversity in the conception and meaning of BDE, epistemological disparity, and technical challenges. He also points out "the new forms of empiricism are featured by emergent research design, shaped by the technological setting, complex and dynamic data, where data science is the fourth research methodology tradition in educational research." A similar target of review is tackled by Fischer et al. [128], who admit the value of digital traces of learners' behavior for more scalable and finer-grained understanding to back learning processes through three educational contexts—micro-level, meso-level, and macro-level—that are stemmed from 370, 175, and 75 papers, respectively. They conclude that "The ubiquity of EBD claims for preparing students in educational graduate programs to utilize data science."

McFarland et al. [218] draw a timeline for EDS, where the kick-off represents the history of data science and the appearance of EDS to affirm, "EDS corpus has grown more than 30-fold in the past decade!" At present, authors analyze 12 papers to highlight the use of natural language processing in large education corpora, the experiments and treatments employed in education research, the question on how data science thinking informs some of education's debates and dualisms, the gathering of new forms of data via new technologies, and the reliability of ML predictions. For the future, they visualize the following chances for EDS: draw on the humanistic and social science traditions that inform research, methods are providing new complementary views on student-level and institutional data, and psychometrics offers modeling venues to be exploited by ML and ways to train students for being involved in EDS. Lastly, three warnings are given: the limits of EDS, the bias in computational approaches, and the need to work toward fairness.

3.3.3 Approach

Concerning the *approaches* that represent an advance for the EDS field, a couple of subcategories are distinguished. The first corresponds to the *logistic* proposals oriented to guide a perspective for EDS endeavor, while the second depicts particular *applications* aimed to solve a given kind of educational target.

3.3.3.1 Logistic

With the purpose of grounding the EDS labor, three bricks have been identified in the literature: *frames* to propose a specific foundation to settle future work, *frameworks* to offer a guideline for leading the development of applications, and *conceptual* to discourse about an interesting matter to enrich EDS duty.

Frames

As for the articles that tailor a frame to leverage an EDS task, a couple of proposals are profiled in this subsection, where the first makes the question, "Who owns educational theory?" As such, Williamson [207] asserts how data science epistemologies and algorithms contribute to theorize education, in addition to warning how knowledge generation is being allocated in data-driven profit organizations as a new fashion of political economy of education. Inspired in two case studies, the author asserts, "EDS is a trans-sector enterprise, using its knowledge and technologies across of variety of schools and Higher Education." Thus, EDS is thought to be as a "field of power" that produces a capital to create its proper identity by the junction of social, economic, and cultural assets led by a sociotechnical imagery of a data science season for knowledge generation and educational research.

In order to deal with the lack of references concerned with the steps of filtering, organizing, and visualizing educational data to boost data-driven decision-making and EDS, Bowers and Krumm [257] introduce a "Data use theory of action." The theory is a continuous cyclical improvement process through five stages: (1) data access and collection; (2) organization, filtering, and analysis; (3) knowledge generation; (4) knowledge application; (5) effectiveness assessment.

Frameworks

In regard to guidelines that define a conceptual setting or process to carry out EDS endeavor, a couple of series are identified in this subsection. The former corresponds to those proposals that shape a given framework, while the latter tailors the method to head diverse activities to achieve a wished goal.

In this sense, Williamson [258] highlights the "Efficacy Framework" tailored by Pearson, which applies the "tried and tested" method to understand how services reach their aimed results by tracking how users interact with data to monitor its efficacy and then optimize its tools to impact on future use. Meanwhile, Klašnja-Milićević et al. [32] shape the "Architecture framework for support higher educational research" that contains five items: (1) data capture and collection module;

(2) extractions, transformations, and loading module; (3) data sources; (4) analysis engine; (5) presentation layer. All of these create an ecosystem to back the analysis of multi-structured datasets. Moreover, Demchenko et al. [38, 259] present the "EDISON Data Science Framework for EDS" that embraces the body of knowledge, model curriculum, etc. as items to design an EDS setting, an education and training directory, and a data science portal to define a data science curriculum and deliver education. As for the claim that "neural networks can discover unexpected deep traits of educational data," Perrotta and Selwyn [260] draw a "Relational Framework" that includes three components: (1) educational dataset and broader digital learning platform; (2) ML method (e.g., a recurrent neural network that owns input, hidden, and output layers) involved in the contrast between "new knowledge" that is tried to be discovered in the data and the "existing knowledge" codified in the data setting; (3) the cultural, discursive, and economic aspects of EDS. Moreover, Sakulwichisintu [41] draws an EDS framework for being used in educational sceneries to foster learners by means of four tiers: *micro* to identify teaching methods that most bias learning, *macro* to enhance learning achievements based on adaptive approaches, *institutional* to focus on the educational organization, *systematic–instructional improvement* to assess teaching methods.

Conceptual

In addition to the prior logistic subjects, there are some efforts that express conceptual topics, for instance, education affairs and datafication tendencies, where bio-datafication enriches EDS and governance also contributes to the field.

In this sense, Wise [261] takes the quest for preparing data scientists, who are able to create ecosystems of data that demand new approaches to education as a result of current changes in the nature, size, scope, and types of data. Hence, the data scientist needs technical and domain expertise as well as skills to play the following roles: *data reader* to make sense of data, *data communicator* to present data for some aims, and *data maker* to monitor their own data contributions.

Regarding *datafication*, defined as "the rendering of social and natural worlds in machine-readable digital format" [51], it takes many forms in education due to the diversity and scale of practices and systems, as well as the influence that is exerted on learners' lives. As data represent the basis of influential endeavors of knowledge production and decision-making, EDS and critical social science try to draw a collaborative interdisciplinary methodology for meaningful analysis of EDS. Additionally, *datafication in education* is seen from a critical and supposed view by Selwyn and Gašević [262], who quote datafication as "the drive to turn vast amounts of activity and human behaviour into data points that are tracked, collected, and analysed to back the data-driven processes in education". Thus, digital trace data is dynamically produced allowing EDS to analyze learning as a developmental process comprised of sequential events.

Concerning bio-datafication, Williamson [263] claims, "With the inclusion of bioinformatics and genetics in EDS, key findings have emerged from complex associations between educational traits and DNA (e.g., test performance, attainment, cognitive skills, achievement. . .), as well as life outcomes, which aim at considering that *precision education* is possible to personalize learning processes." In this sense, Gulson and Webb [264] figure out ideas of life surrounded by augmentation, computation, and intervention to identify their bias for education policy, focusing on new ways of network governance, self-responsibilization and choice, and the move to EDS. In this sense, authors highlight the distinction between computer science and biology, as well as the merging of biological problems and computational problems, which includes the use of forms of algorithmic governance in EDS that relies on biology, neurosciences, etc. Moreover, Perrotta and Williamson [265] notice the rise of public, private, and hybrid forms of data governance in education, which poses a problematic relation between policy, evidence, and profit, as private companies offer technical platforms and EDS methods for the measurement and analysis of datasets are gradually recognized as authoritative sources of evidence and knowledge about education. In this concern, a cognitive infrastructure is designed by Sellar and Gulson [266] to foster practices of automated thinking in education policy and governance that sketch human decision-making to improve learning outcomes in order to explain how data science and automation create new values in education policy contexts. In another vein, Williamson [267] conceives digital laboratories as the assembly of data science and artificial intelligence systems with psychological, biological, and brain sciences that exist inside computer machinery with the goal of instilling neuroimaging, molecular genetic analysis, and learning engineering to boost education research as an experimental data-intensive science.

3.3.3.2   Applications

In regard to the application of EDS in diverse educational affairs, four instances are distinguished: the first concerns *instruction,* the second *apprenticeship*, the third is related to *evaluation,* and the last to *resources.*

Instruction

In relation to teaching of a given domain that considers the support of EDS in some sense, six applications are found to illustrate how to promote an active role for teachers, offer a data-driven approach, use data science techniques, and instruct data science, which are introduced as follows. Firstly, McCoy and Shih [268] assert EDS is able to enable teachers to use educational analytics tools and empower them to become active players in the arena in spite of their shortages related to data literacy. Furthermore, Ndukwe et al. [214] propose an EDS approach to boost the culture of data-informed decision-making in higher education, which is based on the "teaching outcome model" that includes the collection, analysis, and delivery of data-driven evidence to foster teaching analytics. Regarding educational debates as a way of teaching, Doroudi [269] applies the bias–variance tradeoff of ML to debate about learning pedagogy and theories using diverse data science techniques (e.g., regularization, model ensembles) to navigate tradeoffs and generate valuable insights for educational research.

Concerning *data science education* as domain knowledge to be taught, it has also been a target of research as the survey made by Song and Zhu [270] reports, where in the USA four levels of programs consider such a topic: bachelor, certificate, master, and specialization in a doctorate degree. As one outcome of the study, authors propose a "data analytics lifecycle model" composed of eight steps (i.e., business understanding, data understanding, data preparation, model planning, model building, evaluation, deployment, and review and monitoring) as a critical path in data science education and practice. An additional case represents the experience shared by Turek et al. [271], who find that academy and industry have adopted data science, but theoretical criteria for effective tertiary-level data science education remain missed. Hence, they demand suitable components that range from a full end-to-end workflow to technological tools for development and team communications and a dose of motivation and incentives. Another case corresponds to a "data science curriculum for German upper secondary schools" (i.e., aged 15–18) designed by Heinemann et al. [272] that relies on both math and computer science including societal issues. It is inspired by the core of some models of data called *data literacy* to name the process oriented to instill the ability to collect, manage, evaluate, and apply data in a critical way. As an instance of a data science course, four modules are taken into account: from data to information, big data and artificial intelligence, data projects, and data science and society.

Apprenticeship

The application of EDS to learning endeavor is one of the main reasons to study in this field with the purpose of facilitating and enhancing the acquired domain knowledge at classroom and distance learning settings. In this regard, six works share the efforts for enhancing learning, reflection, and motivation as follows.

Baldassarre [273] pursues to reach five goals stated for improving learner outcomes, learning experience, and education guidance, as well as reducing dropouts and creating mass student-centric programs by means of some constructs, such as "data, information, knowledge, and wisdom model," three types of EBD (e.g., structured, unstructured, and semi-structured), three models of EBD (e.g., descriptive, predictive, and prescriptive), and seven phases for a data science approach (i.e., goal definition, data retrieval, data preparation, data exploration, data modeling, outcome visualization, and analysis automation). In relation to social–emotional learning, Liu and Huang [274] apply EDS for the prediction and measurement of achievements, including their relationship with interventions. With this aim, a data analytics thinking framework is used to formalize the knowledge discovery process through six steps (i.e., understanding of the problem domain, understanding of the data, preparation of the data, mining of the data, evaluation of the discovered knowledge, and use of the discovered knowledge).

Meanwhile, Pesonen et al. [40] leverage a student registry data warehouse and EDS techniques to train predictive models with the goal of estimating students' probability of graduation and graduation on time, as well as to predict time-to-degree. Based on the predictions, authors warn of diverse ethical issues, for instance, the danger of unintentionally promoting social injustice and the discrimination of minorities. With the purpose of offering personalized learning support to instill students' reflective skills during their academic progress, Jung et al. [275] make use of theory-informed data science methods to analyze the measure of reflection depth (e.g., no, shallow, deep) and automatically detect six theoretically indicated items of reflection quality (i.e., description, analysis, feeling, perspective, evaluation, and outcome) relying on linguistic features in the reflections.

Concerning the study of learners' motivation to engage in academic practices, a pair of works are briefed, where Douglas et al. [276] seek sources of evidence for the validity of using their motivational scale, in addition to some instruments for measurement, learner performance, and EDS techniques as an argument-based approach to provide stronger levels of evidence to justify claims made about learners' motivation based on the outcomes, whereas Johnson and Olney [277] strive to increase motivation to learn data science by accessing a community-sourced dataset and deal with real-life data science problems, where the students' discourse is analyzed through word count and linguistic inquiry.

Evaluation

In this subsection seven works are introduced to show how EDS fosters assessment duties in general and particularly those related to learners' performance. The aim is to shape a glimpse of the way that EDS deals with evaluation labor. Once again, the pioneer EDS approach built by Gibson and Webb [202] is worthy to be resumed to precise that it was oriented to the assessment of collaborative problem solving, particularly the four already mentioned targets, as a need of the shift from paper-based tests to online learning, and with the aim of progressing on and the calling for the restructuring of training of the current generation. Another early EDS work corresponds to the assessment of data generated by students' writing, which Cope and Kalantzis [278] tackle to identify the range of processes for collecting and interpreting evidence of learning that have the potential to produce unprecedented amounts of data: structured data embedded in learning, machine assessments, and unstructured data collected incidental to learning activity, where the main concerns correspond to data models, data access, and data privacy.

With respect to performance, there are some EDS efforts to predict learners' realization during academic labor. For instance, Waheed et al. [215] aim at predicting performance of students using deep learning, which evolved from ML and is featured by several computational tiers to model the way of learning from examples, patterns, or events with the goal of representing the students' learning perception, facilitating longitudinal interventions by the academia, and tailoring teaching methods to enrich learners' experience. With the idea of improving data pipelines to predict learners' performance, Bertolini et al. [279] harness educational corpora to identify a set of key traits prior to model development and quantify the stability of feature selection techniques. Specifically, aside from pinpointing a subset of suitable attributes, it is possible to reduce the quantity of variables that need to be managed by stakeholders, in addition to making "black-box" procedures more interpretable and guiding faculty to deploy targeted interventions.

Furthermore, Jantakun et al. [280] also try to predict learners' performance by means of a system architecture that includes a data science process, ML methods (e.g., classification through a decision tree, neural networks, naïve Bayes, etc.), and datasets to carry out a context systematic analysis and assessments. What is more, Garmpis et al. [281] are also interested in predicting the length of time needed for students to pass their courses. As such, they define a data science strategy named "automated ML algorithm" to find out the best models and parameters using their historical course performance and academic assessments. Such an approach mixes fuzzy-based active learning methods for predicting students' performance and semi-supervised classification tasks for predicting student dropout. Lastly, Hidalgo et al. [249] strive to predict students' performance too, taking care of clickstream activity to gather huge data about their behavior during the course. Hence, they design a model that deploys deep learning and meta-learning (i.e., techniques that enable the predictive model to "learn to learn") with the idea of automatically optimizing the architecture and hyper-parameters of a deep neural network.

Resources

In relation to the kind of resources that contribute to deploying EDS applications in pursuit of adapting, personalizing, and enhancing the teaching and learning labor, a brief of the gained experience is outlined in this subsection, where the first corresponds to datasets. This valuable resource is indispensable for testing EDS approaches, as Quy et al. [282] access the two variations of a real dataset often used in EDS and a student performance data science dataset gathered at their institute to test a multi-fair-capacitated view that fairly partitions students into non-overlapping groups while ensuring balanced group cardinalities. Concerning education data scientists, Rosenberg et al. [283] analyze their profile and find out they have confidence with mathematics and the substantive aspects of their work oriented to programming, particularly with big data and ML.

Another main resource is the algorithm that exploits EBD to discover valuable knowledge, as Maylawati et al. [284] report the use of the classic *k-means* and exploratory data analysis to examine questionnaires responded to by students to seek a relevant pattern that inspires a recommendation model for strategies to strengthen digital culture in the academic community. A typical software used to deploy EDS procedures is *R*, which is promoted by Estrellado et al. [285] to motivate educators who wonder how to use data better and engage in learning to produce code to deploy EDS approaches that improve their teaching affairs. The implementation of EDS applications demands the use of suitable platforms, for instance, the Data Science

Tutor built by Kotsiopoulos et al. [286] to cover the lack of EDS facilities that deliver automated little-by-little trial solving in a diversity of data science algorithms giving insights into the traits for a given approach.

Lastly, social networks are also a demanding setting that generates a diversity of data that have represented an object of interest for the application of EDS. In this concern, Rosenberg et al. [287] harness Twitter and the next generation of science standards to understand public sentiment about large-scale educational reforms. Particularly, they focus on social media to catch opinion and sentiment expressions to get an overall idea of people's position expressed toward the current system-wide curricular standards-based reform effort in science education.

3.4 Discussion

Now that a landscape of the EDS arena has been outlined and a diversity of related works have illustrated the proposed *taxonomy of EDS*, in this section the discussion of the results is achieved through the following four topics: *analysis* of the results, a *reflection* of the status from four perspectives, a *comparison* of the contribution of the present review to similar ones, and the *responses* to research questions.

3.4.1 Analysis of Results

Essentially this chapter introduces *the taxonomy of EDS* outlined in Sect. 3.2.3, which is the result of the used logistic and available materials already described. As such, in this subsection, three aspects are explained, one related to the number of articles reported, another that analyzes the annual tendency of publications, and the last concerning the classification of works.

With respect to the taxonomy of EDS, it organizes two series of articles, one to embrace the domains that represent the baseline and the related fields considered for EDS, and another to feature the literature that is related with EDS. This is the reason to explain why the accumulation of articles ignores the quantity of papers that have been cited in the instances of the subcategories "1.1 Baseline" and "1.2 Related Fields." Moreover, the summations, $\sum$, edited in Tables 3.1 and 3.2, which, respectively, represent the number of articles annually published and the amount of works classified for each category (i.e., the "1 Context" partially considered), do not coincide because some contributions appear in more than one instance increasing the total of papers for the subcategories shown in Table 3.2.

As for the timeline of publications annually produced, the statistics edited in Table 3.1 is the main proof that demonstrates *EDS is an emergent field*, whose maturation is slowly in progress according to the growing number of papers, which in average corresponds to 8 and 32 during the first and second partial decades of the twenty-first century, respectively. Therefore, it is expected that during the remainder of the 2020s, the number will continue to increase to reach a stable annual quantity.

In relation to the weight that each element of the taxonomy of EDS represents, nearly one-third correspond to papers that superficially allude to EDS without really contributing to extend the domain; as such, they only spread the field, in contrast with the effect that triggers the lack of authentic EDS content. In addition, both the "umbrella" works and those that only offer quite a few mentions about EDS in their text (e.g., instances 2.2.1 and 2.3.1, respectively) accumulate 40 publications that nearly reach 25% of the total, and similarly to the prior case, the contribution to extend EDS is poor, which produces a feeling of disappointment. Lastly, most of the advance for EDS weights on the 75 articles classified as introductory, concrete, depth, review, and approaches that together represent the 45% of the literature.

3.4.2 Reflection of Educational Data Science

According to the pioneer works, EDS is reaching its first decade of existence. So its promotion and expectations demand an analysis of four factors that could contribute to foster its development or restrict its evolution for being worthy to be really considered a science that formally impacts human and academy life.

- *Strengths*: EDS as a subfield of application of data science benefits from a robust baseline to ground its designs and approaches. Furthermore, its related fields are dynamically progressing, opening new venues for deeper and more specialized applications that correspond to EDS expertise.
- *Weaknesses*: Being a newer domain than its allies is a disadvantage for incorporating authentic and loyal researchers who honor the nature and purpose of EDS, because most of them have been working in those related domains for more than a decade. In contrast with those cognate fields, EDS lacks a specialized society, conference, and journal that encourage, lead, organize, foster, and disseminate worldwide labor. So a call for a community, a society, a conference, and a journal are made!
- *Opportunities*: The massive use of diverse sorts of educational systems at classroom and digital spaces; the revolution of distinct environments, as social networks and virtual settings; and the spread of datafication and datasets are some of the main supporters for the development and application of a real EDS that grounds the entire process of data manipulation. What is more, the demand for adaptive and personalized teaching and learning processes as a way to enhance education endeavor is increasing through academic institutions, authorities, scholars, staff, and specially students; as such, this is the main justification to deploy effective and reliable EDS approaches.
- *Threats*: The dual conception of what really EDS is constitutes the principal menace for its evolution, because the assumption of EDS as an umbrella term constrains the maturation process of the domain as a real discipline, which claims

for sketching its own identity that relies on a formal theory, models, frameworks, and methods, as well as tools to differentiate it from related domains. Furthermore, the window opportunity to evolve is limited, and EDS has to take advantage of being a buzzword that meets the market expectations.

3.4.3 Contribution of the Review

Based on the five works introduced in Sect. 3.3.2.4.2, the merit of this chapter for contributing to extend the EDS arena is highlighted by means of contrasting its features against those of such a sample. Hence, according to the profile given for those reviews, some traits are used for comparison aims:

1. *Donaldson et al.* [256]: The scope is limited to primary and secondary education in the UK, with the inclusion of ML as a partner of EDS and asserting EDS as an umbrella for EDM and LA. In contrast, this review offers a broad scope for all educational levels, acknowledges ML as one of its underlying domains, and contributes to define the identity of EDS as a real domain.
2. *Romero and Ventura* [224]: It sees MOOC as a target for EDS application and considers EDS as an umbrella to cover four fields, only 61 related works are considered for the sample, and it only defines four challenges of application for EDS. With a different vision, this chapter is not engaged with a particular educational setting and admits those fields as related ones and not as subordinates of EDS, its sample of related works is nearly three times bigger, and it identifies a wider variety of applications organized as the "Approach" category.
3. *Daniel* [33]: It links big data with data science for educational affairs, identifies some issues to be considered to work with EBD, and claims for giving attention to preparing educational researchers. In this concern the present chapter recognizes big data as part of the baseline for EDS, the main issue considered here corresponds to the dual identity granted to the domain that produces misconception and restricts its development, and it introduces some works oriented to depict a profile to characterize educational data scientists.
4. *Fischer et al.* [128]: The focus is the mining of EBD, considering EDS as a supplementary topic, and it organizes educational datasets into three distinctive levels. From a different outlook, this chapter strives for dignifying EDS labor in addition to recognizing EBD and EDM as partner fields, but preserving EDS as the main and unique target of interest, and introduces some related works that describe datafication as the explosion of digital heterogeneous data.
5. *McFarland et al.* [218]: A report of the background, present, and future for EDS is outlined, acknowledges statistics and computer science as underlying elements for EDS, and recognizes the dual view of being an umbrella and emergent field for EDS. Some degree of similarities are identified with this chapter, particularly those that correspond with the background and current status for EDS, as well as the contradictory views of its nature; but this review also considers other

underlying fields to ground the EDS domain and analyzes a wider collection of related works than just 12 articles considered by cited authors.

3.4.4 Responses to Research Questions

As the last topic for the discussion of results, in this subsection the answers for the research problem and the five questions made in Sect. 3.1 are given. Hence, according to the landscape outlined through Sects. 3.2.3 and 3.3, in the following paragraph, the response to the question "What really is EDS?" is offered, and the reply for the questions is edited in a series of five items.

EDS is an emergent domain of research, development, and application in educational settings that follows a data-driven strategy to adapt and personalize teaching and learning processes, which, in spite of being attached to a pair of contradictory concepts, pursues to define its proper nature as a science, whose object of study is the data and its characterization, management, meaning, and interpretation given an educational context and a specific approach used to discover knowledge.

1. *What about the research being accomplished in EDS?* An account of the labor carried out since 2013 up to date has been illustrated in this chapter through a sample of 153 papers, their classification, and compilation.
2. *What is the baseline that grounds EDS?* Although additional disciplines could be considered (e.g., mathematics, learning science, computer science, and artificial intelligence), eight domains are chosen as the most relevant.
3. *What are the main work lines that distinguish EDS endeavor?* The taxonomy of EDS is proposed to clarify EDS practice and identify its three main research lines, which are composed of several subcategories and thereby diverse instances that label the essence of the accomplished work.
4. *What are the key features that profile EDS labor?* The focus on data, more than the mining or analysis as its related fields do, as well as the essential use of statistics, data analysis, and ML to be engaged in knowledge discovery.
5. *What are the expectations and shortcomings to be taken into account for boosting the progress of EDS?* Essentially, they correspond to gradually evolving as a mature and empirical science that formally uncovers the hidden mysteries of data to improve educational labor, facilitating teaching duties, and enhancing students' learning; and in contrast, gaining a particular identity that distinguishes it from other related domains instead of simply being a term in vogue!

3.5   Conclusions

In this review a broad landscape is tailored to characterize the research fulfilled in the EDS domain and offer a response to depict the nature and concept of EDS. In consequence, the method applied was oriented to gather 153 articles related to EDS and distinguish how they contribute to extend the domain. Surprisingly, 25% of them inspire the concept of EDS as simply an *umbrella* term that gathers some related works under a "buzzword" to facilitate the diffusion of domains that examine EBD under data-driven strategies to lead educational affairs.

Some of the relevant contributions of this review correspond to the glimpse of EDS produced as a result of a wide sample of related works published since 2013 up to date, the acknowledgment of eight underlying disciplines that ground EDS labor, and particularly the design of *the taxonomy of EDS* that highlights the main research lines that identify the achieved labor and facilitate its organization.

However, EDS lacks the essential epistemological items of a discipline, which gather practitioners through a society, a specialized congress that facilitates the meeting of those interested in establishing academic ties to inspire joint research projects, and a dedicated journal to publish the outcomes of mature work that extends the scope and establishes the theoretical constructs of EDS. In addition, the vast amount of papers that only use the EDS term as an umbrella label their work with a term in vogue or scarcely contribute to advance the domain.

As a future work, EDS practitioners should strive for organizing a formal society that develops the conference events oriented to know, analyze, and disseminate the labor, as well as launching the journal that publishes the relevant works that ground the EDS baseline, offer a broad and deep view of the domain, and report the features and achievements of those works that apply EDS to cope with particular issues of the educational environment.

Lastly, *is EDS an umbrella term or an emergent domain*? The answer is, *up to date, both*.

Acknowledgments This research was partially funded by the grants CONACYT-SNI-36453, IPN-SIBE-2023-2024, IPN-SIP-EDI, IPN-SIP 2022-0803, and IPN-SIP 2023-0248. A special mention and gratitude is given to master student José Morales–Ramirez for his valuable contribution to develop this research and Lawrence Whitehill, a British expert reviewer who tuned the manuscript. Last but not least, I acknowledge and express a testimony of the unction and strength given by my Father, Brother Jesus, and Helper, as part of the research projects performed by World Outreach Light to the Nations Ministries (WOLNM).

References

1. Brynjolfsson, E., McAfee, A.: Race Against the Machine: How the Digital Revolution Is Accelerating Innovation, Driving Productivity, and Irreversibly Transforming Employment and the Economy. Brynjolfsson and McAfee (2011)

2. Baldemair, R., Dahlman, E., Fodor, G., Mildh, G., Parkvall, S., Selén, Y., et al.: Evolving wireless communications: addressing the challenges and expectations of the future. IEEE Veh. Technol. Mag. **8**(1), 24–30 (2013)
3. Baecker, R.M.: Computers and Society: Modern Perspectives. Oxford University Press, USA (2019)
4. Mullan, K., Wajcman, J.: Have mobile devices changed working patterns in the 21st century? a time-diary analysis of work extension in the UK. Work Employ. Soc. **33**(1), 3–20 (2019)
5. Pahl, C., Fronza, I., El Ioini, N., Barzegar, H.R.: A review of architectural principles and patterns for distributed Mobile information systems. In: WEBIST, pp. 9–20 (2019)
6. Yan, S., Ramachandran, P.G.: The current status of accessibility in mobile apps. ACM Trans. Access. Comput. **12**(1), 1–31 (2019)
7. Aichner, T., Grünfelder, M., Maurer, O., Jegeni, D.: Twenty-five years of social media: a review of social media applications and definitions from 1994 to 2019. Cyberpsychol. Behav. Soc. Netw. **24**(4), 215–222 (2021)
8. Wang, H., Skau, E., Krim, H., Cervone, G.: Fusing heterogeneous data: a case for remote sensing and social media. IEEE Trans. Geosci. Remote Sens. **56**(12), 6956–6968 (2018)
9. Rosa, L., Silva, F., Analide, C.: Mobile networks and internet of things: contributions to smart human mobility. In: Dong, Y., Herrera-Viedma, E., Matsui, K., Omatsu, S., González Briones, A., Rodríguez González, S. (eds.) Distributed Computing and Artificial Intelligence, 17th International Conference. DCAI 2020 Advances in Intelligent Systems and Computing, vol. 1237. Springer, Cham (2021). https://doi.org/10.1007/978-3-030-53036-5_18
10. Agrawal, N., Tapaswi, S.: Defense mechanisms against DDoS attacks in a cloud computing environment: state-of-the-art and research challenges. IEEE Commun. Surv. Tutor. **21**(4), 3769–3795 (2019)
11. Haber, M.J., Chappell, B., Hills, C.: Cloud computing. In: Cloud Attack Vectors, pp. 9–25. Apress, Berkeley, CA (2022)
12. Yuan, J., Zhang, J., Shen, L., Zhang, D., Yu, W., Han, H.: Massive data management and sharing module for connectome reconstruction. Brain Sci. **10**(5), 314 (2020)
13. Naeem, M., et al.: Trends and future perspective challenges in big data. In: Advances in Intelligent Data Analysis and Applications, pp. 309–325. Springer, Singapore (2022)
14. Toporowicz, F.Z., Souza, J.T.D., Piekarski, C.M.: The knowledge discovery in databases approach: identifying variables that influence ISO 9001 and ISO 14001 certifications. J. Environ. Plan. Manag. **64**(7), 1271–1290 (2021)
15. Guarascio, M., Manco, G., Ritacco, E.: Knowledge discovery in databases. In: Encyclopedia of Bioinformatics and Computational Biology: ABC of Bioinformatics, p. 336 (2018)
16. Han, J., Pei, J., Tong, H.: Data Mining: Concepts and Techniques. Morgan Kaufmann, Burlington, MA (2022)
17. Davenport, T.H.: From analytics to artificial intelligence. J. Bus. Anal. **1**(2), 73–80 (2018)
18. Martinez, I., Viles, E., Olaizola, I.G.: Data science methodologies: current challenges and future approaches. Big Data Res. **24**, 100183 (2021)
19. Zhu, Y., Xiong, Y.: Towards data science. Data Sci. J. **14**(8), 1–7 (2015)
20. Dhar, V.: Data science and prediction. Commun. ACM **56**(12), 64–73 (2013)
21. Waller, M.A., Fawcett, S.E.: Data science, predictive analytics, and big data: a revolution that will transform supply chain design and management. J. Bus. Logist. **34**(2), 77–84 (2013)
22. Kelleher, J.D., Tierney, B.: Data Science. MIT Press, Cambridge (2018)
23. Blei, D.M., Smyth, P.: Science and data science. Proc. Natl. Acad. Sci. **114**(33), 8689–8692 (2017)
24. Sanchez-Pinto, L.N., Luo, Y., Churpek, M.M.: Big data and data science in critical care. Chest. **154**(5), 1239–1248 (2018)
25. Gibert, K., Horsburgh, J.S., Athanasiadis, I.N., Holmes, G.: Environmental data science. Environ. Model Softw. **106**, 4–12 (2018)
26. George, G., Osinga, E.C., Lavie, D., Scott, B.A.: Big data and data science methods for management research. Acad. Manag. J. **59**(5), 1493–1507 (2016)

27. Provost, F., Fawcett, T.: Data Science for Business: What you Need to Know about Data Mining and Data-Analytic Thinking. O'Reilly Media, Inc., Sebastopol (2013)
28. Green, B.: Data science as political action: grounding data science in a politics of justice. J. Soc. Comput. **2**(3), 249–265 (2021)
29. Sarker, I.H., Kayes, A.S.M., Badsha, S., Alqahtani, H., Watters, P., Ng, A.: Cybersecurity data science: an overview from machine learning perspective. J. Big data. **7**(1), 1–29 (2020)
30. Baldi, P.: Deep learning in biomedical data science. Ann. Rev. Biomed. Data Sci. **1**, 181–205 (2018)
31. Giudici, P.: Financial data science. Stat. Prob. Lett. **136**, 160–164 (2018)
32. Klašnja-Milićević, A., Ivanović, M., Budimac, Z.: Data science in education: big data and learning analytics. Comput. Appl. Eng. Educ. **25**(6), 1066–1078 (2017)
33. Daniel, B.K.: Big data and data science: a critical review of issues for educational research. Br. J. Educ. Technol. **50**(1), 101–113 (2019)
34. Kalota, F.: Applications of big data in education. Int. J. Educ. Pedag. Sci. **9**(5), 1607–1612 (2015)
35. Lara, J.A., Lizcano, D., Martínez, M.A., Pazos, J., Riera, T.: A system for knowledge discovery in e-learning environments within the European Higher Education Area–Application to student data from Open University of Madrid, UDIMA. Comput. Educ. **72**, 23–36 (2014)
36. Williamson, B.: Coding the biodigital child: the biopolitics and pedagogic strategies of educational data science. Pedag. Cult. Soc. **24**(3), 401–416 (2016)
37. Rosenberg, J.M., Lawson, M., Anderson, D.J., Jones, R.S., Rutherford, T.: Making data science count in and for education. In: Research Methods in Learning Design and Technology, pp. 94–110. Routledge, London (2020)
38. Demchenko, Y., Belloum, A., de Laat, C., Loomis, C., Wiktorski, T., Spekschoor, E.: Customisable data science educational environment: from competences management and curriculum design to virtual labs on-demand. In: 2017 IEEE International Conference on Cloud Computing Technology and Science (CloudCom), pp. 363–368. IEEE, New York (2017)
39. Williamson, B.: Digital methods and data labs: the redistribution of educational research to education data science. In: The Digital Academic, pp. 140–155. Routledge, London (2017)
40. Pesonen, J., Fomkin, A., Jokipii, L.: Building data science capabilities into university data warehouse to predict graduation. arXiv preprint arXiv:1805.05401 (2018)
41. Sakulwichisintu, S.: The role of data science in online education. TLA Bull. **65**(2), 24–40 (2021)
42. Provost, F., Fawcett, T.: Data science and its relationship to big data and data-driven decision making. Big Data. **1**(1), 51–59 (2013)
43. Irizarry, R.A.: The role of academia in data science education. Harvard Data Sci. Rev. **2**(1) (2020). https://doi.org/10.1162/99608f92.dd363929
44. Van der Laan, M.J., Rose, S.: Targeted Learning in Data Science. Springer International Publishing, Cham (2018)
45. Buckingham-Shum, S., Hawksey, M., Baker, R.S., Jeffery, N., Behrens, J.T., Pea, R.: Educational data scientists: a scarce breed. In: Proceedings of the Third International Conference on Learning Analytics and Knowledge, pp. 278–281 (2013)
46. Zins, C.: Conceptual approaches for defining data, information, and knowledge. J. Am. Soc. Inf. Sci. Technol. **58**(4), 479–493 (2007)
47. Jifa, G., Lingling, Z.: Data, DIKW, big data and data science. Procedia Comput. Sci. **31**, 814–821 (2014)
48. Rider, F.: The Scholar and the Future of the Research Library: A Problem and its Solution, pp. xiv, 236. The Wesleyan University Library. New York: Hadham Press (1944)
49. Lesk, M.: How much information is there in the world? https://courses.cs.washington.edu/courses/cse590s/03au/lesk.pdf (1997)

50. Selwyn, N.: Data entry: towards the critical study of digital data and education. Learn. Media Technol. **40**(1), 64–82 (2015)
51. Williamson, B., Bayne, S., Shay, S.: The datafication of teaching in higher education: critical issues and perspectives. Teach. High. Educ. **25**(4), 351–365 (2020)
52. Manogaran, G., Thota, C., Lopez, D.: Human-computer interaction with big data analytics. In: Research Anthology on Big Data Analytics, Architectures, and Applications, pp. 1578–1596. IGI Global (2022)
53. Cox, M., Ellsworth, D.: Managing big data for scientific visualization. ACM Siggraph. **97**(1), 21–38 (1997)
54. Mashey, J.R.: Big data and the next wave of {InfraStress} problems, solutions, opportunities. In: 1999 USENIX annual technical conference (USENIX ATC 99) (1999)
55. Bryson, S., Kenwright, D., Cox, M., Ellsworth, D., Haimes, R.: Visually exploring gigabyte data sets in real time. Commun. ACM. **42**(8), 82–90 (1999)
56. Gökalp, M.O., Gökalp, E., Kayabay, K., Gökalp, S., Koçyiğit, A., Eren, P.E.: A process assessment model for big data analytics. Comput. Stand. Interface. **80**, 103585 (2022)
57. Saggi, M.K., Jain, S.: A survey towards an integration of big data analytics to big insights for value-creation. Inf. Process. Manag. **54**(5), 758–790 (2018)
58. Rahman, F., Slepian, M.J.: Application of big-data in healthcare analytics—prospects and challenges. In: 2016 IEEE-EMBS International Conference on Biomedical and Health Informatics (BHI), pp. 13–16. IEEE (2016)
59. Von Mises, R.: Mathematical Theory of Probability and Statistics. Academic Press, Cambridge (2014)
60. Hopcroft, J., Kannan, R.: Foundations of data science. Microsoft. https://www.microsoft.com/en-us/research/publication/foundations-of-data-science/ (2014)
61. Johnson, R.R., Kuby, P.J.: Elementary Statistics. Cengage Learning, Bostan (2011)
62. Press, G.: A very short history of data science. Forbes. https://www.forbes.com/sites/gilpress/2013/05/28/a-very-short-history-of-data-science/?sh=753407c755cf (2013)
63. Olhede, S.C., Wolfe, P.J.: The future of statistics and data science. Stat. Prob. Lett. **136**, 46–50 (2018)
64. Jones, M.L.: How we became instrumentalists (again) data positivism since world war II. Hist. Stud. Nat. Sci. **48**(5), 673–684 (2018)
65. Tukey, J.W.: The future of data analysis. Ann. Math. Stat. **33**(1), 1–67 (1962)
66. Donoho, D.: 50 years of data science. J. Comput. Graph. Stat. **26**(4), 745–766 (2017)
67. Tukey, J.W.: Exploratory Data Analysis, 1st edn. Pearson, London (1977)
68. Morgenthaler, S.: Exploratory data analysis. Wiley Interdisc. Rev. Comput. Stat. **1**(1), 33–44 (2009)
69. Rutherford, A.: Anova and Ancova: A GLM Approach. John Wiley & Sons, Inc., New York (2013)
70. Karageorgiou, E.: The logic of exploratory and confirmatory data analysis. Cogn. Crit., 35–48 (2011)
71. Samuel, A.L.: Some studies in machine learning using the game of checkers. IBM J. Res. Dev. **44**, 206–226 (1959)
72. Hormann, A.M.: Programs for machine learning part I. Inf. Control. **5**(4), 347–367 (1962)
73. Hormann, A.M.: Programs for machine learning. Part II. Inform. Control. **7**(1), 55–77 (1964)
74. Jordan, M.I., Mitchell, T.M.: Machine learning: trends, perspectives, and prospects. Science. **349**(6245), 255–260 (2015)
75. Alpaydin, E.: Machine Learning. MIT Press (2021)
76. Janiesch, C., Zschech, P., Heinrich, K.: Machine learning and deep learning. Electron. Mark. **31**(3), 685–695 (2021)
77. Sarker, I.H.: Machine learning: algorithms, real-world applications and research directions. SN Compu. Sci. **2**(3), 1–21 (2021)
78. Piatetsky-Shapiro, G.: The journey of knowledge discovery. In: Journeys to Data Mining, pp. 173–196. Springer, Berlin (2012)

79. Dasgupta, H.: Data mining and statistics: tools for decision making in the age of big data. In: Handbook of Research on Advanced Data Mining Techniques and Applications for Business Intelligence, pp. 15–33. IGI Global (2017)
80. Sumiran, K.: An overview of data mining techniques and their application in industrial engineering. Asian J. Appl. Sci. Technol. **2**(2), 947–953 (2018)
81. Denton, F.T.: Data mining as an industry. Rev. Econ. Stat., 124–127 (1985)
82. Lovell, M.C.: Data mining. Rev. Econ. Stat. **65**(1), 1–12 (1983)
83. Marquez, J., Shack-Marquez, J., Wascher, W.L.: Statistical inference, model selection and research experience: a multinomial model of data mining. Econ. Lett. **18**(1), 39–44 (1985)
84. Mauleón, I.: Stability Testing in Regression Models. Banco de España, Madrid (1985)
85. Mayer, T.: Economics as a hard science: realistic goal or wishful thinking? Econ. Inq. **18**(2), 165 (1980)
86. McCloskey, D.N.: The loss function has been mislaid: the rhetoric of significance tests. Am. Econ. Rev. **75**(2), 201–205 (1985)
87. Aggarwal, C.C.: Data Mining: The Textbook, vol. 1. Springer, New York (2015)
88. Sousa, M.S., Mattoso, M.L.Q., Ebecken, N.F.F.: Data mining: a database perspective. Trans. Inform. Commun. Technol. **19**, 413–431 (1998)
89. Niño, M., Illarramendi, A.: Understanding big data: antecedents, origin and later development. DYNA New Technol. **2**(1), 1–8 (2015)
90. Kantardzic, M.: Data Mining: Concepts, Models, Methods, and Algorithms. John Wiley & Sons, New York (2011)
91. Fayyad, U.: Knowledge discovery in database: an overview. In: Proceedings of Inductive Logic Programming: 7th International Workshop, ILP-97, pp. 3–16 (1997)
92. Maimon, O., Rokach, L. (eds.): Data Mining and Knowledge Discovery Handbook (2005)
93. Frawley, W., Piateski-Shapiro, G., Matheus, C.J.: Knowledge discovery in databases. AI Mag. **13**(3) (1992)
94. Steele, J.A., McDonald, J.R., D'Arcy, C.: Knowledge Discovery in Databases: Applications in the Electrical Power Engineering Domain (1997)
95. Piateski, G., Frawley, W.: Knowledge Discovery in Databases. MIT Press, Cambridge (1991)
96. Rezende, S.O., Oliveira, R.B.T., Felix, L.C.M., Rocha, C.A.J.: Visualization for knowledge discovery in database. In: Transactions on Information and Communication Technologies, vol. 19. WIT Press (1998)
97. Singhal, N., Himanshu: A review on knowledge discovery from databases. In: Electronic Systems and Intelligent Computing, pp. 457–464 (2022)
98. Dong, X., Yu, Z., Cao, W., Shi, Y., Ma, Q.: A survey on ensemble learning. Front. Comp. Sci. **14**(2), 241–258 (2020)
99. Davenport, T.H.: Analytics 3.0. Harv. Bus. Rev. **91**(12), 64–72 (2013)
100. Davenport, T.H.: Competing on analytics. Harv. Bus. Rev. **84**(1), 98 (2006)
101. Davenport, T., Harris, J.: Competing on Analytics: Updated, with a New Introduction: The New Science of Winning. Harvard Business Press, Boston (2017)
102. Cooper, A.: What is analytics? definition and essential characteristics. CETIS Anal. Ser. **1**(5), 1–10 (2012)
103. Turkay, C., Kaya, E., Balcisoy, S., Hauser, H.: Designing progressive and interactive analytics processes for high-dimensional data analysis. IEEE Trans. Vis. Comput. Graph. **23**(1), 131–140 (2016)
104. Greasley, A.: Simulating business processes for descriptive, predictive, and prescriptive analytics. In: Simulating Business Processes for Descriptive, Predictive, and Prescriptive Analytics. de Gruyter, London (2019)
105. Raina, A.: Optimizing Interactive Analytics Engines for Heterogeneous Clusters (2018)
106. Frazzetto, D., Nielsen, T.D., Pedersen, T.B., Šikšnys, L.: Prescriptive analytics: a survey of emerging trends and technologies. VLDB J. **28**(4), 575–595 (2019)
107. Marriott, K., Schreiber, F., Dwyer, T., Klein, K., Riche, N.H., Itoh, T., Thomas, B.H. (eds.): Immersive Analytics, vol. 11190. Springer, Berlin (2018)

108. Runkler, T.A.: Data Analytics. Springer Fachmedien Wiesbaden, Berlin (2020)
109. Ghavami, P.: Big Data Analytics Methods. de Gruyter, Berlin (2019)
110. Evans, J.R.: Business Analytics. Pearson, London (2017)
111. Stieglitz, S., Dang-Xuan, L., Bruns, A., Neuberger, C.: Social media analytics. Bus. Inf. Syst. Eng. **6**(2), 89–96 (2014)
112. El-Nasr, M.S., Drachen, A., Canossa, A.: Game Analytics. Springer, London (2016)
113. Souza, G.C.: Supply chain analytics. Bus. Horiz. **57**(5), 595–605 (2014)
114. Ledford, J.L., Teixeira, J., Tyler, M.E.: Google Analytics. John Wiley and Sons, New York (2011)
115. Naur, P.: The science of datalogy. Commun. ACM. **9**(7), 485 (1966)
116. Naur, P.: Concise Survey of Computer Methods. Petrocelli Books (1974)
117. Hayashi, C., Yajima, K., Bock, H.H., Ohsumi, N., Tanaka, Y., Baba, Y. (eds.): Data Science, Classification, and Related Methods. In: Proceedings of the Fifth Conference of the International Federation of Classification Societies (IFCS-96), Kobe, Japan, Mar 27–30, 1996.. Springer Science & Business Media (2013)
118. Van Dyk, D., Fuentes, M., Jordan, M.I., Newton, M., Ray, B.K., Lang, D.T., Wickham, H.: ASA statement on the role of statistics in data science. Amstat News. **460**(9), 24 (2015)
119. Dobre, C., Xhafa, F.: Intelligent services for big data science. Futur. Gener. Comput. Syst. **37**, 267–281 (2014)
120. Cao, L.: Data science: challenges and directions. Commun. ACM. **60**(8), 59–68 (2017)
121. Sato-Ilic, M.: Part V: data science and analytics preface. Procedia Comput. Sci. **36**, 276–277 (2014)
122. Blum, A., Hopcroft, J., Kannan, R.: Foundations of Data Science. Cambridge University Press, Cambridge (2020)
123. Fan, J., Li, R., Zhang, C.H., Zou, H.: Statistical Foundations of Data Science. Chapman and Hall/CRC, London (2020)
124. Igual, L., Santi, S.: Introduction to data science: a python approach to concepts. Tech. Appl. (2017)
125. NAE, National Academy of Education: Big Data in Education: Balancing the Benefits of Educational Research and Student Privacy. A Workshop Summary. ERIC Clearinghouse (2017)
126. Baig, M.I., Shuib, L., Yadegaridehkordi, E.: Big data in education: a state of the art, limitations, and future research directions. Int. J. Educ. Technol. High. Educ. **17**(1), 1–23 (2020)
127. Ruiz-Palmero, J., Colomo-Magaña, E., Ríos-Ariza, J.M., Gómez-García, M.: Big data in education: perception of training advisors on its use in the educational system. Soc. Sci. **9**(4), 53 (2020)
128. Fischer, C., et al.: Mining big data in education: affordances and challenges. Rev. Res. Educ. **44**(1), 130–160 (2020)
129. Sanjeev, P., Zytkow, J.M.: Discovering enrollment knowledge in university databases. In: KDD, pp. 246–251 (1995)
130. Zaiane, O., Xin, M., Han, J.: Discovering web access patterns and trends by applying OLAP and data mining technology on web logs. In: Advances in Digital Libraries, pp. 19–29 (1998)
131. Ingram, A.: Using web server logs in evaluating instructional web sites. J. Educ. Technol. Syst. **28**(2), 137–157 (1999)
132. Peled, A., Rashty, D.: Logging for success: advancing the use of www logs to improve computer mediated distance learning. J. Educ. Comput. Res. **21**(4), 413–431 (1999)
133. Rahkila, M., Karjalainen, M.: Evaluation of learning in computer based education using log systems. In: ASEE/IEEE Frontiers in Education Conference, San Juan, Puerto Rico, pp. 16–21 (1999)
134. Peña-Ayala, A.: Educational data mining. In: Studies in Computational Intelligence, p. 524. Springer, Berlin (2014)

135. Tsiakmaki, M., Kostopoulos, G., Kotsiantis, S., Ragos, O.: Implementing AutoML in educational data mining for prediction tasks. Appl. Sci. **10**(1), 90 (2019)
136. Du, X., Yang, J., Hung, J.L., Shelton, B.: Educational data mining: a systematic review of research and emerging trends. Inform. Discov. Deliv. **48**(4), 225–236 (2020)
137. Baek, C., Doleck, T.: Educational data mining: a bibliometric analysis of an emerging field. IEEE Access. **10**, 31289–31296 (2022)
138. Yağcı, M.: Educational data mining: prediction of students' academic performance using machine learning algorithms. Smart Learn. Environ. **9**(1), 1–19 (2022)
139. Tinto, V.: Leaving College: Rethinking the Causes and Cures of Student Attrition, 2nd edn. University of Chicago Press, Chicago (1987)
140. Mitchel, J., Costello, S.: Internationale-VET Market Research Report: A Report on International Market Research for Australian VET Online Products and Services. John Mitchell & Associates, Sydney, NSW (2000)
141. Berk, J.: The State of Learning Analytics. Report for American Society for Training & Development (2004)
142. Moore, C.: Measuring Effectiveness with Learning Analytics. Chief Learning Officer. https://www.chieflearningofficer.com/2005/05/03/measuring-effectiveness-with-learning-analytics/ (2005)
143. Retalis, S., Papasalouros, A., Psaromiligkos, Y., Siscos, S., Kargidis, T.: Towards networked learning analytics–a concept and a tool. In: Proceedings of the Fifth International Conference on Networked Learning, pp. 1–8 (2006)
144. Bach, C.: LeaGrning Analytics: Targeting Instruction, Curricula and Student Support. Drexel University, Philadelphia, PA (2010)
145. Peña-Ayala, A., Cárdenas-Robledo, L.A., Sossa, H.: A landscape of learning analytics: an exercise to highlight the nature of an emergent field. In: Learning Analytics: Fundaments, Applications, and Trends, pp. 65–112. Springer, Cham (2017)
146. Peña-Ayala, A.: Learning analytics: a glance of evolution, status, and trends according to a proposed taxonomy. Wiley Interdisc. Rev. Data Min. Knowl. Discov. **8**(3), e1243 (2018)
147. Lang, C., Siemens, G., Wise, A., Gasevic, D., (Eds.).: Handbook of Learning Analytics, p. 23. SOLAR, Society for Learning Analytics and Research, New York (2017)
148. Kew, S.N., Tasir, Z.: Learning analytics in online learning environment: a systematic review on the focuses and the types of student-related analytics data. Technol. Knowl. Learn. 1–23 (2021)
149. Du, X., Yang, J., Shelton, B.E., Hung, J.L., Zhang, M.: A systematic meta-review and analysis of learning analytics research. Behav. Inform. Technol. **40**(1), 49–62 (2021)
150. Paiva, R., Bittencourt, I.I., Lemos, W., Vinicius, A., Dermeval, D.: Visualizing learning analytics and educational data mining outputs. In: International Conference on Artificial Intelligence in Education, pp. 251–256 (2018)
151. Kiss, B., Nagy, M., Molontay, R., Csabay, B.: Predicting dropout using high school and first-semester academic achievement measures. In: 2019 17th International Conference on Emerging eLearning Technologies and Applications (ICETA), pp. 383–389. IEEE (2019)
152. Alonso, J.M., Casalino, G.: Explainable artificial intelligence for human-centric data analysis in virtual learning environments. In: International Workshop on Higher Education Learning Methodologies and Technologies Online, pp. 125–138. Springer, Cham (2019)
153. Qu, S., Li, K., Wu, B., Zhang, S., Wang, Y.: Predicting student achievement based on temporal learning behavior in MOOCs. Appl. Sci. **9**(24), 5539 (2019)
154. Yücel, E., Erol, S.: The gender analysis of enrolled students: a comparison study of Austrian and Turkish higher education. In: Proceedings of the International Symposium for Production Research 2019, pp. 36–47. Springer, Cham (2019)
155. Albó, L., Hernández-Leo, D.: How educators value design analytics for blended learning. In: HLS-D3@ EC-TEL, pp. 53–55 (2019)

156. Zahedi, L., Lunn, S.J., Pouyanfar, S., Ross, M.S., Ohland, M.W.: Leveraging machine learning techniques to analyze computing persistence in undergraduate programs. In: 2020 ASEE Virtual Annual Conference Content Access (2020)
157. Nguyen, Q., Rienties, B., Richardson, J.T.: Learning analytics to uncover inequality in behavioural engagement and academic attainment in a distance learning setting. Assess. Eval. High. Educ. **45**(4), 594–606 (2020)
158. Naseer, M., Zhang, W., Zhu, W.: Early prediction of a team performance in the initial assessment phases of a software project for sustainable software engineering education. Sustainability. **12**(11), 4663 (2020)
159. Marrhich, A., Lafram, I., Berbiche, N., El Alami, J.: A khan framework-based approach to successful MOOCs integration in the academic context. Int. J. Emerg. Technol. Learn. **15**(12), 4–19 (2020)
160. Michos, K., Hernández-Leo, D.: CIDA: a collective inquiry framework to study and support teachers as designers in technological environments. Comput. Educ. **143**, 103679 (2020)
161. Tsai, Y.S., et al.: Learning analytics in European higher education—trends and barriers. Comput. Educ. **155**, 103933 (2020)
162. Baranyi, M., Molontay, R.: Comparing the effectiveness of two remedial mathematics courses using modern regression discontinuity techniques. Interact. Learn. Environ. **29**(2), 247–269 (2021)
163. Sreenivasulu, M.D., Devi, J.S., Arulprakash, P., Venkataramana, S., Kazi, K.S.: Implementation of latest machine learning approaches for students grade prediction. Int. J. Early Child. **14**(3) (2022)
164. Yang, Y., Majumdar, R., Li, H., Flanagan, B., Ogata, H.: Design of a learning dashboard to enhance reading outcomes and self-directed learning behaviors in out-of-class extensive reading. Interact. Learn. Environ., 1–18 (2022). https://doi.org/10.1080/10494820.2022.2101126
165. Perez-Alvarez, R., Jivet, I., Pérez-Sanagustin, M., Scheffel, M., Verbert, K.: Tools designed to support self-regulated learning in online learning environments: a systematic review. IEEE Trans. Learn. Technol. (2022). https://doi.org/10.1109/TLT.2022.3193271
166. Maraza-Quispe, B., Valderrama-Chauca, E.D., Cari-Mogrovejo, L.H., Apaza-Huanca, J.M., Sanchez-Ilabaca, J.: A predictive model implemented in knime based on learning analytics for timely decision making in virtual learning environments. Int. J. Inform. Educ. Technol. **12**(2), 91–99 (2022)
167. Göktepe Körpeoğlu, S., Göktepe Yıldız, S.: Comparative analysis of algorithms with data mining methods for examining attitudes towards STEM fields. Educ. Inf. Technol., 1–36 (2022). https://doi.org/10.1007/s10639-022-11216-z
168. Ince, M.: Automatic and intelligent content visualization system based on deep learning and genetic algorithm. Neural Comput. Appl. **34**(3), 2473–2493 (2022)
169. Buchanan, R.A., Forster, D.J., Douglas, S., Nakar, S., Boon, H.J., Heath, T., Heyward, P., D'Olimpio, L., Ailwood, J., Eacott, S., Smith, S., Peters, M., Tesar, M.: Philosophy of education in a new key: exploring new ways of teaching and doing ethics in education in the 21st century. Educ. Philos. Theory. **54**(8), 1178–1197 (2022)
170. Daud, A., Aljohani, N.R., Abbasi, R.A., Lytras, M.D., Abbas, F., Alowibdi, J.S.: Predicting student performance using advanced learning analytics. In: Proceedings of the 26th International Conference on World Wide Web Companion, pp. 415–421 (2017)
171. Guo, S., Zhang, G.: Analyzing concept complexity, knowledge ageing and diffusion pattern of Mooc. Scientometrics. **112**(1), 413–430 (2017)
172. Selwyn, N.: Data points: exploring data-driven reforms of education. Br. J. Sociol. Educ. **39**(5), 733–741 (2018)
173. Deborah, L., Inger, M., Pat, T.: The digital academic: identities, contexts and politics. In: The Digital Academic, pp. 1–19. Routledge, London (2018)
174. Wise, A.F., Cui, Y.: Envisioning a learning analytics for the learning sciences. In: ICLS 2018 Proceedings, 1799–1806 (2018)

175. Knox, J.: Beyond the "c" and the "x": learning with algorithms in massive open online courses (MOOCs). Int. Rev. Educ. **64**(2), 161–178 (2018)
176. Jarboui, F., et al.: Markov decision process for MOOC users behavioral inference. In: European MOOCs Stakeholders Summit, pp. 70–80. Springer, Cham (2019)
177. Johanes, P., Thille, C.: The heart of educational data infrastructures= conscious humanity and scientific responsibility, not infinite data and limitless experimentation. Br. J. Educ. Technol. **50**(6), 2959–2973 (2019)
178. Kör, H., Erbay, H., Engin, M.: Activity suggestion decision support system design in online learning environment. Electron. Lett. Sci. Eng. **15**(3), 8–22 (2019)
179. Khalid, M.S., Chowdhury, S.A., Parveen, M.A.: A theoretical framework to analyze students' formative feedback on classroom teaching. In: 1st International Conference on Education in the Digital Ecosystem: Blended Learning in Teaching Training–Innovation and Good Practices (2019)
180. Wasson, B., Kirschner, P.A.: Learning design: European approaches. TechTrends. **64**(6), 815–827 (2020)
181. Khatri, P., Raina, K., Wilson, C., Kickmeier-Rust, M.: Towards mapping competencies through learning analytics: real-time competency assessment for career direction through interactive simulation. Assess. Eval. High. Educ. **45**(6), 875–887 (2020)
182. Cappello, G., Rizzuto, F.: Journalism and fake news in the Covid-19 era. Perspectives for media education in Italy. Media. Education. **11**(2), 3–13 (2020). https://doi.org/10.36253/me-9682
183. Baranyi, M., Nagy, M., Molontay, R.: Interpretable deep learning for university dropout prediction. In: Proceedings of the 21st Annual Conference on Information Technology Education, pp. 13–19 (2020)
184. Mor, Y., Dimitriadis, Y., Köppe, C.: Workshop Report: Hybrid Learning Spaces–Data, Design, Didactics (2020)
185. Deshmukh, K.S., Chand, V.S., Shukla, K.D., Laha, A.K.: Exploring associations between participant online content engagement and outcomes in an online professional development Programme. In: International Working Conference on Transfer and Diffusion of IT, pp. 126–136. Springer, Cham (2020)
186. Alrmah, I.A.O., Lokman, A.: Predicting student performance in massive open online courses (MOOCs) using big data analysis and convolutional neural network. Int. J. Innov. Sci. Res. Technol. **5**(3), 778–786 (2020)
187. Lingard, B., Wyatt-Smith, C., Heck, E.: Transforming schooling through digital disruption: big data, policy, teaching, and assessment. In: Wyatt-Smith, C., Lingard, B., E. (eds.) Heck Digital Disruption in Teaching and Testing Assessments, Big Data, and the Transformation of Schooling, pp. 1–44. Routledge, London (2021)
188. de Andrade, T.L., Rigo, S.J., Barbosa, J.L.: Active methodology, educational data mining and learning analytics: a systematic mapping study. Inform. Educ. **20**(2), 171–203 (2021)
189. Parreira do Amaral, M., Hartong, S.: National Education Systems in the Post-national Era: The Territorial and Topological (De-) construction of National Education. The Education Systems of the Americas, pp. 1–22 (2021)
190. Kolber, S., Heggart, K.: Education focused pracademics on twitter: building democratic fora. J. Prof. Capital Commun. (2021)
191. Wyatt-Smith, C., Lingard, B., Heck, E.: Digital disruption in teaching and testing: assessments. In: Big Data, and the Transformation of Schooling, p. 248. Taylor & Francis, Milton Park (2021)
192. Castellanos-Reyes, D.: The dynamics of a MOOC's learner-learner interaction over time: a longitudinal network analysis. Comput. Hum. Behav. **123**, 106880 (2021)
193. Fincham, E., Rózemberczki, B., Kovanović, V., Joksimović, S., Jovanović, J., Gašević, D.: Persistence and performance in co-enrollment network embeddings: an empirical validation of Tinto's student integration model. IEEE Trans. Learn. Technol. **14**(1), 106–121 (2021)

194. Mangina, E., Psyrra, G.: Review of learning analytics and educational data mining applications. In: Proceedings of EDULEARN21 Conference, vol. 5, pp. 949–954 (2021)
195. Nawang, H., Makhtar, M., Hamzah, W.: Comparative analysis of classification algorithm evaluations to predict secondary school students' achievement in core and elective subjects. Int. J. Adv. Technol. Eng. Explor. **9**(89), 430 (2022)
196. Okoye, K., Arrona-Palacios, A., Camacho-Zuñiga, C., Achem, J.A.G., Escamilla, J., Hosseini, S.: Towards teaching analytics: a contextual model for analysis of students' evaluation of teaching through text mining and machine learning classification. Educ. Inf. Technol. **27**(3), 3891–3933 (2022)
197. Uttamchandani, S., Quick, J.: An introduction to fairness, absence of bias, and equity in learning analytics. In: Handbook of Learning Analytics (2022)
198. Gupta, S.L., Mishra, N.: Artificial intelligence and deep learning-based information retrieval framework for assessing student performance. Int. J. Inform. Retriev. Res. **12**(1), 1–27 (2022)
199. Gourlay, L.: Surveillance and datafication in higher education: documentation of the human. In: International Conference on Networked Learning 2022 (2022)
200. Finzer, W.: The data science education dilemma. Technol. Innov. Stat. Educ. **7**(2), 1–9 (2013)
201. Piety, P.J., Hickey, D.T., Bishop, M.J.: Educational data sciences: framing emergent practices for analytics of learning, organizations, and systems. In: Proceedings of the Fourth International Conference on Learning Analytics and Knowledge, pp. 193–202 (2014)
202. Gibson, D.C., Webb, M.E.: Data science in educational assessment. Educ. Inf. Technol. **20**(4), 697–713 (2015)
203. Mitchell, J.C.: MOOCS on and off the farm: MOOCs and technology to advance learning and learning research (ubiquity symposium). Ubiquity. **2014**, 1–10 (2014)
204. Williamson, B.: Smart schools in sentient cities. https://www.storre.stir.ac.uk/bitstream/1893/21500/1/WilliamsonB_Smart%20schools_2014.pdf (2014)
205. Hood, N., Littlejohn, A., Milligan, C.: Context counts: how learners' contexts influence learning in a MOOC. Comput. Educ. **91**, 83–91 (2015)
206. Williamson, B.: Automated knowledge discovery: tracing the Frontiers, infrastructures, and practices of education and data science. In: Digital Disruption in Teaching and Testing, pp. 45–59. Routledge, London (2021)
207. Williamson, B.: Who owns educational theory? Big data, algorithms and the expert power of education data science. E-learn. Dig. Media. **14**(3), 105–122 (2017)
208. Kitto, K., Whitmer, J., Silvers, A., Webb, M.: Creating Data for Learning Analytics Ecosystems. SoLAR Position Papers. The Society for Learning Analytics Research (SoLAR) (2020)
209. Nuankaew, P., Sittiwong, T., Nuankaew, W.S.: Characterization clustering of educational technologists achievement in higher education using machine learning analysis. Int. J. Inform. Educ. Technol. **12**(9) (2022)
210. Martinez-Maldonado, R.: Seeing learning analytics tools as orchestration technologies: towards supporting learning activities across physical and digital spaces. In: CEUR Workshop Proceedings (2016)
211. Wong, C.: Sequence based course recommender for personalized curriculum planning. In: International Conference on Artificial Intelligence in Education, pp. 531–534. Springer, Cham (2018)
212. Echeverria, V., Martinez-Maldonado, R., Buckingham Shum, S.: Towards collaboration translucence: giving meaning to multimodal group data. In: Proceedings of the 2019 Chi Conference on Human Factors in Computing Systems, pp. 1–16 (2019)
213. Selwyn, N.: Re-imagining 'learning analytics'… a case for starting again? Internet High. Educ. **46**, 100745 (2020)
214. Ndukwe, I.G., Daniel, B.K., Butson, R.J.: Data science approach for simulating educational data: towards the development of teaching outcome model (TOM). Big Data Cogn. Comput. **2**(3), 24 (2018)

215. Waheed, H., Hassan, S.U., Aljohani, N.R., Hardman, J., Alelyani, S., Nawaz, R.: Predicting academic performance of students from VLE big data using deep learning models. Comput. Hum. Behav. **104**, 106189 (2020)
216. İnan, E., Ebner, M.: Learning analytics and moocs. In: International Conference on Human-Computer Interaction, pp. 241–254. Springer, Cham (2020)
217. Aljawarneh, S., Lara, J.A.: Data science for analyzing and improving educational processes. J. Comput. High. Educ. **33**(3), 545–550 (2021)
218. McFarland, D.A., Khanna, S., Domingue, B.W., Pardos, Z.A.: Education data science: past, present, future. AERA Open. **7**, 23328584211052055 (2021)
219. Srinivasa, K.G., Kurni, M.: Educational data mining & learning analytics. In: A Beginner's Guide to Learning Analytics, pp. 29–60. Springer, Cham (2021)
220. Gupta, A., Garg, D., Kumar, P.: Mining sequential learning trajectories with hidden Markov models for early prediction of at-risk students in e-learning environments. IEEE Trans. Learn. Technol. (2022)
221. Susnjak, T., Ramaswami, G.S., Mathrani, A.: Learning analytics dashboard: a tool for providing actionable insights to learners. Int. J. Educ. Technol. High. Educ. **19**(1), 1–23 (2022)
222. Wise, A.F.: Learning analytics: using data-informed decision-making to improve teaching and learning. In: Contemporary Technologies in Education, pp. 119–143. Palgrave Macmillan, Cham (2019)
223. Munoz-Najar Galvez, S., Heiberger, R., McFarland, D.: Paradigm wars revisited: a cartography of graduate research in the field of education (1980–2010). Am. Educ. Res. J. **57**(2), 612–652 (2020)
224. Romero, C., Ventura, S.: Educational data science in massive open online courses. Wiley Interdisc. Rev. Data Min. Knowl. Discov. **7**(1), e1187 (2017)
225. Clark, J.A., Liu, Y., Isaias, P.: Critical success factors for implementing learning analytics in higher education: a mixed-method inquiry. Australas. J. Educ. Technol. **36**(6), 89–106 (2020)
226. Jaakonmäki, R., vom Brocke, J., Dietze, S., Drachsler, H., Fortenbacher, A., Helbig, R., Kickmeier-Rust, M., Marenzi, I., Suarez, A., Yun, H.: Learning Analytics Cookbook: How to Support Learning Processes Through Data Analytics and Visualization, pp. 7–14. Springer International Publishing, Berlin (2020)
227. Fancsali, S., Murphy, A., Ritter, S.: "Closing the loop" in educational data science with an open source architecture for large-scale field trials. In: Proceedings of the 15th International Conference on Educational Data Mining, p. 834 (2022)
228. Williamson, B.: Digital education governance: an introduction. Eur. Educ. Res. J. **15**(1), 3–13 (2016)
229. Gibson, D.C., Ifenthaler, D.: Preparing the next generation of education researchers for big data in higher education. In: Daniel, B.K. (ed.) Big Data and Learning Analytics in Higher Education, pp. 29–42. Springer, Cham (2017)
230. Chopade, P., Edwards, D., Khan, S.M., Andrade, A., Pu, S.: CPSX: using AI-machine learning for mapping human-human interaction and measurement of CPS teamwork skills. In: In 2019 IEEE International Symposium on Technologies for Homeland Security (HST), pp. 1–6. IEEE (2019)
231. Martins, V., Oyelere, S.S., Tomczyk, L., Barros, G., Akyar, O., Eliseo, M.A., Albuquerque, C., Silveira, I.F.: A blockchain microsites-based ecosystem for learning and inclusion. In: Brazilian Symposium on Computers in Education, vol. 30(1), p. 229 (2019)
232. Rosé, C.P., McLaughlin, E.A., Liu, R., Koedinger, K.R.: Explanatory learner models: why machine learning is not the answer. Br. J. Educ. Technol. **50**(6), 2943–2958 (2019)
233. Davis, G.M., AbuHashem, A.A., Lang, D., Stevens, M.L.: Identifying preparatory courses that predict student success in quantitative subjects. In: Proceedings of the Seventh ACM Conference on Learning@Scale, pp. 337–340 (2020)
234. Williamson, B., Eynon, R.: Historical threads, missing links, and future directions in AI in education. Learn. Media Technol. **45**(3), 223–235 (2020)

235. Arthurs, N., Alvero, A.J.: Whose Truth Is the "Ground Truth"? College Admissions Essays and Bias in Word Vector Evaluation Methods. International Educational Data Mining Society (2020)
236. Jasim, A.A., Hazim, L.R., Abdullah, W.D.: Characteristics of data mining by classification educational dataset to improve student's evaluation. J. Eng. Sci. Technol. **16**(4), 2825–2844 (2021)
237. Fancsali, S.E., Li, H., Sandbothe, M., Ritter, S.: Targeting design-loop adaptivity. International Educational Data Mining Society. In: 14th International Conference on Educational Data Mining, pp. 323–330 (2021)
238. Madeira, B.C., Tasci, T., Celebi, N.: Prediction of Student Performance Using Rough Set Theory and Backpropagation Neural Networks (2021)
239. Dowell, N.M., McKay, T.A., Perrett, G.: It's not that you said it, it's how you said it: exploring the linguistic mechanisms underlying values affirmation interventions at scale. AERA Open. **7**, 23328584211011611 (2021)
240. Lee, C.A., Tzeng, J.W., Huang, N.F., Su, Y.S.: Prediction of student performance in massive open online courses using deep learning system based on learning behaviors. Educ. Technol. Soc. **24**(3), 130–146 (2021)
241. Zhang, Y., Liu, S., Shang, X.: An MRI study on effects of math education on brain development using multi-instance contrastive learning. Front. Psychol. **12**, 765754 (2021)
242. Fancsali, S.E., Li, H., Ritter, S.: Toward Scalable Improvement of Large Content Portfolios for Adaptive Instruction (2021)
243. Panwong, P., Natthakan, I.O., Mullaney, J.: Improved Cluster Analysis for Graduation Prediction Using Ensemble Approach (2021)
244. Aljawfi, O., Pei, T., Abu-El Humos, A.: Analyzing the associations between educational background factors and problem-solving in technology-rich environments: an investigation of united state Adult's proficiency level in PIAAC. In: International Multi-Conference on Complexity, Informatics and Cybernetics, pp. 144–149 (2021)
245. Maldonado, S., Miranda, J., Olaya, D., Vásquez, J., Verbeke, W.: Redefining profit metrics for boosting student retention in higher education. Decis. Support. Syst. **143**, 113493 (2021)
246. Chaturapruek, S., Dalberg, T., Thompson, M.E., Giebel, S., Harrison, M.H., Johari, R., Stevens, M.L., Kizilcec, R.F.: Studying undergraduate course consideration at scale. AERA Open. **7**, 2332858421991148 (2021)
247. Hou, J., Kylliäinen, I., Katinskaia, A., Furlan, G., Yangarber, R.: Applying gamification incentives in the Revita language-learning system. In: Proceedings of the 9th Workshop on Games and Natural Language Processing within the 13th Language Resources and Evaluation Conference, pp. 7–16 (2022)
248. Yang, B., Tang, H., Hao, L., Rose, J.R.: Untangling chaos in discussion forums: a temporal analysis of topic-relevant forum posts in MOOCs. Comput. Educ. **178**, 104402 (2022)
249. Hidalgo, Á.C., Ger, P.M., Valentín, L.D.L.F.: Using meta-learning to predict student performance in virtual learning environments. Appl. Intell. **52**(3), 3352–3365 (2022)
250. Childs, J., Taylor, Z.: Googling for schools: do K-12 school districts purchase Adwords to drive website traffic? J. Commun. Media Technol. **12**(3), e202215 (2022)
251. Barbosa-Manhães, L.M.B., Zavaleta, J., Cercear, R., Costa, R.J.M., da Cruz, S.M.S.: Investigating STEM courses performance in Brazilians higher education. In: International Conference on Computer Supported Education, pp. 212–231. Springer, Cham (2022)
252. Rus, S.E.V., Fancsali, Jr, P.P., Venugopal, D., Arthur, C.: The Learner Data Institute— conceptualization: a progress report. In: Proceedings of the 2nd Learner Data Institute Workshop in Conjunction with the 14th International Educational Data Mining Conference (2021)
253. Jandrić, P., Knox, J., Macleod, H., Sinclair, C.: Learning in the age of algorithmic cultures. E-learn. Dig. Media. **14**(3), 101–104 (2017)
254. Tsiakmaki, M., Kostopoulos, G., Kotsiantis, S., Ragos, O.: Fuzzy-based active learning for predicting student academic performance using autoML: a step-wise approach. J. Comput. High. Educ. **33**(3), 635–667 (2021)

255. Bekmanova, G., Ongarbayev, Y., Somzhurek, B., Mukatayev, N.: Personalized training model for organizing blended and lifelong distance learning courses and its effectiveness in higher education. J. Comput. High. Educ. **33**(3), 668–683 (2021)

256. Donaldson, P., Ntarmos, N., Portelli, K.: A Systematic Review of the Potential of Machine Learning and Data Science in Primary and Secondary Education (2017)

257. Bowers, A.J., Krumm, A.E.: Supporting the initial work of evidence-based improvement cycles through a data-intensive partnership. Inform. Learn. Sci. **122**, 629–650 (2021)

258. Williamson, B.: Digital methodologies of education governance: Pearson plc and the remediation of methods. Eur. Educ. Res. J. **15**(1), 34–53 (2016)

259. Demchenko, Y., Cuadrado-Gallego, J.J., Brewer, S., Wiktorski, T.: EDISON data science framework (EDSF): addressing demand for data science and analytics competences for the data driven digital economy. In: 2021 IEEE Global Engineering Education Conference, pp. 1682–1687. IEEE (2021)

260. Perrotta, C., Selwyn, N.: Deep learning goes to school: toward a relational understanding of AI in education. Learn. Media Technol. **45**(3), 251–269 (2020)

261. Wise, A.F.: Educating data scientists and data literate citizens for a new generation of data. J. Learn. Sci. **29**(1), 165–181 (2020)

262. Selwyn, N., Gašević, D.: The datafication of higher education: discussing the promises and problems. Teach. High. Educ. **25**(4), 527–540 (2020)

263. Williamson, B.: Bringing up the bio-datafield child: scientific and ethical controversies over computational biology in education. Ethics Educ. **15**(4), 444–463 (2020)

264. Gulson, K.N., Webb, P.T.: 'Life' and education policy: intervention, augmentation and computation. Discour. Stud. Cult. Politics Educ. **39**(2), 276–291 (2018)

265. Perrotta, C., Williamson, B.: The social life of learning analytics: cluster analysis and the 'performance' of algorithmic education. Learn. Media Technol. **43**(1), 3–16 (2018)

266. Sellar, S., Gulson, K.N.: Becoming information centric: the emergence of new cognitive infrastructures in education policy. J. Educ. Policy. **36**(3), 309–326 (2021)

267. Williamson, B.: New digital laboratories of experimental knowledge production: artificial intelligence and education research. Lond. Rev. Educ. **18**(2), 209–220 (2020)

268. McCoy, C., Shih, P.: Teachers as producers of data analytics: a case study of a teacher-focused educational data science program. J. Learn. Anal. **3**(3), 193–214 (2016)

269. Doroudi, S.: The bias-variance tradeoff: how data science can inform educational debates. AERA Open. **6**(4), 2332858420977208 (2020)

270. Song, I.Y., Zhu, Y.: Big data and data science: what should we teach? Expert. Syst. **33**(4), 364–373 (2016)

271. Turek, D., Suen, A., Clark, D.: A project-based case study of data science education. Data Sci. J. **15**, 1–10 (2016)

272. Heinemann, B., Opel, S., Budde, L., Schulte, C., Frischemeier, D., Biehler, R., Podworny, S., Wassong, T.: Drafting a data science curriculum for secondary schools. In: Proceedings of the 18th Koli Calling International Conference on Computing Education Research, pp. 1–5 (2018)

273. Baldassarre, M.: Think big: learning contexts, algorithms and data science. Res. Educ. Media. **8**(2), 69–83 (2016)

274. Liu, M.C., Huang, Y.M.: The use of data science for education: the case of social-emotional learning. Smart Learn. Environ. **4**(1), 1–13 (2017)

275. Jung, Y., Wise, A.F., Allen, K.L.: Using theory-informed data science methods to trace the quality of dental student reflections over time. Adv. Health Sci. Educ. **27**(1), 23–48 (2022)

276. Douglas, K.A., Merzdorf, H.E., Hicks, N.M., Sarfraz, M.I., Bermel, P.: Challenges to assessing motivation in MOOC learners: an application of an argument-based approach. Comput. Educ. **150**, 103829 (2020)

277. Johnson, J.C., Olney, A.M.: Using Community-Based Problems to Increase Motivation in a Data Science Virtual Internship (2022)

278. Cope, B., Kalantzis, M.: Big data comes to school: implications for learning, assessment, and research. Aera Open. **2**(2), 2332858416641907 (2016)

279. Bertolini, R., Finch, S.J., Nehm, R.H.: Enhancing data pipelines for forecasting student performance: integrating feature selection with cross-validation. Int. J. Educ. Technol. High. Educ. **18**(1), 1–23 (2021)
280. Jantakun, K., Jantakun, T., Jantakoon, T.: The architecture of system for predicting student performance based on data science approaches (SPPS-DSA architecture). Int. J. Inf. Educ. Technol. **12**(8), 778–785 (2022)
281. Garmpis, S., Maragoudakis, M., Garmpis, A.: Assisting educational analytics with AutoML functionalities. Computers. **11**(6), 97 (2022)
282. Quy, T.L., Friege, G., Ntoutsi, E.: Multiple fairness and cardinality constraints for students-topics grouping problem. arXiv preprint arXiv:2206.09895 (2022)
283. Rosenberg, J.M., Galas, E., Willet, K.: Who are the data scientists in education? an investigation of the identities and work of individuals in diverse roles. In: Proceedings of the 15th International Conference of the Learning Sciences-ICLS 2021. International Society of the Learning Sciences (2021)
284. Maylawati, D.S.A., Priatna, T., Sugilar, H., Ramdhani, M.A.: Data science for digital culture improvement in higher education using K-means clustering and text analytics. Int. J. Electric. Comput. Eng. **10**(5), 2088–8708 (2020)
285. Estrellado, R.A., Freer, E.A., Mostipak, J., Rosenberg, J.M., Velásquez, I.C.: Data Science in Education Using R. Routledge, London (2020)
286. Kotsiopoulos, C., Doudoumis, I., Raftopoulou, P., Tryfonopoulos, C.: DaST: an online platform for automated exercise generation and solving in the data science domain. In: Proceedings of the 8th Computer Science Education Research Conference, pp. 104–109 (2019)
287. Rosenberg, J.M., Borchers, C., Dyer, E.B., Anderson, D., Fischer, C.: Understanding public sentiment about educational reforms: the next generation science standards on twitter. AERA Open. **7**(1), 1–17 (2021)

Part III
Applications

Chapter 4
Educational Data Science Approach for an End-to-End Quality Assurance Process for Building Creditworthy Online Courses

May Kristine Jonson Carlon, Sasipa Boonyubol, Nopphon Keerativoranan, and Jeffrey S. Cross

Abstract Despite for-credit distance education existing since the latter half of the nineteenth century, the creditworthiness of massive open online courses (MOOCs), a form of distance education, remains debatable outside a few domains. This may be due to the perceived lack of quality in MOOCs, either due to the wide variety of MOOCs or the difficulty in translating quality expectations from brick-and-mortar courses to online courses. In this chapter, the quality expectations for stereotypical university credit are explored and compared with emerging patterns in MOOCs. Based on this analysis, a data-driven end-to-end quality assurance (QA) process to bridge the gap is proposed. Portions of this process are demonstrated using the Tokyo Institute of Technology's (Tokyo Tech) MOOCs as case studies. This demonstration has shown how a modified Deming cycle can provide actionable quality improvement measures at different stages of the MOOC cycle. In particular, each stage has showcased how the most commonly used approaches in educational data science (EDS), namely, educational data mining (EDM) and learning analytics (LA), are applied to gather information, course contents, and learner feedback data from various interested parties as support evidence for for-credit MOOCs. Henceforth, for improving quality assurance, this demonstration helps clarify the concerns regarding the worthiness of the online learning experience.

M. K. J. Carlon · J. S. Cross (✉)
Department of Transdisciplinary Science and Engineering, School of Environment and Society, Tokyo Institute of Technology, Ookayama, Meguro-ku, Tokyo, Japan

Online Content Research and Development Section, Center for Innovative Teaching and Learning, Tokyo Institute of Technology, Ookayama, Meguro-ku, Tokyo, Japan
e-mail: mcarlon@acm.org; cross.j.aa@m.titech.ac.jp

S. Boonyubol · N. Keerativoranan
Department of Transdisciplinary Science and Engineering, School of Environment and Society, Tokyo Institute of Technology, Ookayama, Meguro-ku, Tokyo, Japan
e-mail: boonyubol.s.aa@m.titech.ac.jp; keerativoranan.n.aa@m.titech.ac.jp

A. Peña-Ayala (ed.), *Educational Data Science: Essentials, Approaches, and Tendencies*, Big Data Management,
https://doi.org/10.1007/978-981-99-0026-8_4

Keywords Educational data science · Massive open online courses · Educational quality assurance · Learning analytics · Natural language processing · Educational data mining

Abbreviations

ABET	Accreditation Board for Engineering and Technology Inc.
AI	Artificial intelligence
ALS	Alternative learning system
CATS	Credit Accumulation and Transfer Scheme
ECTS	European Credit Transfer and Accumulation System
EDM	Educational data mining
EDS	Educational data science
ERT	Emergency remote teaching
LA	Learning analytics
LMS	Learning management system
MOOC	Massive open online course
NLP	Natural language processing
OCRD	Online Content Research and Development
OCW Initiative	OpenCourseWare Initiative
OERs	Open Educational Resources
OMS	Online Master of Science
PDCA	Plan-Do-Check-Act
PLATO	Programmed Logic for Automated Teaching Operations
QA	Quality assurance
STEM	Science, Technology, Engineering, and Mathematics

4.1 Introduction

Massive open online courses (MOOCs) are courses targeted to be large-scale and made accessible through the internet. Since most MOOCs are produced by reputable universities, they have been instrumental in improving marketability for individuals with no postsecondary credentials from the perspective of employers [1]. Unfortunately, MOOCs did not end up replacing universities as some of their evangelists touted earlier on in 2012 when they emerged. Most learners see online learning either as an augmentation of their other learning pursuits or simply for entertainment purposes. Thus, there is no motivation to dive into deep learning [2].

Despite initial disappointments, MOOCs still appear standing strong after more than a decade since their inception. Because of their online nature, researchers can gather significant amounts of data from MOOCs, which can be a fertile landscape for educational data science (EDS) research. EDS has been instrumental in better understanding the potential of MOOCs, such as improving prospects for justice, equity, inclusion, and diversity [3] and dispelling incorrect notions—online counterparts cannibalizing enrollment in traditional degree programs or being of lower quality [4]. Several other researchers see EDS as a transdisciplinary amalgamation of educational research with mathematical and technological research such as analytics and data mining [5–7]. EDS can be viewed in the following still-developing foundational areas [8]:

- *Holistic disciplinary practice*: Different disciplines will require, to some extent, different approaches to instruction. For instance, computer science courses will benefit from more well-defined exercises as opposed to ethics courses where learners can develop their ideas better from open discussions.
- *Technology-mediated psychometrics*: Learning is a very humanistic experience; as such, it is important to understand dispositions such as sentiments or engagement levels, especially in environments where these factors would not be easily observable.
- *Analyses of complex performance*: Which sequence of actions would result in optimal learning outcomes? Which activities appear to be bottlenecks for the learners? These are some questions that can potentially be answered by tracking complex learner interactions.
- *Formative and situated or classroom assessment*: With little opportunity for foresight, academic dishonesty is a formidable challenge in online education.
- *Self-regulated learning and metacognition*: Metacognition is an individual's awareness and regulation of their knowledge. Self-regulated learning is a learner's ability to adjust avoiding depletion of learning resources such as time and motivation, among others. Self-regulated learning is critical for success in autonomous learning, which is the typical case for online learning.

Among MOOC platforms, edX is an educational portal for MOOCs founded by Harvard University and the Massachusetts Institute of Technology in 2012, and Open edX is the open source platform that powers edX. Aside from MOOCs, edX also enables the delivery of small private online courses or SPOCs, which can be used locally on-campus. As one of the leading academic institutions in Japan, the Tokyo Institute of Technology (Tokyo Tech) recognizes the importance of MOOCs, and online classes in general, in the formal education landscape. Tokyo Tech's Online Content Research and Development (OCRD) Section[1] has started to offer MOOCs on edX since 2015 under the TokyoTechX brand, as well as edX SPOCs to deliver blended learning courses or fully online non-credit courses for satisfying ethics training requirements for its students. Tokyo Tech has also used its online courses for virtual exchanges with several partner universities [9]. Because of the importance of its online courses, Tokyo Tech has been conducting extensive

---

[1]This was formerly called the Online Education Development Office (OEDO). For inquiries regarding the tools and processes here, please feel free to reach out through the website, https://www.oedo.citl.titech.ac.jp/contact.

research and development on improving MOOC quality despite not offering online degrees now, mostly applying the EDS fundamentals. With a consistent focus on quality improvement, MOOCs might just be able to achieve their potential instead of falling into obsoletion as what happened with educational radio. Even though educational radio was cheap and accessible, funding to support it eventually dried up as educators grappled with its proper place in education [10].

What is preventing Tokyo Tech and several other institutions from providing credits for their MOOCs? Credentials signify that the credentialing institution assures that the learners indeed acquired the knowledge afforded to them. Having received credentials enables other entities such as employers to assess whether the learners can execute their intended tasks. Herein lies the difficulty that lifelong learners using MOOCs for career growth face: some MOOCs appear to be not on par with online for-credit courses. This is on top of other concerns related to ensuring the identity of the students: proving the individuals who complete associated activities with online courses can be a challenge, and existing solutions such as proctoring software can have questionable privacy handling.

This chapter first looks at how distance education, a predecessor to MOOCs, was developed in Sect. 4.2. Then it explores the quality expectations of for-credit courses by studying prevailing conventions in providing credits in Sect. 4.3. Then, the origins of the notion that MOOCs are of a lower quality are investigated to formulate an end-to-end quality assurance (QA) process not just for MOOCs but for online courses in general in Sect. 4.4. Tokyo Tech has conducted research relevant to the EDS foundational areas mentioned above, from which a modified Deming cycle is formulated and discussed in Sect. 4.5. Other efforts demonstrating what else can be done are discussed in Sect. 4.6 before concluding in Sect. 4.7. The goal is to provide various interested parties with ideas on how to use EDS to derive actionable quality assurance measures from the course and the learners' data within and outside one's institutions. For instance, instructors can be more receptive to the different ways the learners could be expressing their sentiments as can be deduced from Sect. 4.5.3.2. Administrators from non-native English-speaking institutions can take inspiration from Sect. 4.6.1 in rallying their faculty to take a more active role in creating accessible courses. In the end, this chapter is intended to be a starting point for iterative processes involving probing what concerns interested parties have that prevent them from embracing MOOC-like courses for credentialing.

4.2 Credentialing in Distance Education

Aside from current examples of MOOC-based degrees, giving credits for work done outside traditional instruction methods is not unprecedented. Long before MOOCs were introduced, and even before the internet was invented, a similar mode has always been used: distance education.

4.2.1 *Background on Distance Education*

Distance learning, or learning for students who cannot be physically present in the school, has been an opportunity enabler for those facing constraints including having to fit schooling with their jobs or those with mobility issues [11]. While self-study or studying with learning groups has always been possible, distance learning had been in many cases the only learning opportunity for would-be students in the latter half of the nineteenth century, when education and training had become important social concerns. Distinguishing marks of distance education include mediated subject-matter presentation and mediated student-tutor interaction, with media playing a huge role in serving as a substitute for face-to-face interaction between students and tutors.

Pregowska et al. [11] provided a comprehensive overview of the history of distance education. Formerly known as correspondence education, distance education was introduced in 1728 by Caleb Phillips who sent training materials on shorthand via postal services to his mostly female students. It took more than an entire century before the notion of feedback in distance education was introduced by Sir Isaac Pitman in 1840, and the practice was adopted by the Phonographic Correspondence Society 3 years later. Forty years later, the first correspondence school was established in the United States.

As new communication technologies are developed, so are new media for distance education adopted. The University of Wisconsin-Extension used an amateur radio wireless station for broadcasting in 1906, and the University of Iowa was the first scientific unit to provide education via television in 1934. Later, storage media such as cassettes and disks were used for educational content as well, forming the earlier forms of on-demand distance education. Systems such as the Programmed Logic for Automated Teaching Operations (PLATO) became available in the 1960s. This is the precursor to the online education that we know today.

With the advent of the internet, educational institutions were more compelled to make their learning resources more accessible. Open Educational Resources (OERs) have become more commonplace, with the Massachusetts Institute of Technology spearheading the effort through their OpenCourseWare Initiative (OCW Initiative) in 2001. Tokyo Tech also started its OpenCourseWare website in 2005, as well as a number of other Japanese institutions. However, like how the presence of books cannot always guarantee to learn, being able to access distance education resources cannot guarantee recognition of learning from the providing institutions. As such, it is important to look at how credentials are granted in modern distance education programs.

4.2.2   Credentials in Distance Education

The most prolific institution with regard to research on OERs from 2002 to 2020 is the Open University, an institution in the United Kingdom granting degrees through distance learning [12]. Being the only UK university dedicated to distance learning, having accessible educational resources is paramount. Open University's public-facing OERs are aligned with their core student provision; thus, their OERs serve a dual role of student support and public advocacy. Another institution that considers degree-granting distance learning as its main advocacy is the University of the People. Touting itself as a "tuition-free university," the University of the People is accredited by the US Distance Education Accrediting Commission. On a nationwide level, SWAYAM is the Government of India's platform for hosting online course content whose quality is ascertained by national coordinators, thus allowing learners on the platform to receive degrees should they choose to take proctored examinations.

Common to all the mentioned degree-granting distance learning platforms is tracking student achievement through the notion of credits. Since credit management is a familiar concept for traditional higher education institutions, several universities also have distance learning components, usually targeted at adult learners. For instance, Harvard University has the Harvard Extension School that offers flexible learning arrangements such as evening or online courses. The University of the Philippines System has the University of the Philippines Open University as an independent body to offer distance education. In other cases, institutions offer online counterparts to their regular offerings. This is the case for the Georgia Institute of Technology's Online Master of Science (OMS) programs.

4.3   The Notion of Credits

Credits serve as the currency indicating learner achievement for most educational institutions. They can be earned at least in two different ways: through dedicating hours to meet defined learning outcomes or through gathering experience that matches learning outcomes as certified by a credentialing body. This is especially important with the emergence of the "credential creep" phenomenon, where minimum job requirements, typically in terms of degrees, continue to rise [13]. While credential creep is another problem that must be tackled albeit with more interested parties, allowing learners to keep up with future job demands through more accessible credentialing is where the academic community can have more influence. Programs such as OMS are already taking strides in this direction, but in order to influence more institutions to join in the revolution, there is a need for a better understanding of the notion of credits. In this section, the focus will be specifically on credits in higher education.

Table 4.1 Approximate number of hours needed to earn one university credit

Credit method	Hours
The United Kingdom's credit accumulation and transfer scheme (CATS) [15]	10
European Credit Transfer and Accumulation System (ECTS) [16]	25–30
California Code of Regulations [17]	48 for semester basis, 33 for quarter basis
Quality Assurance Framework of Higher Education in Japan [18]	45 learning hours (contact hours can be as little as 15)

4.3.1 Credit Hours

In 1906, the Carnegie Foundation used credit hours, or the perceived amount of time a faculty needs to deliver a course to determine workloads [14]. This workload calculation was then used to assess the pension of faculty who did not invest their earnings and thus might not be able to retire. While the way faculty pension is assessed has changed since then, the use of credit hours has stuck around. Over the years, credit hours have evolved from calculating retirement funds to a pseudo-measurement of the rigor of student learning outcomes. Table 4.1 shows the associated number of hours to receive a credit across different regions. From available information such as this, it is evident that every region, or even every degree-granting requirement, may have different notions of what deserves credit.

Converting from credit to hours can be a convenient way of measuring whether students have exerted sufficient effort to merit credit from one institution to another. However, due to credit hours' financial roots, some institutions were found to commit credit-related malpractices to secure more federal funding allocations [19]. Additionally, while credits allow students to get recognition for their efforts through credit transfer from one institution to another, this only works seamlessly for the traditionally linear approach where students who graduate from high school immediately transfer to a postsecondary institution and graduate within the pre-scribed amount of time. Credit transfer for students who took nontraditional path-ways (e.g., exploring community colleges or vocational schools), moved from 4-year degree programs to 2-year degree programs and vice versa, and among other approaches is not as seamless as policymakers intended it to be [20].

4.3.2 Quality Worthy of Credit

To address the shortcomings of credit hours due to their usage other than their initial intention and to accommodate changing learning paths, several policymakers are looking into measuring learning competencies in giving credits. For instance, the Philippines' Department of Education has instituted the alternative learning system

(ALS) where experts design target competencies for which individuals who may have achieved those through other means, such as through their jobs, can take examinations and receive credits that can count for a future degree [21]. On a more macro scale, the Bologna Process has enabled European and other participating countries to have a sense of credit equivalency by setting expected learning outcomes that are detached from specific content [22]. Accreditation programs, which are typically used to signal quality, such as the Accreditation Board for Engineering and Technology Inc. (ABET), also emphasize generic student outcomes that can be tailored to a more specific curriculum [23].

Central to these new ways of looking at credits through the student learning outcome perspective is the importance of quality. Naturally, the quality of a program depends on several factors such as the curriculum, the qualifications of its faculty, available facilities, and institutional support, among others [23]. In this chapter though, the focus will be given on a narrower scale that is within the locus of control of course developers and instructors: MOOC quality.

4.4 Massive Open Online Courses (MOOCs) for Credit?

MOOCs have led to better labor market outcomes, not just as perceived by students [24], but as evidenced by income potentials as well [25]. Credentialing from MOOCs can help MOOC learners realize these benefits.

4.4.1 Background on MOOCs

In March 2011, Sebastian Thrun of Stanford University presented his work on self-driving cars at TED Conference in Long Beach, California. Presenting at the same conference was Sal Khan of Khan Academy, a wildly popular internet platform for instructional videos that tutors millions [26]. Inspired by a new challenge after listening to Sal Khan, Thrun started teaching lessons on artificial intelligence (AI) written on napkins and recording them with an inexpensive digital camera in his living room. In that year, Stanford University offered three courses free to the public: Jennifer Widom's Databases and Andrew Ng's Machine Learning on a student-developed platform that eventually became Coursera and Thrun and Peter Norvig's Artificial Intelligence on a different platform that evolved to become Udacity [27]. Each of these classes enrolled hundreds of thousands of students, and in some cases, online learners outperformed (i.e., getting better scores) their Stanford University counterparts. By delivering the lessons via pre-recorded lectures and having machine-graded assessments, MOOCs are seen as potential disruptors to education with their scalability and availability. With this positive reception, the *New York Times* labeled 2012 as the Year of the MOOCs [28].

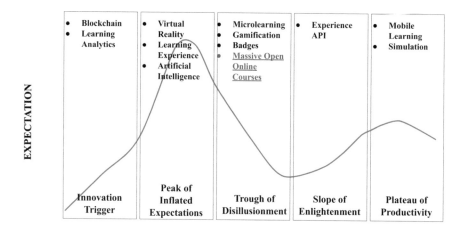

Fig. 4.1 e-Learning Gartner Hype Curve from Web Courseworks' analysis [76]

It did not take long for a critical problem with MOOCs to become evident: student withdrawal rate is too high, sometimes even reaching about 100% [29]. While some of the dropout reasons can be attributed to the students' circumstances such as lack of time, motivation, and background knowledge, other reasons can be traced back to the course quality such as complaints on course design (e.g., too many modules, overly complex or technical, etc.) and lack of interactivity leading the learners to feelings of isolation. It does not help that most MOOC platforms moved from being completely free to charging fees for complete access to materials and certifications. In 2019, the e-learning company Web Courseworks placed MOOCs on the Trough of Disillusionment on its Gartner Hype Curve [2] as can be seen in Fig. 4.1.

A renewed interest in online learning, in general, resurfaced when educational institutions were forced to shift to emergency remote teaching (ERT) with the onset of the COVID-19 pandemic. Like MOOCs, ERT led to feelings of isolation due to its online nature [31]. Unlike MOOCs though, ERT was characterized by a quick transition from face-to-face methods to online, leading to several barriers such as both instructors and learners not being comfortably ready to use technology extensively in their classes. Quick improvisations are made to deliver a class online instead of intentionally making content that would lead to learning despite the lack of direct human interaction. More widespread use of MOOCs might have made the easing from one format to another smoother. This notion becomes even more pertinent as the COVID-19 pandemic situation remains volatile even after more than 2 years since its onset, with some countries such as the Philippines choosing to

---

[2] The Gartner hype cycle is a marketing tool used to assess the public perception of a technology trend [30]. The Trough of Disillusionment refers to the point where public interest is starting to wane as it becomes evident that technology implementation fails to deliver.

be more cautious in reopening brick-and-mortar universities [32]. Since online-only instruction has been going on for an extended time in several locations, depending on how things unfold further, legitimate interest in granting degrees through fully online delivery might grow.

Nowadays, MOOCs have been used for various purposes such as training future teachers [33], providing short-term training for developing various competencies such as research self-efficacy [34], and pushing for better social inclusion [35]. These aspirational applications of MOOCs can have significant individual, societal, and vocational relevance to their learners when quality expectations are met [36].

4.4.2 Assessing MOOC Quality

When gauging MOOC quality, learners and potential employers could be looking for these factors [37]:

- *Context*: The platform where the course is offered, the institution providing the course, the recommended hours to complete, the price for the certification, the language and transcript the course provided, the level of the course, the pre-requisites the learners required before enrolling in the course, and whether it belongs to a micro-credentialing program
- *Course volume*: Contents in the course that may be quantified by the totals of minutes of videos presented in the course and/or the total number of words in the transcripts as well as the amount of reading material
- *Academic integrity*: Availability of integrity-preserving measures such as proctoring or plagiarism checks being in place
- *Course content and evaluation*: The number and kinds of assessments, whether students are required to engage or not, and whether assessments are auto-graded or not

While the context factor is a characteristic that can only be improved over an extended period, the other three factors could be actively measured and tracked through learning analytics (LA). Recent innovations in this have tapped into natural language processing (NLP) techniques to gleam through different types of data such as course content and interaction sequences on top of typical tabular data such as student grades. Some examples include an alerting system to inform novice instructors of potential quality issues in online classes [38] and an analytics dashboard retrieving quality indicators from open responses in prompts designed to develop metacognitive skills [39].

Table 4.2 Micro-credentials from various MOOC platforms [40]

Platform	Micro-credentials
Coursera	Specialization, MasterTrack, professional certificate
edX	XSeries, MicroBachelors, MicroMasters, professional certificate, professional education
Udacity	Nanodegree
FutureLearn	Program, ExpertTracks, Microcredential
Kadenze	Program

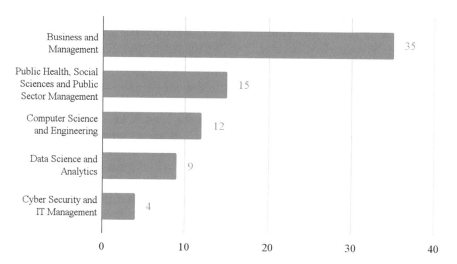

Fig. 4.2 MOOC-based master's degrees on various domains. (Data from Class Central [41])

4.4.3   MOOC-Based Degrees and Credits

To support learners in their career development activities, MOOC platforms have created various micro-credentials that provide certifications to learners. Table 4.2 shows the micro-credentials available as of 2021 according to Class Central, the online education aggregation company.

Ever since the Georgia Institute of Technology announced its Online Master of Science in Computer Science delivered via Udacity in 2014, the number of full-fledged degrees on MOOC platforms has increased. Figure 4.2 shows the master's degrees categorized according to their domains from data collected by Class Central as of January 2022. Aside from more than 70 master's degrees, there were also 7 bachelor's degrees offered fully online.

4.5 End-to-End Quality Assurance Process for Online Courses

What is then creditworthy in the online learning context? From the above discussion on MOOCs and credits, the following criteria may be useful in examining whether a MOOC is creditworthy:

- Sufficient amount of content to approximately equal the effort required for traditional programs or sufficient experience opportunities to match the expectations of alternative learning system certification programs
- Stronger focus on learning outcomes, potentially those that can be generalizable beyond the course itself, instead of on specific content details
- Credible assessment of both content and learning outcomes

While there are several advancements in assessing content based on expected outcomes, which arguably can be modality-independent (i.e., could be applicable for both on-site and distance education), getting the learners to meet the expected outcomes can be a challenge. To encourage continued participation in online courses, assuring the quality of information (i.e., content), the system (i.e., the platform itself), and the service (i.e., responsiveness to learner concerns) is important [42]. While system quality might be of reasonable maturity for most platforms having been continuously developed for more than a decade at this point, information and service quality can further be improved using various quality tools [43]. That being said, creditworthy MOOCs can have a better chance of success if the implementing institution has in place a process to ensure the following:

- There are no or very minimal barriers to learners accessing the content—may it be to limitations or preferences.
- There is an infrastructure to enable timely detection and support of learner needs.
- There is a consistent effort to improve the associated learning systems.

These criteria, both for the course and the institution, can be solidified and transformed into an end-to-end quality assurance process as demonstrated in this section.

4.5.1 Process Overview

Quality assurance (QA) processes are not unprecedented in the educational setting, and several of these processes applied in traditional educational platforms have found their way to the online setting. For instance, Quality Matters, a framework initially developed for postsecondary education by the US Department of Education, has been used to analyze online education quality by assessing faculty professional development, online course design, and course reviews [44]. The problem with adhering to most established frameworks is that compliance is given more

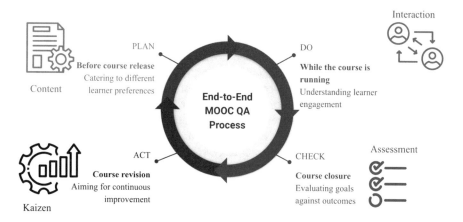

Fig. 4.3 An overview of the proposed end-to-end quality assurance process for MOOCs

importance over other critical factors such as learning outcomes [45]. As it is impossible to create frameworks that will cover all possible scenarios such as transcending cultural and economic differences, it is important to give leeway for QA implementers to adjust the frameworks for those to be meaningful to them [46].

Faithful to the Japanese quality traditions, we are looking at a MOOC QA process that banks on *monozukuri* as shown in Fig. 4.3. *Monozukuri* (物作り or making things) is the Japanese craftsmanship philosophy of aiming to deliver the best possible quality to customers. It is closely related to several other concepts: first is dual-aspect monism or the necessary presence of complementary forces (e.g., Yin and Yang and thinking and feeling); second is *hitozukuri* (人作り or person building), where the character is also built as one strives for excellence; and third is *kaizen* (改善 or continuous improvement). These *monozukuri* concepts are set within a Deming cycle, also commonly known as the Plan-Do-Check-Act (PDCA) cycle.

During the plan and do stages, the dual-aspect monism of content and interaction is put under scrutiny. Learning happens when learners interact in the classroom: may it be with their fellow learners, the instructors, or the instructional materials. In an asynchronous online classroom such as a MOOC, the main interface is through content; thus, well-designed content is instrumental to fuel interactions. The quality of content and interaction can be understood through the approaches of the two most prevalent EDS communities: *educational data mining* (EDM) to address educational questions through large data coming from educational environments and *learning analytics* (LA) to analyze learning context through data and act accordingly to optimize the learning experience [47]. When linking back to the EDS foundational areas discussed in Sect. 4.1, EDS uses technology-mediated psychometrics for understanding interaction and caters to promoting self-regulated learning and meta-cognition as well as a holistic disciplinary practice. Is there sufficient content in different modalities that can allow learners of different abilities and preferences to learn on their own in the first place? When interacting with the course, do learners exhibit positive engagement and sentiments?

If an online course is to be anthropomorphized, then the check stage is dedicated to the course's character building. In self-actualization theory, a person has a sense of fulfillment when their idealized self is in congruence with their actual self. While this is grossly simplifying the concept, the notion of self-actualization in courses can be manifested through assessments: the goals set out during course creation are indeed achieved at course closure. Another EDS facet, educational statistics, can be used to check if intended course improvements are effective, such as through A/B tests. Finally, continuous improvements can be targeted during the act stage. During the *kaizen* opportunity discovery activities, two considerations must be kept in mind: how can the product (i.e., the online course) be made better and how to ensure that the improvement efforts do not result in a degraded experience in the future. This alludes to the EDS foundational area pertaining to analyzing complex performance: improvement efforts can potentially have complexities that can be detrimental in some aspects while legitimately solving an issue. Just like in most decision-making endeavors, balancing the merits and demerits is crucial.

How can this proposed framework, an end-to-end QA process, produce credit-worthy online courses? And how is it different from currently existing course QA processes such as those mentioned in Sect. 4.4.2? The proposed process deviates from the traditional manufacturing PDCA cycle where each phase is intended to evaluate a single aspect of a product. Instead, the proposed process is more akin to the phase-exit review process used in project management, where each phase has its deliverables to be inspected. Each stage in the PDCA cycle is set to correspond to different phases in the online course lifecycle: creation, deployment, closure, and revision. Unlike existing QA frameworks that tend to be punitive due to the focus on compliance, this process focuses on discovery. Through the constant discovery process, it is hoped that learners and instructors will be flexible, as well as other interested parties such as administrators, parents, and employers, to give their inputs about course quality that can be acted upon accordingly.

In the following subsections, the usage of the proposed process is illustrated through different sample cases. The results of these activities are still being utilized and continuously being improved up to the time of writing. Some of the cases mentioned are yet to produce results, while some ideas presented are yet to be implemented. These cases are aptly labeled as samples; they are intended to demonstrate how the end-to-end QA process can be used and is not necessarily prescriptive. Because of the unique context that TokyoTechX might be in (e.g., language and cultural differences and organizational goals, among others), some of the measures applied might not apply to another context. Nevertheless, the readers are encouraged to get inspiration from the problem-solving approaches applied.

4.5.2 Plan: Before Course Release—Catering to Different Learner Preferences

When creating content, typical approaches include the traditional approach where the desired topics are known ahead of time from which content and assessments are created or the backward design approach where the goals are first decided from which assessments and finally contents are created [48]. While these approaches help ensure that there is sufficient content for learners to achieve the desired learning outcomes, these may be lacking when viewed from the lens of dual-aspect monism. The Yin or the thinking aspect can be determined from the learner assessment results, but how do we know if it is possible for the learners to "feel" the content? In this subsection, the question to be answered in the planning phase is what types of content learners can resonate with. To answer this question, a custom web crawler was created to download data from online courses as shown in Fig. 4.4. Since web crawlers can be dependent on how websites are displayed, it was decided to consolidate effort for the creation and maintenance of the tool by focusing on a single online learning platform. edX was the selected platform since it is where TokyoTechX is hosted.

For content-related quality assurance activities, the focus is not just on TokyoTechX. Data from 304 courses offered by 77 institutions covering 28 subjects downloaded using the abovementioned crawler in March 2018 were used for EDM. The goal is to answer questions by uncovering hidden patterns in large amounts of data. This overcomes the risk of evaluating courses on the limited aspirations of an institution or faculty unit and opens the possibility of learning new goals based on widespread yet latent best practices. The analyses from these downloaded data are heavily based on simple word counts in the text material, assessments, and video

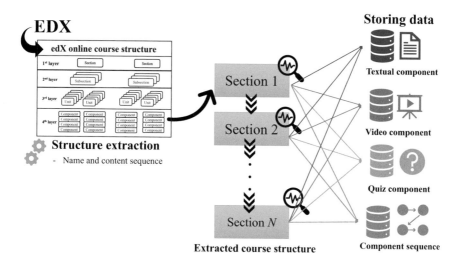

Fig. 4.4 An overview of the custom content web crawler for edX courses

transcripts. The selected samples to be discussed include measuring content type distribution and analyzing content type sequence.

4.5.2.1 Sample 1: Measuring Content Type Distribution

There is a need to present content in different formats to cater to different learner needs. For instance, while learners may prefer video over text content, the text is easier to consume on the go (e.g., while on a commute or waiting in line) and is easier to render in areas with limited resources (i.e., text loads faster than videos when the internet speed is low). To understand the hidden consensus among edX courses on what is the ideal content type distribution, the word counts gathered from the crawled data are categorized into text, video, and assessment components [49].

Figure 4.5 shows the content type distribution for the 304 downloaded courses. Table 4.3 shows the word count statistics summarized from Fig. 4.5. This information can help course developers judge whether the size of their courses is reasonable or not.

Aside from the course's size per se, it is also crucial to understand the right mix of content types. After looking at raw word counts, the crawled courses are represented as vectors each containing the percentages of text, video, and assessment according to their word counts. A K-means clustering as illustrated in Fig. 4.6 was conducted to have a sense of what might be the hidden consensus of course developers on the ideal

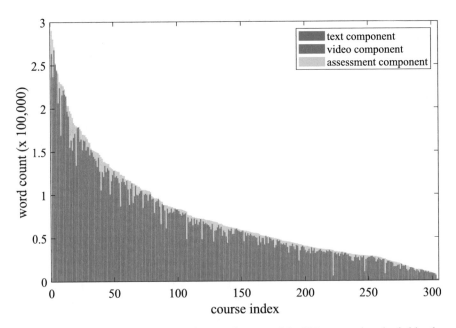

Fig. 4.5 Content type distribution based on word counts of the 304 courses downloaded by the custom content web crawler [49]

Table 4.3 Word count statistics for the 304 crawled courses [49]

Statistic	Text	Video	Assessment	Total
Minimum	0	346	0	1728
First quartile	3312	18,314	1618	33,062
Median	9514	40,406	3656	58,030
Mean	14,464	54,289	6516	75,268
Third quartile	18,556	75,037	7404	104,808
Maximum	129,456	264,406	59,342	290,693

Note that each row corresponds to the statistic and not to a specific course (i.e., the course with the minimum word count for Text may not be the same course with the minimum word count for Video). Likewise, Total refers to the statistic associated with the total word counts in a course and not the sum of Text, Video, and Assessment in their associated row.

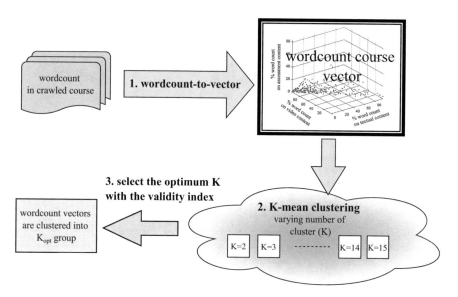

Fig. 4.6 K-means clustering process for word count data

content type distribution. The K-means clustering algorithm starts by randomly initializing K cluster centroids. In this scenario, a centroid is just a three-dimensional vector representing common word count ratio vectors (course grouping). Every vector is assigned to one K cluster whose distance to its centroid is minimized. The cluster centroids are then recalculated based on all members until reaching an optimum value. In this K-means determination, the number of K clusters is needed before performing clustering. Therefore, this data-driven technique as a validity index is necessary to determine the best K that fits the dataset.

Figure 4.7 shows the result of the K-means clustering. The K-means clustering resulted in seven clusters, with members of most clusters appearing to be well congregated except for clusters 6 and 7.

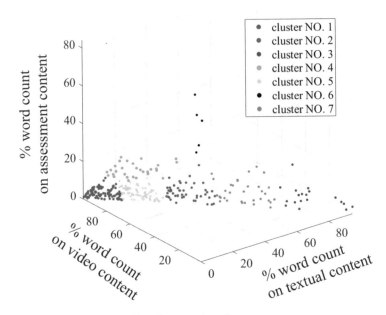

Fig. 4.7 Visualization of the resulting K-means clustering

Table 4.4 Average content type distribution according to word count in percentages per resulting cluster [49]

Cluster	Course count	Average percentages		
		Text	Video	Assessment
1	100	5.47	89.73	4.70
2	49	33.42	57.94	8.64
3	19	77.26	13.26	9.47
4	30	8.85	67.22	23.92
5	68	17.66	75.88	6.46
6	5	15.72	24.05	60.23
7	33	52.96	35.92	11.10

Table 4.4 shows the details of each cluster. More than 70% of the data are within three clusters, clusters 1, 5, and 2. Each of these major clusters has videos comprising more than 50% of the content. This indicates that despite some of the advantages of text content, most developers choose to use videos instead, possibly due to their better ability to engage learners.

From this measurement utilizing data from 304 crawled courses, it is shown to have a high tendency of having video content as the main component. How many videos should be created? How much text should be added? What should be the ratio of video, text, and assessment? There is no exact and correct answer to respond to these questions. However, the results in this section could provide the direction, as a guideline, to the course developers to begin the planning stage of how their intended online courses would be.

4.5.2.2 Sample 2: Analyzing Content Type Sequence

edX recommends that assessments, text, and videos should be interleaved to improve learner engagement based on their experience in assisting in the creation of hundreds of MOOCs. To easily check for compliance with this criterion, a tool to visualize the content type sequence was created. A sample visualization of two MOOCs with different course structures is shown in Fig. 4.8. While one course follows the style of an alternate video followed by a quiz in one section, another presents a video followed by multiple assessments. This is part of OCRD's effort to put research results of other institutions into action. Different MOOCs may have various interleaved methods and sequences depending on many reasons such as subjects and preferences of the MOOC development teams. Understanding such structure may lead to the relationship between certain type sequences and learning engagement for domain subjects.

Knowing the sequence of components (text, assessment, video) would allow the course developers to plan and design the course. Creating the assessment following the video segment to test the learners' understanding right after watching and placing the assessment after a series of video lectures to allow learners to connect the knowledge along the journey before checking their mastery at the end are both common patterns in online courses. The course developers are able to organize either style to meet their final goal.

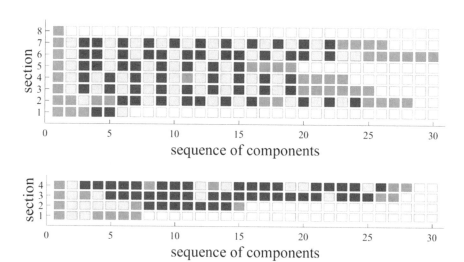

Fig. 4.8 Visualization of content type sequences of two different courses, where blue is for text, red is for assessments, and yellow is for videos [43]

4.5.3 Do: While the Course Is Running—Understanding Learner Engagement

Learner engagement refers to the learner's cognitive and emotional energy dedicated to achieving a learning task. In online classrooms, indicators of learner engagement can include various data such as clickstreams, video interactions, or participation in discussion forums. Technology-mediated psychometrics can be formulated with these multimodal data. This is especially advantageous with MOOCs as their digital nature allows more convenient ways to collect huge amounts of data. Advancements in AI-related fields, may it be through the creation of models based on big data or the invention of more robust algorithms, also enable richer analysis of learner behaviors beyond what traditional tabular data such as attendance records and assessment results can offer. For the do, check, and act stages, the focus will be on learner-specific indicators. As such, the samples to be discussed are based on specific courses offered on TokyoTechX.

4.5.3.1 Sample 1: Analyzing Activity for Different Video Types

Interactions with video content are a method for understanding learner engagement, which can be rather unique to the online learning environment. These interactions can be part of two different cognitive functions: first is meaning processing where learners perceive, decode, and process information, and second is memory where learners encode, store, and retrieve previously known information. Therefore, more clicks could be more indicative of more cognitive energy invested and thus more learner engagement.

Clickstream data associated with video interactions were analyzed for two versions of the *Computer Science and Programming* MOOC: one offered in Japanese[3] and another offered in English.[4] The videos of these two courses can be categorized as lecture videos and videos related to exercises. The result of the analysis is shown in Fig. 4.9. The initial goal of the inspection was to understand whether the relatively unfamiliar Japanese language would make the learners devote more cognitive energy. Upon closer inspection of the videos, it became apparent that these were not directly comparable. Therefore, the types of videos generating engagement were investigated instead. It was found that videos related to exercises generate more engagement than lecture videos. The exercises though are direct applications of the lectures; hence, it is possible that learners were spending more time deciphering instructions rather than understanding the concepts being taught. This is one caveat when analyzing learner engagement: more engagement may not necessarily mean that the course is better.

---

[3] https://www.edx.org/course/introduction-to-computer-science-and-programming-3

[4] https://www.edx.org/course/introduction-to-computer-science-and-programming

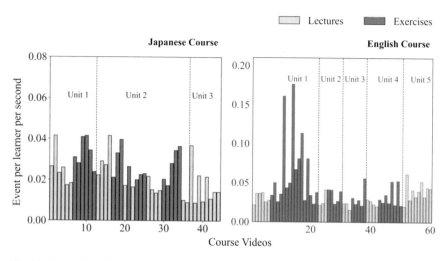

Fig. 4.9 Interactions in video content on the Computer Science and Programming MOOC [50]

The learners are the most important interested party in MOOCs; but in the absence of visible, synchronous interaction, it can be difficult to gauge the learners' behaviors. While the course is running, checking the learners' behavior occasionally is also an important aspect to analyze the course. This sample has illustrated the effect of a relatively unfamiliar language (Japanese) on the learner's cognition, which shows no difference compared to the course provided in English, while the impact could be observed with the higher learner engagement (more clicks) on the videos related to exercises than the lecture videos. This data could also be further analyzed to investigate the reason behind the high cognitive energy to understand the learners more whether it tends to be positive (the content is interesting) or negative (the content is difficult).

4.5.3.2 Sample 2: Measuring Learner Sentiment on Discussion Forums

The lack of face-to-face interaction in online courses is a significant hurdle to learner satisfaction: learners can rate an online course to be perfect, but still prefer to attend the course in person if given the chance [51]. All is not lost, though; even though challenging, once the engagement is spurred in online classrooms, particularly in discussion boards, cognitive presence can be established [52]. To understand how interactions in discussion boards look like, a web crawler for discussion boards as shown in Fig. 4.10 was created.

One of the most common analyses done on discussion board data is sentiment analysis or understanding whether the data have positive, negative, or neutral outlooks. Using the sentiment trained model available on Python's TextBlob library, Fig. 4.11 shows visualizations of sentiment analysis results on two TokyoTechX courses, *Autophagy* and *Modern Japanese Architecture*. The mean sentiment

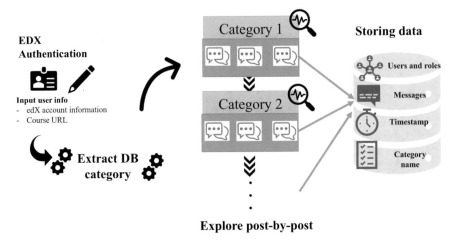

Fig. 4.10 An overview of the custom discussion board web crawler for edX courses

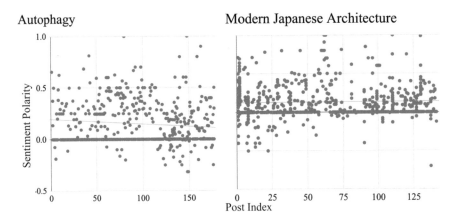

Fig. 4.11 Sentiment polarity scores for discussion board posts [43]

polarity score is indicated by a horizontal line. The sentiment polarity scores for the analysis method used range from -1 to 1 indicating sentimental levels from negative to positive. Just based on the horizontal line, it appears that the overall sentiment in both courses appears somewhat neutral. However, the scatterplots show that the sentiments in the *Modern Japanese Architecture* course are more varied, with several negative sentiments throughout. These negative comments are mostly complaints about ambiguous text on quizzes. From this result, it is evident that quality decisions should not be based on aggregate end values alone but more careful analysis of hidden trends.

This section shows another example of what can be analyzed while the course is running. The discussion board data could help the course developers understand more of learners' behavior through the text that was input by the learners directly giving their own impression of the course or the confusion in some specific content presented in the course. The implementation of the sentiment analysis technique would help not only in understanding the learners' feedback but also in enabling the course developers to evaluate their own courses. In courses where supervision or facilitation of an instructor is maintained (which is ideal but can be limited by human resources), regular sentiment analysis can also help in online classroom management.

4.5.4 Check: Course Closure—Evaluating Goals Against Outcomes

In an educational setting, the purpose of an evaluation is typically to see whether actual learning outcomes match the preset learning goals. Several researchers working on the EDS fundamental area of formative and situated or classroom assessments attempt to reduce threats to the integrity of assessment methods. However, recent research and discussions on "ungrading"—or using alternatives to assignments, quizzes, and tests for assessment—indicate that decoupling from conventional evaluation methods can better support a learner's humanity: that can commit mistakes and learn from those [53]. Should traditional assessments be removed from MOOCs, multiple viable sources for assessment can be used instead, such as, once again, discussion forums. These new approaches can also enable new ways of thinking about assessments. It is not just the learner performance that we can provide feedback to: instructors can learn from these assessments as well. Additionally, in online course development, the goal-outcome comparison can also take a different dimension. For example, the topics can be analyzed and compared between in the course (provided by the course developers) and in the discussion boards (created by the learners), which is discussed in Sect. 4.5.4.1, or innovations may be introduced into a course, such as having automatically dubbed videos as described in Sect. 4.5.4.2. Checking whether these innovations bring forth their desired positive effect is critical to avoid burdening learners with unnecessary learning environment complexity.

4.5.4.1 Sample 1: Comparing Topics Discussed in the Course-Provided Materials and the Discussion Boards

For evaluating target learning outcomes with actual learning outcomes, the second iteration of the MOOC titled *Autophagy: Research Behind the 2016 Nobel Prize in Physiology or Medicine*[5] that was released in September 2019 was used for demonstration. The data were crawled on March 24, 2021. This course is designed to be

---

[5] https://www.edx.org/course/autophagy-research-behind-the-2016-nobel-prize-in

completed by learners in 3 weeks with 1–2 h of effort per week. The following questions are probed using the data crawled from both the course contents and the discussion board posts:

- Did the instructor discuss what they planned for the learners to learn?
- What did the learners end up talking about in discussion boards?

To answer these questions, topic modeling was conducted on the course contents and the discussion board posts. Since the course is 3 weeks, it was decided to derive five topics: one for each week, one extra for introductions, and another for other topics such as complaints or expressions of gratitude if any. Arguably, complaints and gratitude will not be derived from the content topics. In that case, it can be expected that the additional content topics will correspond to subtopics within the planned content. Functions on Python's Natural Language Toolkit (NLTK) were used to vectorize and train topic models using Latent Dirichlet Allocation. The resulting words associated with the models created were inspected to look for words that did not contribute much meaning. These non-contributing words were put on the list of stop words, and topic modeling was iteratively conducted. In the end, the models were analyzed by manually mapping them to the syllabus topics and visualizing the results to see the connections. Figure 4.12 shows the resulting topics for the course content and discussion board posts.

Table 4.5 shows the mapping of the resulting models to the course week goals. The second week received considerable emphasis on the course content, which is not surprising since it is the week related to the discussion on the Nobel prize research. It is noticeable though that there is a content topic that is not mappable to any of the weeks, so a future improvement that might be considered is adding a target learning outcome for this extra topic that appeared to be important to the instructor during course creation. It also appears that there is not enough content for a topic to be modeled related to the first week. Nevertheless, the learners did discuss a lot about the topics associated with this week. Hence, the instructor can decide whether they want to enhance the first week's contents or leave it as it is since it is enough to encourage learners to have discussions. Finally, it appears that learners discussed the third week's goals, which are about the role of autophagy. Depending on the instructor's judgment, this can be a negative sign since the learners should be equally motivated to discuss other weeks, especially the second week, but it can be a positive sign as well since the learners are enthusiastic about the course takeaways relevant to them.

Topic modeling can serve as an applicable tool for course quality assurance. In this example, the evaluation is performed at the course closure to evaluate the goals against the outcomes. However, the application is not fixed to the course closure only; it could also be executed even before the course release or while the course is running. The course developers might be able to notice some particular items that need attention in the course content and could take immediate action to respond to learners' needs.

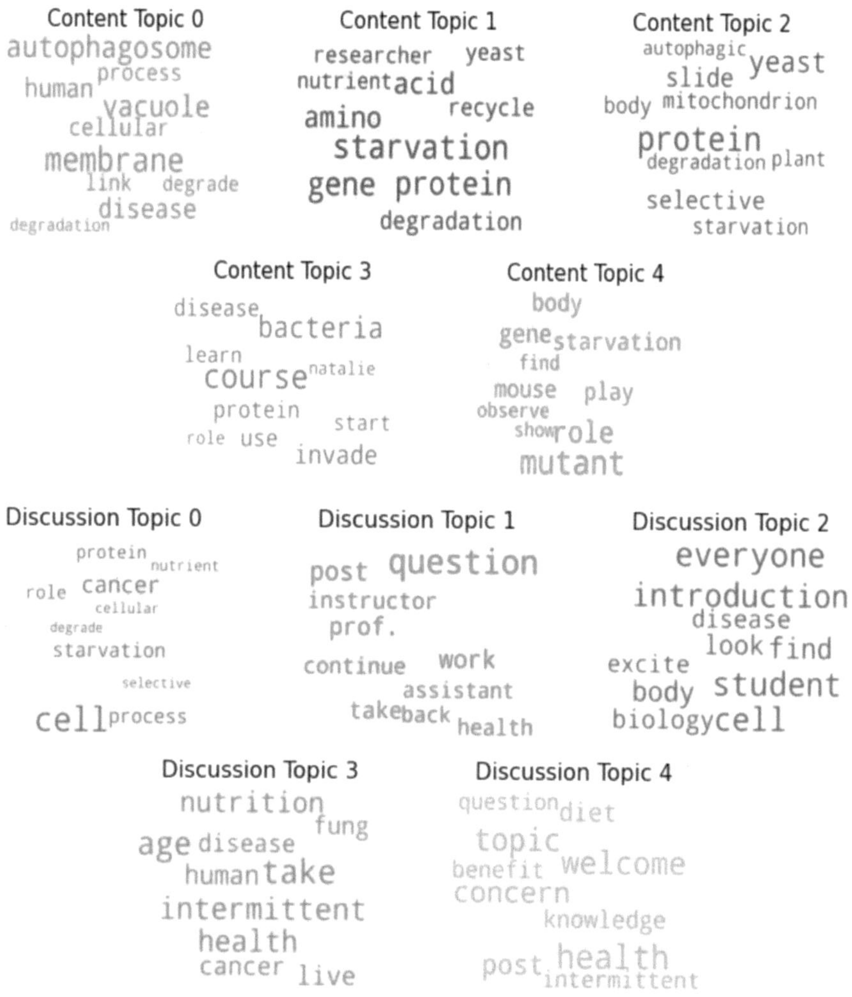

Fig. 4.12 Topic modeling results on the content and discussion board posts in the Autophagy course [54]

Table 4.5 Topic and course outline mappings [54]

Course week	Content topic	Discussion topic
1		0, 1
2	1, 2, 3	2
3	4	0, 3, 4
Not mappable	0	

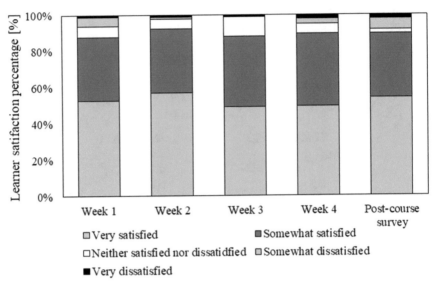

Fig. 4.13 Learner satisfaction ratings on automatically dubbed English videos [55]

4.5.4.2 Sample 2: Gauging the Acceptance of Automated Text-to-Speech Dubbing on Videos

Creating the transcript in a language other than the original audio source is an easy and efficient method for learners to understand their preferred and comfortable languages better. However, it is still ineffective for visually impaired learners, and the presence of the transcript also tends to draw the learner's attention away from the visual content of the video. One of the possible solutions is to perform video dubbing, which can be categorized into two main methods. The first one is conventional dubbing, which typically requires experienced voice-over artists and a significant amount of time, effort, and human resources to make a good-quality dubbed video. This could be solved by the second method: automatic dubbing with a synthetic voice created from the textual speech powered by AI technology such as Google text-to-speech, which requires less amount of time and a fixed cost independent of the number and length of the videos.

The *Introduction to Electrical and Electronic Engineering* MOOC[6] was originally taught in Japanese and was then revised and released with English videos dubbed with computer-generated synthetic voice [55]. A straightforward way to evaluate the effectiveness of this approach is to ask the learners directly. Figure 4.13 shows the satisfaction ratings provided by the learners on the automatically dubbed English videos. About 90% of the learners are either very satisfied or somewhat

---

[6]https://www.edx.org/course/introduction-to-electrical-and-electronic-engineer

satisfied with the dubbing. While the entire process was challenging, this approach can be recommended for online courses with concerns related to multilingualism.

This section showed one of the examples that emphasize the quality assurance approach. After the course closure of the first original course in Japanese, the Tokyo Tech course developers carefully reviewed the learners' responses from the post-course surveys, which showed the demand for having the course run in English instead of Japanese. The course developers then investigated the method and responded to learners' needs by creating the new course with automatically dubbed videos. The course developers could consider this approach if they plan to release an online course that provides videos in multiple languages to attract learners from around the world.

4.5.5 Act: Course Revision—Aiming for Continuous Improvement

Kaizen is typically taken to mean small incremental changes that are applied regularly to eliminate wastes, may those be procedural or material. The word itself though does not give any indication of size; the more important point is the notion of making changes for the good of the product. How can these target changes be identified, and how do we know if they are effective when implemented? A typical source for improvement points in a traditional classroom setting is the course feedback survey conducted at the end of the term. Unfortunately, these evaluations suffer from validity as they are more likely to measure likability, may it be of the instructor or the course itself, than quality [56]. The focus on likability might not be a huge concern for MOOCs as several learners enroll for edutainment; however, MOOC retention rates are extremely low (around 5%), and dropout reasons can be as mundane as other friends are dropping out too [57]. Thus, learners answering the end-of-course surveys may be a very small minority, risking the responses be not representative of the experience of the majority of the learners. That is not to say that these surveys are useless, but they need deeper scrutiny such as dedicating time with a team to qualitatively derive improvement points from the surveys that can be used in the same course and for other courses as well. Another approach, especially if the course is recurring, is to quantitatively track key metrics from one course cycle to another.

4.5.5.1 Sample 1: Identifying Metric Differences Across Course Revisions

Character building through creation is central to Tokyo Tech education. From bachelor's to doctorate degrees, Tokyo Tech does not only cover specialized education based on the students' course majors but also emphasizes that students should

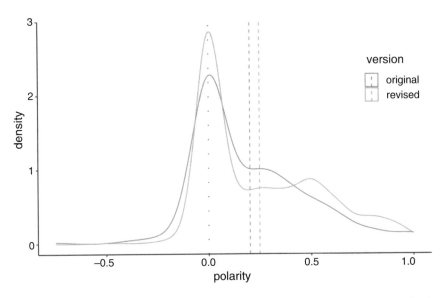

Fig. 4.14 Density distribution of sentiment polarity scores of discussion board posts in the *monozukuri* courses [58]

always be mindful of their intended careers and roles in their societies through career development and liberal arts education. As such, having a *monozukuri* MOOC is very apt to introduce Tokyo Tech to the rest of the world.

The first TokyoTechX *monozukuri* MOOC[7] was offered in 2018 [58]. It was a four-week mechanical engineering course with a week dedicated to each stage of the PDCA cycle. Throughout the course, the lectures on the science and engineering aspects of steam engine boats were presented, followed by a workshop where learners created their own toy steam engine boat, and finally, the philosophical aspects of *monozukuri* were discussed. While there were learners who enjoyed the course, feedback on course surveys indicated that it was hard to keep the focus on the course. The boat-making project took too long to finish, the philosophy lessons sounded a little bit too sentimental, and several learners craved more real-life and relatable examples of *monozukuri*.

To address the course survey feedback, it appeared that revamping the course was necessary. Similar topics were grouped and optimized. Additionally, modules on experiential learning and *monozukuri* practices were introduced. This revised course was offered from 2020. The main goal during course revision was to increase learner satisfaction. Taking a page from the do stage, sentiment analysis was conducted on discussion board posts from both course versions. Figure 4.14 shows the density distribution and mean of the sentiment polarity scores. There was a statistically

---

[7] https://www.edx.org/course/monozukuri-making-things

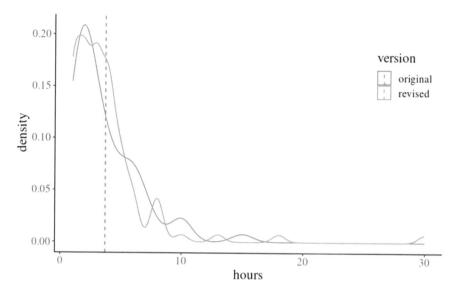

Fig. 4.15 Density distribution of learner self-reported hours spent on the *monozukuri* courses [58]

significant difference between the two course versions, with the newer course having more positive sentiment. This indicates that, indeed, improvement was introduced.

When introducing improvements, equally important to eliminating waste is not introducing new waste. With the introduction of new modules, there is a possibility that tasks are increased unnecessarily, thus negatively burdening the learners' effort. To check if this unwanted effect was present, self-reported estimated hours from course review surveys for each version were compared similarly to what was done to the sentiment polarity scores. Figure 4.15 shows the result of this comparison. Hypothesis testing indicated that there was no statistical difference between the reported hours in the old version and the new version of the course. As such, the course revision was a success.

This type of analysis, comparing two different versions of the same course, can provide value to different interested parties. Platform developers and researchers had been using this A/B test-like approach for significant software changes and experiments. Instructors and course developers will know concretely whether the changes they introduced in their courses are effective. Administrators will be able to see if new investments translate to actual benefits without inadvertent cost.

4.5.5.2   Sample 2: Brainstorming from Course Feedback Survey

Brainstorming is a group creativity technique typically done in a relaxed manner to encourage participants to share their ideas, sometimes irrespective of the ideas' feasibility. The significant changes introduced to TokyoTechX courses such as the creation of the English version of the programming course, the introduction of

Fig. 4.16 A screenshot of one of the 360° videos in the Japanese Architecture and Structural Design MOOC [59]

automated dubbing in the electrical engineering course, and the *monozukuri* course revamp were all from careful brainstorming activities.

Another course that saw a significant change in revision after brainstorming is the Japanese Architecture and Structural Design MOOC. Learners expressed interest in experiencing architecture more intimately. In its revision, three-dimensional 360° videos were introduced in the course to hopefully magnify the learners' spatial awareness and for them to have better architecture appreciation if they have suitable equipment (i.e., head-mounted virtual reality display). Figure 4.16 shows a frame from one of such videos. The benefit of this addition is currently being investigated using a human-computer interaction research approach. Other approaches that may also be used include technical evaluations such as those laid out by the International Organization for Standardization and International Electrotechnical Commission [60]. This includes assessing for adverse physiological effects such as discomfort (e.g., nausea or dizziness), eyesight problems (e.g., fatigue or blur), and musculo-skeletal concerns (strain due to the extended carrying of the device), among others.

Just like the previously discussed course comparisons, brainstorming can provide value to different interested parties. As the adage goes, two heads are better than one: when approached with an open mind, interested parties can gather and provide input on how they imagine the learning experience can be improved in a more communi-cative manner. This approach is also not new: at least in some social science research fields, offline measures (i.e., measuring the target construct outside the target activity, such as in surveys) are augmented with online measures (i.e., measuring the target construct while the target activity is happening, such as in computer logs) the same way quantitative methods are augmented with qualitative methods [61]. Brainstorming based on a course feedback survey is akin to combining offline and online measures as well as quantitative and qualitative methods.

4.6 Other Quality Indicators

While the EDS techniques introduced thus far revolve around data that can be harvested from MOOCs, the problems tackled in this chapter are not strictly confined to the online education setting except for automated dubbing. The principles behind—providing materials catering to different learner preferences, checking for learner engagement, evaluating learning outcomes, and aiming for continuous improvement—can likewise be applied to most learning scenarios. The concern relating to the automated dubbing, i.e., making content available in multiple languages, gives us a glimpse of problems that may be unique, at least to some extent, to MOOCs: interested parties such as instructors and learners, as well as their means of interacting with MOOCs, can be very diverse; protecting academic integrity can be more finicky; the asynchronous nature can mean learners needing support when there is no staff available to respond. In this section, other quality indicators that may be critical to other interested parties are presented. For instance, parents will most likely want to know if the special needs of their children can be catered to in the online learning environment. Employers, as mentioned earlier, may be concerned about knowing whether they should trust that potential employees who took online courses for credit are indeed competent.

4.6.1 Accessibility: Readability and Listenability

In software engineering, dealing with a user base having diverse backgrounds is done by improving the software's accessibility. Arguably, MOOCs are also software products; more importantly, since MOOCs are typically accessed through web browsers, the devices from which they will be accessed are also diverse. MOOC accessibility can be improved by targeting it to be mobile-friendly. Some measures that can be applied include minimizing typical readability issues such as illegible fonts and situational impairments such as hard-to-see or hard-to-hear videos [62].

Outside software aspects, notions of readability and listenability are also important in creating MOOCs for a different reason: while most MOOCs are delivered in English, most MOOC learners do not have English as their first language as can be seen in Fig. 4.17. More importantly for Tokyo Tech, the students who will be using the SPOCs that are derived from TokyoTechX MOOCs are mostly Japanese, as well as the course instructors. To address this concern, Tokyo Tech pays extra attention to making all the text in the online courses as readable as possible.

Using the same downloaded courses for analyzing content type distribution, Flesch-Kincaid readability tests were administered on all extracted content. A pushback that is encountered for promoting more readable content is that having well-written materials may not be always easy. Sometimes, the nature of the courses renders themselves to be less readable. It is also possible that the instructor's English ability may be a hindrance. To loosely disprove this point, readability scores for each

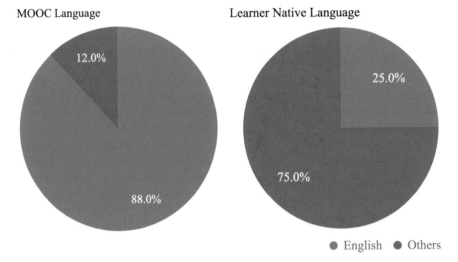

Fig. 4.17 Disparity between MOOC languages and learner native languages. (Data from [63])

Table 4.6 Top rankings in terms of readability [49]

Courses that are		Institutions with courses that are	
Easiest to read	Hardest to read	Easiest to read	Hardest to read
Biology and life sciences	Philosophy and ethics	DavidsonX	Amnesty InternationalX
Engineering	Chemistry	BrownX	ColumbiaX
History	Law	USMx	UC3Mx
		PekingX	UC BerkeleyX
		DartmouthX	UTPermianBasinX

course domain and each institution are averaged and ranked. It can be seen in Table 4.6 that Science, Technology, Engineering, and Mathematics (STEM) and liberal arts courses can be both readable. Likewise, institutions from native English-speaking countries as well as non-native English countries can produce readable content. Thus, improving readability is an actionable measure to improve the accessibility, and thus quality, of MOOCs.

Aside from advanced scripting, careful instructional design can also be employed to make adjacent sections in a course more connected with each other. This connectedness can be measured by checking for the similarity of the content. A sample visualization aggregating the readability and similarity of sections in a MOOC is shown in Fig. 4.18.

Even after a course is well-scripted, pushbacks may still be encountered during video recording. Instructors may be concerned that their pronunciation may be hard to understand. To help instructors feel at ease with their speech, a means to measure listenability was developed. Figure 4.19 sample output shows that the recording is reasonably understandable. To achieve these good results, instructors who are less

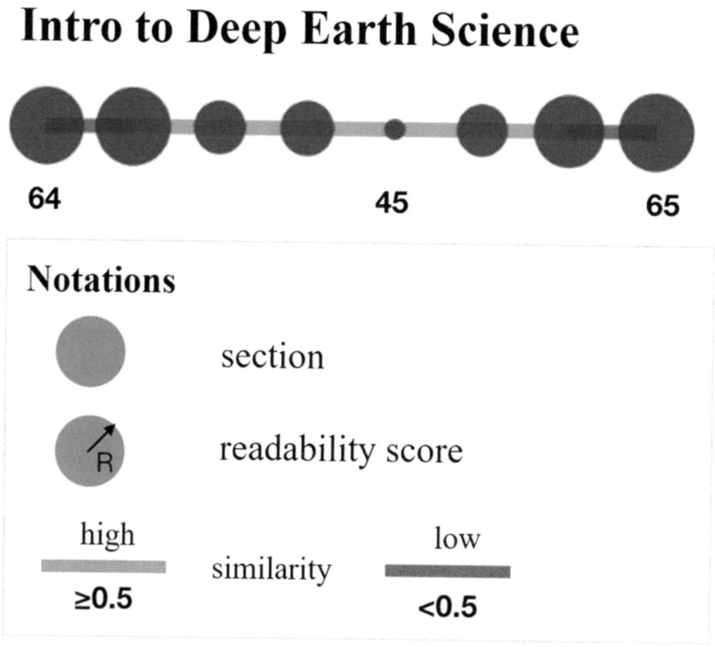

Intro to Deep Earth Science

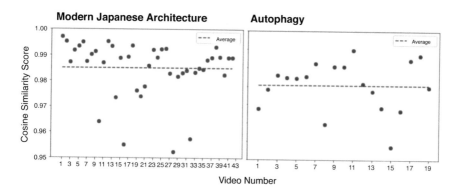

Fig. 4.18 Visualization of readability scores and similarity between adjacent sections

Fig. 4.19 Listenability rating by comparing transcripts prepared by course staff and transcripts autogenerated using Google Docs' Voice Typing feature [43]

confident with their speaking ability can speak more slowly during video recording. While some research indicates that faster speaking rates can lead to better engagement as slow speakers can be perceived to be boring, it was also found that artificially speeding up videos does not negatively reduce learning ability [64].

While much of Tokyo Tech's research is on using readability and listenability to support non-native speakers, it is important though to remember that these measures are also intended for different populations, such as individuals with visual or hearing impairment. Sometimes, accommodations are made such as adding screen reader-friendly alternative texts for images or video insets of sign language translators. In some cases, because these accommodations add more information, they lead to higher cognitive loads for their intended audience [65]. Going back to the basics of instructional material creation may be more critical here than the add-ons. The same as how elementary pupils are reminded to have a single topic for a paragraph, instructional units such as a video slide should also focus on one topic at a time to help students focus. Additionally, add-ons such as alternative texts should be checked for accuracy as well as consistency with the rest of the content just like how adjacent sections are checked for similarity.

Aside from course developers and instructors, it is also important for educational policymakers to take interest in the accessibility of online courses if those will be offered for credit. Accessibility is an important aspect that benefits from a top-down approach with consultation as it requires a cultural paradigm: inclusivity must be integrated at all stages of an endeavor, and feedback from both end users (in our case, learners) and experts must be sought [66]. The highest levels may be at a global scale such as the Web Accessibility Initiative[8] or at the national level in the form of legislation.[9] In the end, heads of institutions are sufficiently positioned to enact inclusivity measures that are less temporal than individual faculty members can affect.

4.6.2 Trustworthiness: Competency-Based Assessments That Are Hard to Cheat

Since Tokyo Tech has not yet offered MOOC-based degrees, the trustworthiness of assessments is an area largely unexplored in TokyoTechX offerings. Nevertheless, there is a wealth of research on academic dishonesty in online learning that we can derive inspiration from. Several of the proposed solutions are technology-based: some are for tackling academic dishonesty head-on such as the use of plagiarism checking software, and others are in support of further learning such as adaptive tests or delivering assessments in line with what an AI agent deems to be the most useful for the learner or less palatable solutions such as using proctoring software, which can violate privacy [67].

A lot of these findings though are already long codified as best practices for creating assessments regardless of online or on-site classes. These include redesigning assessments to prevent cheating such as staying away from multiple-choice questions; not using textbook questions to which learners might have

---

[8] https://www.w3.org/WAI/

[9] An example of such legislation is Japan's Basic Act on the Formation of an Advanced Information and Telecommunications Network Society enacted in 2000 https://japan.kantei.go.jp/it/it_basiclaw/it_basiclaw.html.

acquired answer keys somewhere else; requiring learners to show their work; creating real-world requirements, especially for programming courses; and favoring smaller and frequent quizzes over high-stakes examinations [68]. Pedagogically, these approaches tap into higher-order thinking skills as opposed to recall, enable appreciation of the value of what was learned, and ease anxiety, thus making cheating less enticing.

Still, some recommendations are not specific to online courses but are easier to implement when the course has some digital component. This includes understanding the right amount of time to be allocated for exercises, possibly through previous learner data [69]. Another approach is to randomize questions, which can be conveniently done in most learning management systems (LMSs) such as Canvas and Moodle [70].

Just like accessibility, trustworthiness can benefit from a strong initiative from educational policymakers. While instructors may have leeway on what is commensurate as academic dishonesty, institutional policies are what eventually dictate the corresponding disciplinary measures. It is also at the institutional level where the fairness of academic honesty allegations is decided.

4.6.3 Quality of Service: Timeliness of Feedback and Support

Tokyo Tech has further developed its discussion board web crawler to automatically download forum data from its online courses regularly and alert course staff about potential learner concerns. Additionally, dedicated teaching assistants are also assigned to grade open-response assessments. While traditional pedagogical knowledge points to the benefit of peer feedback in enabling learners to improve their understanding through learning by teaching, research on peer feedback points to several disadvantages such as low reliability and validity of peer feedback [71], on top of the possibility of some exercises ending up not being assessed by any peer. Also, tools for inspecting simple but hard-to-locate issues such as broken links were created to help prevent minor yet annoying inconveniences. Admittedly, the timeliness of these actions is yet to be evaluated.

One approach to improve the timeliness of feedback and support that Tokyo Tech is yet to try but can already be attested to by other MOOC research is enabling collaboration among learners. For instance, having a collaborative video caption editing feature can augment video-based learning on top of improving caption quality [72]. Not only are the learners compelled to pay more attention to the videos and the captions, but they are once again able to learn by teaching.

As can be deduced from earlier sections, quality of service is an iterative aspect that affects and involves multiple interested parties. Online course staff or even platform developers can continue to innovate by improving the timeliness of feedback and support to both learners and instructors. Learners can take a more active role by learning how to provide meaningful feedback both to the instructors and their peers. Instructors can then dedicate more of their time to mentorship and facilitation.

4.7   Conclusions

This chapter discussed an end-to-end quality assurance process to ensure that a MOOC can be creditworthy. This involved using different EDS techniques spanning from NLP-based approaches, detailed analytics derived from interaction data, and comparisons across different course versions to assess the goal of course improvement. It must be noted though that having creditworthy MOOCs is not sufficient for an institution to grant degrees via distance learning: they should equally be prepared to provide technological infrastructure, human infrastructure, and governance to help distance learners succeed [73]. Additionally, the process described here is from the actual practice done at Tokyo Tech and thus was planned and executed to solve the problems encountered that may be unique in that specific context. Nevertheless, the motivation behind should not be discounted. For instance, while institutions in the Anglo region may not have a concern with instructors hesitating to deliver courses in English, the issue that the learners' cognitive burden might be coming from hard-to-read content must still be considered. Eventually, it is recommended that those who intend to follow this framework remember the combined *monozukuri* and *kaizen* concepts. While continuous improvement is most often desired, there are typically at least two sides in a given situation.

It is hoped that this work can inspire the use of EDS for improving and spreading teaching and learning duties. Ensuring that learning happens has traditionally been the teacher's responsibility. MOOCs, with reduced learner-teacher contact, shift some of the burdens to the learners to take control of their learning. Providing credits via MOOCs will lead administrators to be also involved, as well as parents interested in quality indicators of courses targeted at younger learners. EDS techniques can provide evidence of the conduciveness of online courses to learning (e.g., content scaffolding, learner engagement, etc.) and the trustworthiness of the perceived outcomes. In this chapter, a discovery approach to quality assurance was presented as opposed to audit-style approaches. Samples that hopefully will inclusively resonate with several interested parties were presented. Through these samples, ways on how the said interested parties can observe and have a say on the quality of an online course were shown. Because of the inclusivity and discovery aspects, it is further aspired that this work can enhance EDS research in general. Continuous tracking of multimodal data throughout the course lifecycle is crucial in promoting more holistic applications of EDS. Additionally, attributing value to courses via credits, thus inducing economic repercussions, can force EDS practitioners to present information in a manner digestible for different kinds of interested parties.

Just like in other research studies, there are several opportunities for future work here. For one, large language models are continuously improving; thus, NLP-based approaches are only expected to get better. On a more cautious note, with robust tracking of learner activities, privacy becomes a concern. Measures must be in place to ensure the ethical use of EDS techniques: that they are utilized for support and not for other nefarious surveillance activities [74]. On a more speculative side, recently, there have been active discussions about Web 3.0 or decentralized internet. As such, the use of blockchain technologies for education, especially with the economics of credentials, is also worthy of further study [75]. Finally, at least in consideration of Japan's goal to reach Society 5.0, assuring the quality of courses using emerging technologies such as extended realities should also be considered.

Acknowledgments This work was supported by the Center for Innovative Teaching and Learning (CITL) at Tokyo Tech and partially supported by a Japan Society for the Promotion of Science (JSPS) Grants-in-Aid for Scientific Research (Kakenhi) Grant Number JP20H01719. We are also grateful to numerous teaching assistants who have conducted research results to support this work as well as the reviewers and editor for their constructive input.

References

1. Rivas, M.J., Baker, R.B., Evans, B.J.: Do MOOCs make you more marketable? An experimental analysis of the value of MOOCs relative to traditional credentials and experience. AERA Open. **6**(4), 2332858420973577 (2020)
2. Adnin, R., Afroz, S., Majid Taseen, M., Sharmin, S.: Students' adoption of online platforms for learning purposes in Bangladesh International Conference on Human-Computer Interaction, pp. 3–11. Springer, Cham (2022)
3. Williamson, K., Kizilcec, R.: A review of learning analytics dashboard research in higher education: implications for justice, equity, diversity, and inclusion. In: LAK22: 12th International Learning Analytics and Knowledge Conference, pp. 260–270 (2022)
4. Joyner, D.A., Isbell, C., Starner, T., Goel, A.: Five years of graduate CS education online and at scale. In: Proceedings of the ACM Conference on Global Computing Education, pp. 16–22 (2019)
5. Daniel, B.K.: Big data and data science: a critical review of issues for educational research. Br. J. Educ. Technol. **50**(1), 101–113 (2019)
6. McFarland, D.A., Khanna, S., Domingue, B.W., Pardos, Z.A.: Education data science: past, present, future. AERA Open. **7**, 23328584211052055 (2021)
7. Rosenberg, J.M., Lawson, M., Anderson, D.J., Jones, R.S., Rutherford, T.: Making data science count in and for education. In: Research Methods in Learning Design and Technology, pp. 94–110. Routledge (2020)
8. Romero, C., Ventura, S.: Educational data science in massive open online courses. Wiley Interdisc. Rev. Data Min. Knowl. Discov. **7**(1), e1187 (2017)
9. Joyner, D.A., Carlon, M., Cross, J., Corpeño, E., Hernández Rizzardini, R., Rodas, O., Cortes-Mendez, M., Staubitz, T., Ruipérez-Valiente, J.A.: Global learning@ scale. In: Proceedings of the Seventh ACM Conference on Learning@ Scale, pp. 229–232 (2020)
10. Dousay, T.A., Janak, E.: All things considered: educational radio as the first MOOCs. TechTrends. **62**(6), 555–562 (2018)
11. Pregowska, A., Masztalerz, K., Garlińska, M., Osial, M.: A worldwide journey through distance education—from the post office to virtual, augmented and mixed realities, and education during the COVID-19 pandemic. Educ. Sci. **11**(3), 118 (2021)
12. Mishra, M., Dash, M.K., Sudarsan, D., Santos, C.A.G., Mishra, S.K., Kar, D., Bhat, I.A., Panda, B.K., Sethy, M., da Silva, R.M.: Assessment of trend and current pattern of open educational resources: a bibliometric analysis. J. Acad. Librariansh. **48**(3), 102520 (2022)
13. Veillard, L.: Alternance training as a way to improve the attractivity of vocational education programmes in France. In: The Standing of Vocational Education and the Occupations it Serves, pp. 139–158. Springer, Cham (2022)

14. McMillan, A., Barber III, D.: Credit hour to contact hour: using the Carnegie unit to measure student learning in the United States. J. High. Educ. Theory Pract. **20**(2), 88–99 (2020)
15. The Open University: What are CATS points or credits? https://www.open.ac.uk/study/credit-transfer/faqs/what-are-cats-points-or-credits (n.d.)
16. Directorate-General for Education: ECTS Users' Guide 2015. Photo of Publications Office of the European Union. https://op.europa.eu/en/publication-detail/-/publication/da7467e6-8450-11e5-b8b7-01aa75ed71a1 (2017)
17. Thomson Reuters Westlaw: 55002.5. Credit hour definition. https://govt.westlaw.com/calregs/Document/I10E4210986BA4C93A057CD9CBC951F0B (n.d.)
18. Higher Education Bureau: Quality Assurance Framework of Higher Education in Japan. Ministry of Education, Culture, Sports, Science and Technology. https://www.mext.go.jp/component/english/__icsFiles/afieldfile/2011/06/20/1307397_1.pdf (2009)
19. Noda, A.: How do credit hours assure the quality of higher education? In: CEAFJP Discussion Paper Series 16-01 CEAFJPDP—FFJ.EHESS.FR. http://ffj.ehess.fr/upload/Discussion/CEAFJPDP16-01.pdf (2016)
20. Giani, M.S.: The correlates of credit loss: how demographics, pre-transfer academics, and institutions relate to the loss of credits for vertical transfer students. Res. High. Educ. **60**(8), 1113–1141 (2019)
21. Alternative Learning System: About the Program. Department of Education. https://www.deped.gov.ph/about-als/ (n.d.)
22. Brøgger, K.: How education standards gain hegemonic power and become international: the case of higher education and the Bologna process. Eur. Educ. Res. J. **18**(2), 158–180 (2019)
23. ABET: Criterion 3. Student outcomes. In: Criteria for Accrediting Engineering Programs, 2021–2022. https://www.abet.org/accreditation/accreditation-criteria/criteria-for-accrediting-engineering-programs-2021-2022/ (2021)
24. Castaño-Muñoz, J., Rodrigues, M.: Open to MOOCs? Evidence of their impact on labour market outcomes. Comput. Educ. **173**, 104289 (2021)
25. Heinrich, C.J., Cheng, H.: Does the labor market give credit for learning online? Online course-taking in high school and later labor market outcomes. In: 2020 APPAM Fall Research Conference. APPAM (2020)
26. Chafkin, M.: Udacity's Sebastian Thrun, Godfather of Free Online Education, Changes Course. Fast Company. https://www.fastcompany.com/3021473/udacity-sebastian-thrun-uphill-climb (2016)
27. Ng, A., Widom, J.: Origins of the modern MOOC (xMOOC). In: Hollands, F.M., Tirthali, D. (eds.) MOOCs: Expectations and Reality: Full Report, pp. 34–47 (2014)
28. Pappano, L.: The year of the MOOC. The New York Times. https://www.nytimes.com/2012/11/04/education/edlife/massive-open-online-courses-are-multiplying-at-a-rapid-pace.html (2012)
29. Dalipi, F., Imran, A.S., Kastrati, Z.: MOOC dropout prediction using machine learning techniques: review and research challenges. In: 2018 IEEE Global Engineering Education Conference (EDUCON), pp. 1007–1014. IEEE (2018)
30. Gartner Hype Cycle Research Methodology: Gartner. https://www.gartner.com/en/research/methodologies/gartner-hype-cycle (n.d.)
31. Donham, C., Barron, H.A., Alkhouri, J.S., Changaran Kumarath, M., Alejandro, W., Menke, E., Kranzfelder, P.: I will teach you here or there, I will try to teach you anywhere: perceived supports and barriers for emergency remote teaching during the COVID-19 pandemic. Int. J. STEM Educ. **9**(1), 1–25 (2022)
32. Magsambol, B.: CHED Allows Colleges, Universities to Decide on Face-to-Face Classes. RAPPLER. https://www.rappler.com/nation/ched-lets-colleges-universities-decide-face-to-face-classes-academic-year-2022-2023/ (2022)
33. Cross, J.S., Nagahama, T., Murota, M., Goto, S.: Tokyo Tech graduate student teaching assistant online course development program. J. JSEE. **69**(6), 6_59–6_64 (2021)

34. Seng, C., Carlon, M.K.J., Gayed, J.M., Cross, J.S.: Long-term effects of short-term intervention using MOOCs for developing Cambodian undergraduate research skills. EMOOCs. **2021**(2021), 49–62 (2021)
35. Lambert, S.R.: Do MOOCs contribute to student equity and social inclusion? A systematic review 2014–18. Comput. Educ. **145**, 103693 (2020)
36. Herranen, J.K., Aksela, M.K., Kaul, M., Lehto, S.: Teachers' expectations and perceptions of the relevance of professional development MOOCs. Educ. Sci. **11**(5), 240 (2021)
37. Lohse, J.J., Altoe, F., Jose, J., Nowotarski, A.M., Rahman, F., Tuck, R.C., Joyner, D.A.: The search for the MOOC credit hour. In: 2020 IEEE Learning with MOOCS (LWMOOCS), pp. 124–130. IEEE (2020)
38. Chen, J., Li, H., Wang, W., Ding, W., Huang, G.Y., Liu, Z.: A multimodal alerting system for online class quality assurance. In: International Conference on Artificial Intelligence in Education, pp. 381–385. Springer, Cham (2019)
39. Carlon, M.K.J., Cross, J.S.: Learning analytics dashboard prototype for implicit feedback from metacognitive prompt responses. In: Proc. 29th Int. Conf. on Computers in Education (2021)
40. Shah, D.: Massive list of MOOC-based microcredentials. The Report by Class Central. https://www.classcentral.com/report/list-of-mooc-based-microcredentials/ (2021)
41. Ledwon, H., Ma, R.: [2022] 70+ Legit online master's degrees. The Report by Class Central. https://www.classcentral.com/report/mooc-based-masters-degree/ (2022)
42. Gu, W., Xu, Y., Sun, Z.J.: Does MOOC quality affect users' continuance intention? Based on an integrated model. Sustainability. **13**(22), 12536 (2021)
43. Cross, J.S., Keerativoranan, N., Carlon, M.K.J., Tan, Y.H., Rakhimberdina, Z., Mori, H.: Improving MOOC quality using learning analytics and tools. In: 2019 IEEE Learning with MOOCS (LWMOOCS), pp. 174–179. IEEE (2019)
44. Zimmerman, W., Altman, B., Simunich, B., Shattuck, K., Burch, B.: Evaluating online course quality: a study on implementation of course quality standards. Online Learn. **24**(4), 147–163 (2020)
45. Lowenthal, P.R., Lomellini, A., Smith, C., Greear, K.: Accessible online learning: an analysis of online quality assurance frameworks. Q. Rev. Distance Educ. **22**, 15–29 (2021)
46. Edge, C., Monske, E., Boyer-Davis, S., VandenAvond, S., Hamel, B.: Leading university change: a case study of meaning-making and implementing online learning quality standards. Am. J. Dist. Educ. **36**(1), 53–69 (2022)
47. Romero, C., Ventura, S.: Educational data mining and learning analytics: an updated survey. Wiley Interdisc. Rev. Data Min. Knowl. Discov. **10**(3), e1355 (2020)
48. Cheng, Z., Anderson, T.R., Pelaez, N.J.: Backward designing a lab course to promote authentic research experience according to students' gains in research abilities. In: Trends in Teaching Experimentation in the Life Sciences, pp. 91–104 (2022)
49. Carlon, M.K.J., Keerativoranan, N., Cross, J.S.: Content type distribution and readability of MOOCs. In: Proceedings of the Seventh ACM Conference on Learning@ Scale, pp. 401–404 (2020)
50. Gaddem, M.R., Aouadi, L., Nagahama, T., Cross, J.S.: Exploring learners' video interaction in edX computer science MOOCs. In: Proceedings of the 2022 Annual Spring Conference of the Japan Society for Educational Technology, pp. 251–252 (2022)
51. Lodi, M., Sbaraglia, M., Zingaro, S.P., Martini, S.: The online course was great: I would attend it face-to-face: the good, the bad, and the ugly of IT in emergency remote teaching of CS1. In: Proceedings of the Conference on Information Technology for Social Good, pp. 242–247 (2021)
52. Lee, J., Soleimani, F., Hosmer IV, J., Soylu, M.Y., Finkelberg, R., Chatterjee, S.: Predicting cognitive presence in at-scale online learning: MOOC and for-credit online course environments. Online Learn. **26**(1) (2022)
53. Gorichanaz, T.: "It made me feel like it was okay to be wrong": student experiences with ungrading. In: Active Learning in Higher Education, 14697874221093640. (2022)

54. Carlon, M.K.J., Asa, A.D.D., Keerativoranan, N., Nagahama, T., Cross, J.S.: Topic modeling in MOOCs: what was to be discussed, what the instructor discussed, and what the learners discussed. In: 2021 IEEE International Conference on Engineering, Technology & Education (TALE), pp. 849–853. IEEE (2021)
55. Boonyubol, S., Kabir, S., Cross, J.S.: Comparing MOOC learners engagement with Japanese videos and text to speech generated English videos. In: Proceedings of the Ninth ACM Conference on Learning@ Scale, pp. 317–320 (2022)
56. Clayson, D.: The student evaluation of teaching and likability: what the evaluations actually measure. Assess. Eval. High. Educ. **47**(2), 313–326 (2022)
57. Feng, W., Tang, J., Liu, T.X.: Understanding dropouts in MOOCs. In: Proceedings of the AAAI Conference on Artificial Intelligence, vol. 33(01), pp. 517–524 (2019)
58. Carlon, M.K.J., Gaddem, M.R., Augusto, C., Reyes, H., Nagahama, T., Cross, J.S.: Investigating mechanical engineering Learners' satisfaction with a revised Monozukuri MOOC. EMOOCs. **2021**, 237 (2021)
59. Liu, F., Carlon, M.K.J., Cross, J.S.: Trial assessment of online learners' engagement with 360° videos using a head-mounted display. In: Proceedings of the 2022 Annual Fall Conference of the Japan Society for Educational Technology. (2022)
60. Cordero-Guridi, J.D.J., Cuautle-Gutiérrez, L., Alvarez-Tamayo, R.I., Caballero-Morales, S.O.: Design and development of a I4. 0 engineering education laboratory with virtual and digital technologies based on ISO/IEC TR 23842-1 standard guidelines. Appl. Sci. **12**(12), 5993 (2022)
61. Tempelaar, D., Rienties, B., Nguyen, Q.: Subjective data, objective data and the role of bias in predictive modelling: lessons from a dispositional learning analytics application. PLoS One. **15**(6), e0233977 (2020)
62. Kim, J., Choi, Y., Xia, M., Kim, J.: Mobile-friendly content design for MOOCs: challenges, requirements, and design opportunities. In: CHI Conference on Human Factors in Computing Systems, pp. 1–16 (2022)
63. Stratton, C., Grace, R.: Exploring linguistic diversity of MOOCs: implications for international development. Proc. Assoc. Inf. Sci. Technol. **53**(1), 1–10 (2016)
64. Nagahama, T., Makino, M., Morita, Y.: Effect analysis of high-speed presentations of educational video utilizing synthetic speech. Int. J. **13**(1), 66–74 (2019)
65. Rodrigues, F.M., Abreu, A.M., Holmström, I., Mineiro, A.: E-learning is a burden for the deaf and hard of hearing. Sci. Rep. **12**(1), 1–10 (2022)
66. Gies, T.: The ScienceDirect accessibility journey: a case study. Learn. Pub. **31**(1) (2018)
67. Surahman, E., Wang, T.H.: Academic dishonesty and trustworthy assessment in online learning: a systematic literature review. J. Comput. Assist. Learn. **38**, 1507–1819 (2022)
68. Goldberg, D.: Programming in a pandemic: attaining academic integrity in online coding courses. Commun. Assoc. Inf. Syst. **48**(1), 6 (2021)
69. Sloan-Lynch, J., Gay, N., Watkins, R.: Too fast for their own good: analyzing a decade of student exercise responses to explore the impact of math solving photo apps. In: LAK22: 12th International Learning Analytics and Knowledge Conference, pp. 67–76 (2022)
70. Reátegui, J.L., Herrera, P.C.: Artificial intelligence in the assessment process of MOOCs using a cloud-computing ecosystem. In: 2021 IEEE International Conference on Engineering, Technology & Education (TALE), pp. 487–493. IEEE (2021)
71. Garcia-Loro, F., Martin, S., Ruiperez-Valiente, J.A., Sancristobal, E., Castro, M.: Reviewing and analyzing peer review inter-rater reliability in a MOOC platform. Comput. Educ. **154**, 103894 (2020)
72. Bhavya, B., Chen, S., Zhang, Z., Li, W., Zhai, C., Angrave, L., Huang, Y.: Exploring collaborative caption editing to augment video-based learning. In: Educational Technology Research and Development, pp. 1–25 (2022)

73. Li, S., Craig, S.D., Schroeder, N.L.: Lessons learned from online learning at scale: a study of exemplar learning organizations. TechTrends. 1–14 (2022)
74. Geisel, J., Warkentin, H., Snow, J.: Ethical use of learning analytics for student support, not surveillance. Canadian Perspect. Acad. Integ. **5**(1), 28 (2022)
75. Koul, S., Singh, S., Verma, R.: Decentralised content creation in digital learning: a blockchain concept. In: ICT with Intelligent Applications, pp. 583–591. Springer, Singapore (2022)
76. Hicken, A.: Elearning Hype Curve: our predictions for 2019. Web Courseworks. https://webcourseworks.com/elearning-predictions-hype-curve-2019/ (2021)

Chapter 5
Understanding the Effect of Cohesion in Academic Writing Clarity Using Education Data Science

Jinnie Shin and Carrie Demmans Epp

Abstract Presenting complex information in a clear and concise manner in an academic paper is often considered a daunting task for many students. In particular, writing an abstract that could effectively convey the highlights of research findings is important to capture the reviewers' attention. While cohesion has been considered a critical feature for supporting comprehension in academic papers, a recent study has introduced a possible reverse effect that depends on reader knowledge. The purpose of the present study is to explore the effect of text cohesion on writing clarity using the data science methods of text mining and a hybrid classifier that employs fuzzy reasoning to consider a broader range of potential labels. We focused on using educational data science to automatically detect and investigate how referential cohesion features affect writing clarity in the abstracts of academic papers. We introduce a neuro-fuzzy approach to investigate these effects. Referential cohesion features were extracted from the abstract of 224 research articles using Coh-Metrix. We used these features to predict writing clarity scores. Accurate classification results, above 70%, were achieved using the hybrid neuro-fuzzy inference system. Subsequent analyses indicated that text cohesion features negatively impacted writing clarity for peer reviewers.

Keywords Educational text analysis · Educational data science · Natural language processing · Artificial intelligence in education · Reverse-cohesion effect · Coh-Metrix · Hybrid neural fuzzy inference · Writing clarity

J. Shin (✉)
College of Education, University of Florida, Gainesville, FL, USA
e-mail: Jinnie.shin@coe.ufl.edu

C. D. Epp
Department of Computing Science, University of Alberta, Edmonton, AB, Canada
e-mail: demmanse@ualberta.ca

193

Abbreviations

AI   Artificial intelligence
EDM   Educational data mining
EDS   Educational data science
HyFIS   Hybrid neuro-fuzzy inference system
NLP   Natural language processing
SVM   Support vector machines

5.1   Introduction

Many students consider the academic writing process to be daunting and somewhat intimidating. Unlike other types of writing, academic writing requires a formal structure with a logical flow of ideas. Thus, some consider academic writing as a process of helping readers to learn abstract ideas by adopting the role of a good teacher to convey a well-organized synthesis of ideas [1–3]. However, presenting highly complex information in a concise and clear manner requires a tremendous amount of experience and practice. Moreover, the burden that students bear when writing academic papers is heavy, because the consequences of failing to communicate clearly with readers often lead to hardships in publishing research findings. This burden is exacerbated by students' frustration regarding insufficient support for helping them learn how to write clearly [4–7].

Cohesion is one of the most referenced writing features that researchers have used to investigate text comprehension. This is partly because educational data science methods based in natural language processing can be used to identify the elements of a text that contribute to cohesion. Through the use of these automated methods, an increasing amount of evidence has identified cohesion building as one of the key components to improving academic writing [8–10]. Cohesion refers to the presence of explicit cues in a text that allows the reader to make connections between ideas [11]. Cohesion facilitates concise and clear communication between readers and a writer by reducing the cognitive burden required for readers to comprehend the text effectively [12–14]. For example, readers with limited background knowledge about the content would generally prefer cohesive texts that present information with a more thorough and logical flow of ideas. Thus, many consider cohesion as an effective tool for helping readers create a personal experience by making sense of content in a text [15].

While the strong positive connection between text comprehension and text cohesion has been supported with abundant empirical evidence [12, 14, 16, 17], some mixed results have been noted on how the knowledge of readers could change the impact of cohesion features [18–21]. Previous studies have identified a possible interaction effect between the reader's prior knowledge and text cohesion. That is, readers who are highly knowledgeable about the content may not benefit from cohesion devices in a text.

Crossley and McNamara [14] described text comprehension as a product of the interactive communication between the writer and readers, where careful consideration of the reader's background knowledge is essential. The negative impact of text cohesion, when presented to a highly knowledgeable audience, was introduced as the "reverse-cohesion effect", because text cohesion reversely affected text comprehension or writing clarity [18–21]. They explained that text cohesion devices interfered with the reader's experience and were reported to be "redundant" and "distracting" to readers and, in particular, high-knowledge readers.

In short, clarity in writing is achieved as a result of interaction between the writer and the readers with different levels of background knowledge. Therefore, the level of text cohesion should be carefully selected to provide the optimal experience of understanding for readers. This outcome may be particularly important in academic writing, where the primary audience is often well-informed professionals and experts in related fields whom the authors intend to persuade and inform about the significance of the findings. For example, reviewers of conference submissions are typically selected for their expertise in the field and are expected to provide fair and insightful recommendations and feedback. While text cohesion has been highly encouraged and even considered critical for clear communication in academic writing, it is important to understand whether text cohesion could adversely impact the text comprehension of academic articles due to the highly knowledgeable audience who evaluates these submissions, namely the peer reviewers. A better understanding of this effect could inform writing instruction for graduate students, which is an important focus for many higher education programs.

The recent introduction of data science and advanced computational analysis tools, such as natural language processing, in education has increased the possibility of systematically analyzing and informing effective instruction in academic writing [22–25]. Artificial intelligence (AI)-powered writing analytic methods could help systematically and efficiently extract evidence of high-quality writing using advanced psycholinguistic features from a large number of educational texts [26, 27]. Natural language processing combined with machine learning approaches may allow us to locate the complex patterns that connect psycholinguistic features with students' writing outcomes [28–31].

Hence, in the current study, we will introduce an application of artificial-intelligence-based method for the purposes of conducting educational data science work. This work evaluates reverse-cohesion effects in the context of academic writing using a neuro-fuzzy system (e.g., [32, 33]). We provide systematic insight to advance and inform instruction in academic writing, focusing on clarity using a core paradigm from educational data science: educational data mining (EDM) [34]. Educational data science (EDS) adopts data science approaches to solve challenges or resolve issues in education [34]. Piety et al. [35] identifies four general communities that conduct educational data science research, which include those who work on academic and institutional analysis, big data analysis using learning analytics and educational data mining, learner analytics and personalization, and systemic and instructional improvement. The approach we adopted and introduced in the current study follows this traditional paradigm of EDS by adopting the educational data mining approach, where the focus is in discovering structure and mining relationships using computational models to help and benefit learners [36]. For instance, increasing educational data mining research demonstrated success in effectively analyzing students' academic writing, such as conference proceedings (e.g., [37]). The studies shed lights on the importance of using educational data mining approaches to provide more effective and efficient instructional feedback (e.g., [38]). Still, to our knowledge, limited studies have been conducted to understand and evaluate the specific aspects of students' academic writing to help promote clear and effective communication.

Using our neuro-fuzzy approach, we automatically analyze academic writing using abstracts to understand the interactive dynamics among various cohesion indicators in a text that could impact writing clarity. More specifically, we focused on analyzing article abstracts to gather evidence regarding text cohesion and investigated the associations between text cohesion devices and writing clarity. Abstracts were selected because they play a catalytic role in an article, making an important first impression regarding the quality of writing to reviewers [39, 40]. Both a hybrid neuro-fuzzy inference system (HyFIS; [41]) and other classifier algorithms were then used to investigate the complex associations while providing adequate linguistic labeling to represent the level of cohesion present in the analyzed text.

5.2 Related Work

5.2.1 Text Cohesion and Writing Clarity

Clear writing has often been considered a process of representing ideas in a concise manner to effectively communicate with readers. In his classic text, *The Technique of Clear Writing*, Gunning [42] introduced the concept of clear writing by emphasizing the importance of concrete conceptualization, stating that clear and effortless communication with readers can only be established through a clear ideation of concepts by writers. Similarly, Couture [43] defined effective ideation as a representation of a range of elliptical to explicit linguistic choices made by a writer, which consequently encourages successful parallel communication between a writer and a reader. Later, the discourse-processing theory introduced by Gernsbacher [44] provided more concrete guidelines to encourage clear communication using text cohesion.

Gernsbacher [44] introduced three cognitive structure-building processes to explain how readers make sense of a text. Readers generate an information tree in the process of understanding the text and laying a foundation for future information. As the reader processes more information in a text, readers can map the subsequent information to the growing tree by mapping or by shifting to create a new subinformation tree to expand their understanding. The decision between mapping and shifting is based on the coherence of the information. In other words, the reader can map the information onto the existing structure and enhance the existing information tree if the text represents coherent information with familiar contents. Otherwise, the reader should shift to a new structure. Because the mapping process involves the integration of familiar content within existing structures, this process is easier and allows quicker reading and shorter reading times.

Therefore, a reader can comprehend highly cohesive text without producing many inferences and attempting to reconnect the information stored in distinctive trees [12]. Thus, writers are encouraged to use a sufficient number of cohesion devices to facilitate clear communication with readers and to improve text comprehension [14]. Furthermore, previous studies have identified significant positive associations of text cohesion with overall writing quality [16–18, 45]. For example, Chanyoo [46] located a moderate positive correlation between the number of cohesion devices with the overall writing score ($r = 0.44$, $p < 0.05$). In particular, the study identified that the frequency of using referential cohesion devices, such as it, a, and, they, was significantly positively correlated with the writing score in Thai undergraduate student's academic essays ([46]; $r = 0.51$, $p < 0.01$).

In addition, text cohesion devices were commonly identified as critical elements that explain the overall judgements of writing quality of L2 learners [45, 47–50] and learners at different education levels [51–53]. More specifically, Struthers et al. [53] proposed a checklist to evaluate cohesion in writing for elementary students and identified that text cohesion could significantly distinguish students who are high- and low-scoring writers with fair to moderate correlations. Similarly, Crossley et al. [45] asserted that the use of text cohesion devices in L2 learner writing contributed to explaining the majority of the variability in students' overall writing score or quality.

5.2.2   Reverse-Cohesion Effect

Despite the empirical evidence of the strong positive association between text cohesion and the ease of text comprehension and overall writing quality, a few studies have introduced rather contradictory results, indicating a negative effect of cohesion in clear writing, particularly for highly knowledgeable readers [18–21]. See Jones-Mensah and Tabiri's [54] comprehensive literature review on coherence and cohesion for a deeper explanation of cohesion in text.

The interactive effects of reader's prior knowledge and text cohesion on text comprehension was first introduced by McNamara and Kintsch [19] in an attempt to understand the learning effect of text cohesion for adults. They identified that adult learners with higher prior knowledge could perform better on given tasks (e.g., sorting task, open-ended questions) after reading the low-coherence text using a traditional analysis of variance (ANOVA) research methodology. They explained that low-coherence text facilitates more inference processes for high-knowledge readers, which works as a benefit for them to comprehend and process the content in a text.

O'Reily and McNamara [21] introduced that the effect of reverse cohesion occurs when students with low knowledge show better understanding and learning with more cohesive texts, whereas high-knowledge readers benefit more from low-cohesion text. The study included 143 college students, and an ANOVA test was conducted to detect the possible reverse-cohesion effect. They emphasized that increased cohesion gaps in a low-cohesion text facilitates and encourages readers to actively process the text and generate the inferences to comprehend the content [21].

Interestingly, O'Reily and McNamara [21] argued that such reverse-cohesion effect only occurs for less-skilled and high-knowledge readers. High-knowledge readers who are less skilled tend to generate a false sense of understanding without comprehending much detail due to the partial overlap between the displayed content and their knowledge. In other words, high-knowledge and less-skilled readers tend to skim the text as they possess a false impression of understanding, which leads to less comprehension achievement in the high-cohesion text. Moreover, McNamara et al. [14] reported that cohesion indices could not be used to significantly distinguish college students' writing quality.

Consistent with these previous studies, McNamara et al. used the Coh-Metrix analysis tool to extract cohesion devices from a text and then ANOVA to understand the significance of cohesion evidence in distinguishing low- and high-knowledge readers. The findings indicated that the impact of cohesion might depend on the knowledge of readers. In fact, high-knowledge readers can successfully make the inferences required to bridge the gaps in the low-cohesion text, as it forces the reader to make connections that are not explicitly available [21, 55].

Furthermore, cohesion cues might appear unnecessary and disruptive for high-knowledge readers, so cohesion cues may be inversely related to the essay scores if the readers have more knowledge about the topics. Similar to the previous findings, Crossley and McNamara [14] demonstrated that cohesion devices could not be associated with highly coherent texts, which were evaluated by human judges. In other words, the absence of cohesive devices was associated with a more coherent mental representation of a text.

5.2.3 The Role of Abstracts in Academic Writing Clarity

Abstracts in journal articles help readers understand the essence of the articles by encapsulating the important findings of the content [56]. Abstracts form initial impressions of the findings presented in the study. Hence, they "sell" the idea presented in the article [57]. Increasing research evidence compared and associated the readability of abstract and the subsequent part of the journal articles [58–60]. The findings from the previous literature consistently indicate that the contents provided in the abstract and the initial paragraphs of the articles were, oftentimes, significantly more difficult to comprehend compared to latter content. From this conclusion, Hartley and Trueman [60] highlighted the importance of taking the "abstract" into account when evaluating the clarity and the readability of article (e.g., Hartley [61]).

Hence, more recent studies introduced abstracts and the linguistic components extracted from the abstracts to evaluate the overall readability and clarity of research and scientific articles (e.g., Plavén-Sigray et al. [62]). Plavén-Sigray et al. [62] reported a strong positive relationship between the readability of the abstract and the full article, which indicates the complexity of abstracts generalizes to the full text of scientific and academic articles. Given these findings, we focused on understanding the abstract from the scientific article dataset and evaluated the reverse-cohesion effect using the content provided in the article associated with the overall writing clarity.

5.2.4 Measuring Text Cohesion Using Coh-Metrix

Coh-Metrix [8] is one of the most prevalent corpus-analysis tools that is used to investigate the cohesion and the coherence of text [63, 64]. Coh-Metrix 3.0 includes 11 categories of indices and a total of 110 scales that attempt to measure deep language features. The 11 groups include descriptive, text easability principal component scores, referential cohesion, latent semantic analysis, lexical diversity, connectives, situation models, syntactic complexity, syntactic pattern density, word information, and readability.

Referential cohesion is one of the categories that explicitly measures cohesion in text. Coh-Metrix defines cohesion as text characteristics that help readers mentally connect ideas in the text. Referential cohesion attempts to capture the semantic relationships among the parts of a text focusing on the overlaps and co-references in content words [11].

Referential cohesion includes five indices that measure the degree of overlap in different dimensions, such as noun, argument, stem, content word, and anaphor overlap between pairs or all of the sentences. More specifically, noun overlap (CRFNO1 and CRFNOa) and content word overlap refer to co-references and the exact overlap between the adjunct and all sentences. Specifically, the adjacent noun overlap (or local overlap) score (CRFNO1) identifies the number of sentences, on average, that present noun overlaps with the previous sentence. The global noun overlap (CRFNOa) evaluates the noun overlap between all pairs of sentences in the text. Unlike noun overlap, content word overlap (CRFCWO1 and CRFCWOa) evaluates the explicit overlap between the content words.

Stem overlap (CRFSO1 and CRFSOa) focuses on capturing deviational forms of the same word by considering overlaps in the lemma. Anaphor overlap (CRFANP1 and CRFANPa) attempts to account for pronominal references: whether the next sentence included a pronoun that refers to a pronoun or noun in the current sentence. Finally, argument overlap (CRFAO1 and CFRFAOa) evaluates the overlap between the singular and plural forms of the same noun in the previous and the current sentence. Similar to the noun overlap measures, the rest of the four overlap measures produce both the global and local stem overlap scores.

5.2.5   *Hybrid Neural Fuzzy Inference System (HyFIS)*

Data science and its subject matter is driven by large collections of data from which knowledge and information can be extracted [65]. Educational data science (EDS) concerns the extraction of knowledge and information from the data that is gathered in various educational settings and environments with the typical aim of solving educational problems [34, 66]. Various types of analytic approaches and algorithms are introduced, such as statistics, to effectively and accurately extract knowledge from educational datasets. Among the various analytic techniques, machine learning (ML) is a core disciplinary and research area within AI. In order to effectively disambiguate the decisions that involve clarity of academic writing, our study implemented one of the advanced machine learning algorithms, neuro-fuzzy systems.

Neuro-fuzzy systems combine the benefit of fuzzy logic and neural networks to improve performance in various data science tasks across areas. The theory of fuzzy logic was first introduced by Zadeh in the early 1960s to investigate the possibility of capturing and emulating uncertainties in the human cognitive process using linguistic labels. Fuzzy logic is intended to reflect human judgment and reasoning within computational modeling frameworks. These frameworks have often demonstrated many benefits, such as fast and accurate learning and good generalizability in addition to their being semantically understandable and forming rules that communicate how the classification decisions are made [67].

Neural networks employ a data-driven approach to map the expected output with observed features using a multilayer perceptron [68]. These mappings enable fuzzy logic and neural network systems to be considered as representation of two extreme ends of rule-based and machine-learning modeling approaches. The neuro-fuzzy approach was initially proposed to create a synergetic integration of the two paradigms to achieve greater learning capabilities while preserving the ability for the system to capture uncertainty within the labeling process [41]. They can contain various preidentified expert-driven rules that combine human judgment with machine learning. In this study, we focus on introducing one of the more successful neuro-fuzzy frameworks called the Hybrid neural Fuzzy Inference System, or HyFIS.

HyFIS has a unique architecture with two stages, structure learning and parameter learning. These two phases can be repeated on any dataset, making the model most suitable for incremental and online learning in dynamic systems. In other words, the system can change and adapt the rules initially introduced by the experts according to the training data, thus making the learning process more accurate and efficient [41]. More specifically, in the structure learning stage, a set of fuzzy rules are generated with predefined labels. Fuzzy rules refer to "if-then" rules that are used in conjunction with linguistic labels. The basic idea is that we can capture human decision-making and reasoning with fuzzy if-then rules, such as "If A is high and B is low, then, x is y". These rules can then be used to map the relationship between the input (i.e., x) and output (i.e., y) with linguistic labels (e.g., high and low), thus

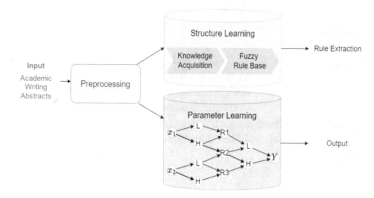

Fig. 5.1 Two-phase learning framework of HyFIS

supporting the decision-making and interpretation stages that are common to data science projects in educational settings.

One question that follows from the above approach is how we determine cutoff values to successfully define linguistic labels and apply fuzzy reasoning. The probability of each point in the input belonging to a particular label is determined using a membership function. A membership function consists of a curve from 0 to 1 that defines the mapping between a variable's value and its label (e.g., high $= 1$ and low $= 0$ in our study). HyFIS uses a Gaussian membership function, which represents the degree of membership of input data in a bell-curve shape, where the shape is determined by the mean and variance of data. After the rules are identified, the system attempts to adjust parameters of the membership functions to optimally map the input and output data. The gradient-based learning approach is used to optimize the parameters while updating the weights (or significance) of each fuzzy rule in mapping the input and output variable. Figure 5.1 provides a representation of the model learning structure of HyFIS.

5.2.6   The Current Study

To provide more effective instructions on academic writing, it is important to understand the strategies that could yield effective comprehension for readers. While some empirical evidence supports the positive associations between text cohesion and text comprehension, a few studies have introduced a potential interactive effect of the reader's knowledge and text cohesion on text comprehension [18–21]. Therefore, we focused on understanding the relationship between text cohesion and writing clarity in academic articles submitted for a conference publication peer-review process. More specifically, we focused on investigating the "reverse-cohesion effect", which was demonstrated to impact writing clarity judgment of high-knowledge readers. To evaluate this effect, we adopted a Hybrid

Neural Fuzzy method of writing analysis, thereby providing more comprehensive and interpretable results while improving the classification performance. More specifically, neuro-fuzzy inference systems can detect both linear and nonlinear relationships between the dependent and independent variables, thus increasing the prediction or classification accuracy significantly. In addition, the systems have an ability to optimize linguistic labels to numeric values to increase the interpretability of the classification rules. We introduced a neuro-fuzzy inference system approach, called HyFIS, to understand the complex dynamics between various cohesion variables while providing linguistic labels to increase the interpretability of the results.

5.3 Methodological Framework

In this section, we describe the dataset used in the current study, the model development and architecture, and the evaluation measures used for model validation.

5.3.1 Data

A subset of 224 items from the PeerRead dataset was used in the study (see Table 5.1). PeerRead [69] is an open-source dataset that contains 14,700 academic articles and their review decisions submitted in top-tier conference venues in computing science: ACL, NIPS, CoNLL, and ICLR . Apart from the acceptance decision, the dataset included several sub-scores with an overall writing clarity score that were analyzed.

The current study focused on the clarity score, which indicated whether the paper was clearly structured to promote easier understanding. Writing clarity was scored based on one of the peer-review questions, "For the reasonably well-prepared reader, is it clear what was done and why? Is the paper well-written and structured?" The score ranged from 1 to 5, each indicating "Much of the paper is confusing", "Important questions were hard to resolve", "Mostly understandable to me", "Understandable by most readers", and "Very clear", respectively.

Table 5.1 A subset of PeerRead dataset with clarity scores

Year	Section	Total N	N accepted	Average clarity scores
2016	CoNLL	221	11	3.97
2017	ACL	22	88	4.19
2017	ICLR	80	NA	4.26
Total		224	NA	4.14

Note. The number of accepted submissions in ICLR could not be identified due to lack of information in the original dataset

Most articles showed relatively high clarity scores with an average score range from 3.72 to 4.26. To avoid any potential issues stemming from a highly skewed score distribution, we decided to categorize the responses into two groups based on their clarity score: low clarity ($N = 61$) and high clarity ($N = 163$) group using 4 as a cutoff score. In addition, we implemented an oversampling approach to avoid having our model simply remember the solution due to the imbalanced representation of classes. The naive random oversampling approach was chosen to introduce a relatively equivalent number of instances from each class in model training. The naive random oversampling approach attempts to remedy the issue of imbalanced data by generating new cases in the classes that are underrepresented.

This approach repeats the original examples of the cases in the underrepresented class without increasing the variety of training cases the model could learn from. While this approach is commonly used, it is associated with concerns about increased risk of over-fitting or limited generalizability [70, 71]. Despite such concerns, several previous studies have also demonstrated the effectiveness of using the random oversampling method to significantly increase model performance (e.g., [72]). We used 20% and 10% of the data for testing and validation purposes, respectively. The remaining 70% was used for training.

5.3.2 Model Development

To effectively explore writing attributes that influence how clearly information is conveyed in academic papers, our framework consisted of four stages—data preprocessing, feature extraction, classification model development to understand the clear associations between text cohesion and writing clarity, and performance evaluation. For the main classification model, we experimented with two classification algorithms: support vector machines (SVMs) and HyFIS.

5.3.2.1 Data Preprocessing

Prior to feature extraction, critical preprocessing steps were conducted to decrease noise and bias. In order to extract clean and representative features from Coh-Metrix, we focused on eliminating potentially troubling symbols and representations in the document that could not be processed by the Coh-Metrix analysis tool. For example, we removed all the citations (e.g., [73]) along with quotation marks and we removed the equations and other nonalphabetic words (e.g., @, #, %). Then, we saved each document into a separate text file to run the Coh-Metrix analysis at the document level. Figure 5.2 provides an example of preprocessing of one of the abstracts analyzed in our study.

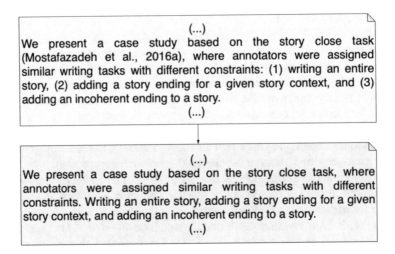

Fig. 5.2 Example of data preprocessing results

5.3.2.2 Feature Extraction and Reduction

We used Coh-Metrix 3.0 text analyzer [8] to extract text features from our dataset. We ran the analysis on the abstract for each article, which resulted in 110 Coh-Metrix features. While most of the Coh-Metrix indices are designed to investigate text cohesion and coherence, they also include 11 subcategories: descriptive indices, text easability principal component scores, referential cohesion, latent semantic analysis scores, lexical diversity, connectives, situation model, syntactic complexity, syntactic pattern density, word information, and readability.

To increase the interpretability of our classification algorithms, we experimented with reducing the number of features using stepwise regression without compromising performance accuracy. Stepwise regression modeling is commonly implemented to automatically perform feature selection using the forward elimination, backward elimination, and combined elimination approach. Stepwise regression performs a randomized search on a possible set of feature combinations while searching for a locally optimal solution for each step to eventually reach a globally optimal solution [74]. The model performance in each step is evaluated based on various goodness-of-fit criteria including Akaika Information Criterion (AIC), Bayesian Information Criterion (BIC), and adjusted R-squared [75]. In this study, we used the forward elimination strategy to return a small number of deterministic features to build our reduced-feature model for classification.

5.3.2.3 Clarity Score Classification Model

Two classification algorithms were implemented as our main models. First, we used the HyFIS model to classify the low-clarity articles from the high-clarity articles based on the reduced Coh-Metrix features. The hybrid neural fuzzy inference system (HyFIS; [41]) mainly consists of a two-phase learning scheme, which is the knowledge acquisition and the parameter learning phase. Following the knowledge acquisition model introduced by Wang and Mendel [76], HyFIS extracts fuzzy rules with linguistic properties.

For example, for a binary classification model with two variables (e.g., $x1$, $x2$), we could expect to locate a rule such as "If $x1$ is low and $x2$ is high, then the class is 0" when using low and high as linguistic labels. The initial sets of rules are provided to the parameter learning stage to extract final sets of rules using a gradient descent learning algorithm, while tuning the Gaussian membership function parameters (i.e., mean and variance) for each linguistic category.

The HyFIS model was chosen for this study, as it demonstrated highly accurate and efficient results in prediction tasks [77] and the Gaussian membership function was appropriate to represent the distribution of our Coh-Metrix dataset as the distribution of the output generally followed a normal distribution. We also experimented with using various combinations of hyperparameters, such as the number of linguistic labels (2–5), step sizes (0.1 and 0.01), and the number of maximum interactions (100–1000) to locate our optimal models.

We used the support vector machine (SVM) algorithm to classify the low-clarity group and the high-clarity group for comparison purposes. SVM is one of the most used machine learning classification algorithms; it has demonstrated highly accurate performance in both linear and nonlinear classification tasks [78]. We used a linear SVC model, which uses a linear kernel as a basis function for classification while minimizing the squared hinge loss. The squared hinge loss is a commonly used loss function in training classification algorithms [79]. The squared hinge loss measures the accuracy of the classification by evaluating and quantifying the difference between the expected outcome and the prediction results. We experimented with various combinations of hyperparameters to locate the optimal classification model. The penalty parameters C ranged from 0.1 to 100, using L1 or L2 regularization and a batch size of 128–256.

5.3.2.4 Performance Evaluation

To thoroughly evaluate the classification accuracy, we used commonly employed classification accuracy metrics, such as the average accuracy, precision, recall, and F1 score. We used human-raters' agreement score as a gold standard and compared the performance of our SVM and HyFIS models to that standard. In addition, we implemented a random guesser algorithm, which randomly guesses the output (0 or 1) to remedy potential bias stemming from imbalanced data. Comparing our results with the random guesser's performance helped prevent the model from simply outputting the overrepresented class to achieve a high classification accuracy.

Precision, recall, and the F1 score were also reported as the evaluation metrics for the same purposes. Precision and recall are commonly used accuracy measures, especially when the classes are imbalanced. While precision represents the relevancy of the result, recall measures the proportion of truly relevant results returned. Precision is computed as a ratio of true-positives over all the positives, and recall is computed as a true-positives over the sum of true-positives and false-negatives. The F1 score refers to the harmonic mean of the precision and recall scores. As such, it represents a trade-off between the types of errors that a classifier can make. F1 scores range from 0 to 1, in which 1 represents perfect precision and recall; it is commonly applied to evaluate classification task performance.

5.4 Analysis and Interpretation of the Results

We compare and discuss the findings to understand whether text cohesion depicts negative relationships with text comprehension for high-knowledge readers. To begin, the preliminary analysis results are presented to provide more insights into the cohesion features used for score classification in the current study. Then, we present more detailed information regarding the feature reduction process using stepwise logistic regression analysis. Finally, model performance is presented and compared to our gold-standard, human-raters' agreement score, and our baseline, a random guesser.

5.4.1 Descriptive Statistics and Feature Selection Results

The lengths of abstracts were relatively short with an average word length of 139 across six sentences in one paragraph. However, there were a few cases where the abstract alone consisted of close to 30 sentences. The average writing clarity score was higher than initially expected (4.14), leaving our output data relatively skewed. Thus, we categorized the final output into two classes, low clarity ($N = 61$) and high clarity ($N = 162$), for more effective classification model development (see Table 5.2). We used stepwise logistic regression to reduce the number of features. Feature selection was conducted to focus on investigating the indicative linguistic features that are highly associated with writing clarity. In addition, this could greatly improve the interpretability of the final classification model and its performance. We used forward conditional binary logistic regression to select deterministic (i.e.,

Table 5.2 Descriptive statistics of the current dataset

Features	M (SD)	Min	Max	N responses
Number of words	139 (60.10)	34	531	
Number of sentences	6.08 (2.72)	1	28	
Average clarity scores	4.14 (0.80)	2	5	224
Low-clarity group		2	3.9	61
High-clarity group		4	5	163

Table 5.3 Stepwise regression analysis results

Variable	Unstandardized coefficients		
	B	S.E.	P-value
DESWLsyd	−3.691**	1.322	0.005
CRFNO1	−1.985*	0.914	0.030
CRFSO1	3.355**	1.082	0.002
CRFCWOa	−9.759*	3.831	0.011

Note. $**p < 0.01$, $*p < 0.05$; Nagelkerke R-squared ($R^2 = 0.162$ and 0.160)

Table 5.4 Reduced feature set in the abstract dataset

	Feature	M (SD)	Min	Max
Descriptive	DESWLsy	1.16 (0.12)	0.81	1.49
Referential cohesion	CRFNO1	0.67 (0.26)	0	1
	CRFSO1	0.82 (0.21)	0	1
	CRFCWOa	0.11 (0.05)	0	0.31

Note. DESWLsy: Word length, Average number of syllables; Computes the number of syllables on average for every word that appears in the text. Word that are shorter are tend to be easy to comprehend. CRFNO1: Noun overlap, Adjacent sentences; Identifies the number of sentences on average that shows noun overlap between the adjacent sentences. Only the nouns that matches the form and polarity are counted as exact matching. CRFSO1: Stem overlap, Adjacent sentences; A local overlap measure where a number of matchings between content words (e.g., nouns, verbs, adjectives, and adverbs) and a noun that shares the common lemma. CFRCWOa: Content word overlap, All sentences; A global overlap measure that computes the proportion of explicit content word overlaps between the all sentences

features that depict significant contributions in classification decisions) Coh-Metrix features. The following four features were extracted from the abstract dataset: DESWLsy, CRFNO1, CRFSO1, and CRFCWOa. While the backward elimination process is often preferred and recommended by statisticians for dimensionality reduction [80], a forward elimination process commonly returns a smaller number of features. As a result, we selected this procedure to focus on a few highly indicative writing features. The findings indicated that three out of four features selected from the abstract dataset belong to the referential cohesion category, measuring various types of overlap (i.e., CRFNO1, CRFSO1, and CRFCWOa). The fourth one is the word length feature from the descriptive category (DESWLsy). Tables 5.3 and 5.4 present more detailed information regarding the analysis results with the descriptions of the Coh-Metrix features. Figure 5.3 presents the distribution of the final four Coh-Metrix features.

5.4.2 Findings from the Classification Model

Next, we used the reduced-feature sets (i.e., DESWLsy, CRFNO1, CRFSO1, and CRFCWOa) from the abstract to predict writing clarity scores. In model training, we

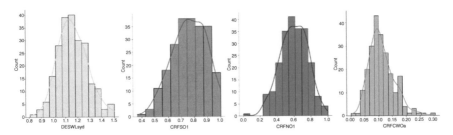

Fig. 5.3 Data distribution of the final Coh-Metrix features

Table 5.5 Classification performance results comparisons

	Random	SVM	HyFIS	Human rater
Accuracy	0.521	0.742	0.717	0.820
Precision	0.650	0.711	0.701	0.805
Recall	0.488	0.825	0.979	0.959
F1 score	0.558	0.764	0.817	0.881

used the naive random oversampling approach to remedy potential problems of the imbalanced dataset, while promoting more effective learning. This sampling technique helps address the issue of imbalanced data by generating new samples in the underrepresented class. Table 5.5 presents the final classification accuracies of our models using the accuracy, precision, recall, and F1 scores. The results indicated that our HyFIS and SVM models could achieve relatively high accuracies in predicting writing clarity scores.

The HyFIS model could also produce a relatively high classification performance when compared to the SVM model performance. Our final SVM model was tuned based on a selection of hyperparameters using a grid search and we found that our linear support vector machine could achieve the best accuracies with the following settings: $C = 1.0$, using a squared hinge loss function, and L2 penalization. Our HyFIS model included two linguistic labels for each variable (e.g., long, short or high, low) to represent the writing features in a more meaningful and interpretable manner. Our final model demonstrated the best performance with a parameter setting of two linguistic labels, a step size of 0.01, and the maximum iteration of 200. Figure 5.4 provides a conceptual representation of our model framework with five layers following the original HyFIS framework by Kim and Kasabov [41]. More specifically, by setting two linguistic or term labels for each variable, HyFIS used the Gaussian membership function to provide the membership values for each input variable.

Figure 5.5 presents the final membership function for the input variables—word length, noun overlap, stem overlap, and content word overlap—with respect to the linguistic labels. The cross point of the two membership functions indicated a higher magnitude of belongingness to the other label. For example, if the standardized score of the word length is below 0.58, then a document would have a higher membership value to be labeled "short" than that for "long". Similarly, the membership functions

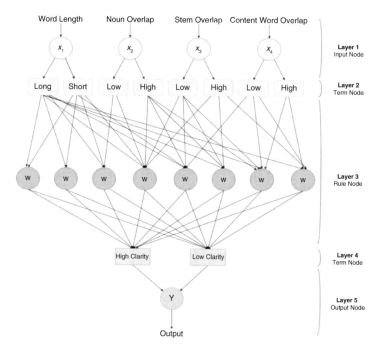

Fig. 5.4 Conceptual representation of the final HyFIS model

for noun, stem, and content overlap suggest that if a document had standardized scores below 0.59, 0.71, and 0.43, then the document should be labeled to have "low" cohesion for the noun, stem, and content overlap.

5.4.3 Findings from the Feature Weights and the Fuzzy Inference System

Out of the total 16 combinations of rules (i.e., $2^4 = 16$), HyFIS could extract the following eight rules in the parameter learning phase. Figure 5.6 depicts the final rules that were extracted for the classification and the findings from the rules that indicated the significance of the noun, stem, and content overlap variables in determining the final class, high or low writing clarity. In addition, we could confirm the "reverse-cohesion effect" based on the rules that are highly associated with low clarity. More specifically, the fuzzy rules that were associated with low clarity consistently corresponded to high values in noun overlap and stem overlap.

This indicated that when the abstract showed a high degree of overlap in noun and stem words, the peer reviewers associated the writing with low clarity. In addition, the most significant fuzzy rule, with the highest function degree, indicated abstracts that showed high values in word length as well as noun, stem, and content word

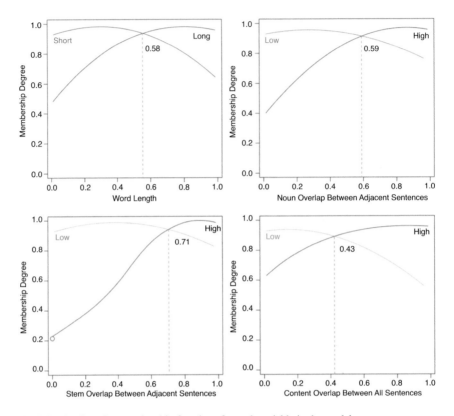

Fig. 5.5 The Gaussian membership functions for each variable in the model

Word Length	Noun Overlap	Stem Overlap	Content Word Overlap	Function Degree	Clarity
long	high	high	high	0.771	
short	high	high	low	0.750	Low Clarity
long	high	high	low	0.732	
short	high	high	low	0.715	
long	low	low	low	0.636	
short	low	low	low	0.595	High Clarity
short	low	high	low	0.586	
long	low	high	low	0.582	

Fig. 5.6 Final rules extracted from the HyFIS model (*Note*. The contrast between the cells is used to denote "high" and "low" membership)

overlap were highly associated with low clarity. In other words, if the word length is long (above 0.58), noun overlap is high (above 0.59), stem overlap is high (above 0.71), and content word overlap is high (above 0.43), then the writing was associated with low clarity (0.771). Likewise, if the word length is short, noun overlap is high, stem overlap is high, and content word overlap is low, then the writing is associated with low clarity. Still, we noticed some inconsistencies where the two consistent rules resulted in different function degrees (lines 2 and 4 of Table 5.5). Overall, our findings indicate the importance of noun overlap in determining the clarity of a text, in which high-cohesion texts were associated with low clarity and less comprehension.

5.5 Discussion, Interpretations, and Findings

Text cohesion facilitates explicit connections in the content to reduce the cognitive burdens imposed on readers to understand the text [12]. Text cohesion has been actively evaluated and referenced in previous research to understand its impact on text comprehension. While text cohesion was commonly thought to have positive association with text comprehension and writing clarity, a few studies have indicated a possible interaction effect between text cohesion and reader knowledge on text comprehension [18–21]. They indicated that high-knowledge readers tend to benefit less from high-cohesion text and referred to such effects as the reverse-cohesion effect. However, little research has been conducted to understand the effect in academic articles that are submitted for conference or journal publication peer reviews. The peer-review process often involves a highly knowledgeable and skilled audience, who have expertise in the related field to provide adequate feedback.

Therefore, the purpose of the study was to investigate the relationship between writing clarity and text cohesion in academic articles, while focusing on detecting potential reverse-cohesion effects. In addition, previous academic writing analysis often involved relatively traditional and linear methods, such as ANOVA and regression analysis, rather than more advanced data science approaches. To provide a more comprehensive understanding of the associations between various text cohesion devices and their impact on writing clarity, we adopted a hybrid neuro-fuzzy inference system (HyFIS) approach to automatically score the clarity of writing and to uncover the underlying rules with interpretable linguistic labels.

The outcomes of the study yield several interesting findings that support the presence of reverse-cohesion effects in a peer-review process. First, we identified that our system could achieve close to 70% accuracy in predicting the writing clarity scores denoted by the peer reviewers using the presence or absence of text cohesion in the abstract alone. More specifically, the co-references in noun, stem, and content words between sentences were identified as critical writing features that determined writing clarity.

Second, we located the presence of reverse-cohesion effect in peer-reviewed academic articles. The results from our HyFIS model indicated that "high" standardized scores in the overlap in noun (≥ 0.59), stem (≥ 0.71), and content words (≥ 0.43) were highly associated with lower writing clarity. While noun overlap and content overlap refer to the exact noun and content overlap between the sentences, stem overlap attempts to account for such deviations by considering overlaps in the lemma. Therefore, all helps, helping, helper, helpless, and helpful could be counted to overlap with the lemma, help.

The findings in the current study provide several important contributions to expand the understanding of current literature with methodological and practical implications. First, the study adds to the extent of previous literature, emphasizing the importance of cohesive reference in academic articles by providing empirical evidence of a reverse-cohesion effect in academic papers [18–21, 81, 82]. Moreover, the study introduced the reverse-cohesion effect in a unique writing setting, peer-review processes of academic papers so that we could expand our understanding of how the reverse-cohesion effect could be applied in high-stakes writing settings.

Unlike the findings of O'Reilly and McNamara [21], we located the reverse-cohesion effects among high-knowledge and high-skilled readers as the clarity judgement scores in our study were collected for peer reviewers who evaluated themselves as established experts in the related fields. O'Reilly and McNamara [21] asserted that high-knowledge and less-skilled readers tend to generate a false impression of complete understanding of the context in a text as they commonly skim the text. The reverse-cohesion effect identified from our findings could be due to current peer-review processes, where reviewers often face difficulties due to lack of time and resources to carefully review papers.

In terms of methodological contribution, our study successfully demonstrated the applicability of educational text analysis using neuro-fuzzy inference systems. Our analytic framework contributes to the educational data sciences community by providing robust methodological guidelines in academic writing analysis research. Specifically, our hybrid neuro-fuzzy inference system (HyFIS) approach could demonstrate highly accurate classification performance while providing interpretable and comprehensive rules that were associated with their classification performance. Moreover, the HyFIS approach could yield and adapt linguistic labels to each variable (e.g., high and low) to improve the interpretability of the associated rules.

Finally, in terms of practical implications for teaching English for academic purposes, our findings provide interesting insights into potential focal points for providing instruction to students so they can learn to communicate effectively with reviewers. Issues with lack of appropriate use of cohesive reference in academic writing have been identified as highly problematic in previous literature [83, 84]. Previous studies have consistently reported that students find writing abstracts in a coherent manner difficult [85–87]. More specifically, Suwandi [87] indicated that most undergraduate students find locating appropriate cohesive devices to connect the sentence together highly challenging. Therefore, the common recommendation centered on emphasizing teaching cohesive devices. However, our findings emphasized that the previous focus on cohesion features could be reduced as the overuse of cohesion devices, indeed, impacted the reviewer's understanding negatively.

5.6 Limitations and Future Work

While the current study was carefully designed and implemented to reduce potential sources of bias and limitations, we acknowledge that there are few methodological and empirical consideration to be further studied. First, the study only considers the cohesion effect devices that were quantified and extracted from the abstract. The importance and the challenging nature of writing abstracts in academic papers are well-establish in the previous literature [62]. Expanding the findings to the other parts of an academic or scientific paper (e.g., conclusion and discussion) would be important to draw a deeper practical conclusion from the findings. Second, our study adopted the Hybrid Neural Fuzzy model to evaluate the cohesion effect from the abstracts of academic writing. While introducing a novel architecture of fuzzy neural system was out of scope, we encourage future research to understand expanded architectures and algorithms to better evaluate the effect of cohesion devices in academic writing. Third, the current analysis did not account for various confounding factors such as the competencies of the peer reviewers (e.g., confidence level, total time spent to review). In addition, only a small set of the subset data is selected based on the availability of the overall writing clarity score. Hence, we encourage future research to focus on how the current findings generalize when mitigated by other textual or human-rater relevant variables.

The findings from the current study suggest that the previous focus on text cohesion should be reduced in academic writing instruction. Specifically, we observed the importance of noun overlap in determining the clarity of a text, in which high-cohesion texts were associated with low clarity and less comprehension.

The findings and the analytic framework demonstrated in this study contributes to the educational data sciences by providing methodological guidelines to automatically analyze and characterize academic writing from a large number of texts. More study is required to understand whether the current findings could be applied to academic articles from different disciplines. Previous studies have reported various disciplinary differences in academic writing [88, 89].

Interestingly, text cohesion was one of the features that could significantly predict the differences in various subdisciplines within engineering and science. Therefore, future studies should carefully investigate whether reverse-cohesion effects can be located in distinct disciplines, such as education. In addition, the present study focused on analyzing abstracts of articles to gather evidence regarding text cohesion and its impact on writing clarity. We believe that including other parts of articles, such as the introduction and conclusion, could help increase the generalizability of our findings through improved classification accuracies that are based on a larger sample of the text.

References

1. Ferretti, R.P., Graham, S.: Argumentative writing: theory, assessment, and instruction. Read. Writ. **32**(6), 1345–1357 (2019). https://doi.org/10.1007/s11145-019-09950-x
2. French, A.: Academic writing as identity-work in higher education: forming a 'professional writing in higher education habitus'. Stud. High. Educ. **45**(8), 1605–1617 (2020). https://doi.org/10.1080/03075079.2019.1572735
3. Sage, S.: Formulaic sequences: importance to academic English writing. Gakuen. **948**, 22–33 (2019)
4. Canadian University Survey Consortium: 2015 Graduating University Student Survey Master Report (2015)
5. Huisman, B., Saab, N., van den Broek, P., van Driel, J.: The impact of formative peer feedback on higher education students' academic writing: a meta-analysis. Assess. Eval. High. Educ. **44**(6), 863–880 (2019). https://doi.org/10.1080/02602938.2018.1545896
6. Rahman, M., Hasan, K.: Academic writing difficulties of Bangladeshi students at a higher institution in Malaysia. J. Res. Multidisc. **2**(2), 145–171 (2019). https://doi.org/10.5281/jrm.v2i2.19
7. Singh, M.K.M.: International graduate students' academic writing practices in Malaysia: challenges and solutions. J. Int. Stud. **5**(1), 12–22 (2019). https://files.eric.ed.gov/fulltext/EJ1052831.pdf
8. Graesser, A.C., McNamara, D.S., Louwerse, M.M., Cai, Z.: Coh-Metrix: analysis of text on cohesion and language. Behav. Res. Methods Inst. Comput. **36**(2), 193–202 (2004). https://doi.org/10.3758/BF03195564
9. He, Z.: Cohesion in academic writing: a comparison of essays in English written by L1 and L2 university students. Theory Pract. Lang. Stud. **10**(7), 761–770 (2020)
10. Weston-Sementelli, J.L., Allen, L.K., McNamara, D.S.: Comprehension and writing strategy training improves performance on content-specific source-based writing tasks. Int. J. Artif. Intell. Educ. **28**(1), 106–137 (2018). https://doi.org/10.1007/s40593-016-0127-7
11. Halliday, M.A.K., Hasan, R.: Cohesion in English. Routledge, London (1976)
12. Best, R.M., Rowe, M.P., Ozuru, Y., McNamara, D.S.: Deep-level comprehension of science texts: the role of the reader and the text. Top. Lang. Disorders. **25**(1), 65–83 (2005). https://journals.lww.com/topicsinlanguagedisorders/fulltext/2005/01000/deep_level_comprehension_of_science_texts__the.7.aspx
13. Crossley, S.A., Kyle, K., Dascalu, M.: The tool for the automatic analysis of cohesion 2.0: integrating semantic similarity and text overlap. Behav. Res. Methods. **51**(1), 14–27 (2019). https://doi.org/10.3758/s13428-018-1142-4
14. McNamara, D.S., Louwerse, M.M., McCarthy, P.M., Graesser, A.C.: Coh-Metrix: capturing linguistic features of cohesion. Discour. Process. **47**(4), 292–330 (2010). https://doi.org/10.1080/01638530902959943
15. Williams, J.M., Nadel, I.B.: Style: Ten Lessons in Clarity and Grace, pp. 22–23. Scott, Foresman, Glenview, IL (1989)
16. Collins, J.L.: Strategies for Struggling Writers. The Guilford Press, New York, NY (1998)
17. DeVillez, R.: Writing: Step by Step. Kendall Hunt, Dubuque, IO (2003)
18. Crossley, S., & McNamara, D. (2010). Cohesion, coherence, and expert evaluations of writing proficiency. In Proceedings of the Annual Meeting of the Cognitive Science Society, 32(32)

19. McNamara, D.S., Kintsch, W.: Learning from texts: effects of prior knowledge and text coherence. Discour. Process. **22**(3), 247–288 (1996). https://doi.org/10.1080/01638539609544975
20. McNamara, D.S., Kintsch, E., Songer, N.B., Kintsch, W.: Are good texts always better? Interactions of text coherence, background knowledge, and levels of understanding in learning from text. Cogn. Instr. **14**(1), 1–43 (1996). https://doi.org/10.1207/s1532690xci1401_1
21. O'Reilly, T., McNamara, D.S.: The impact of science knowledge, reading skill, and reading strategy knowledge on more traditional "high-stakes" measures of high school students' science achievement. Am. Educ. Res. J. **44**, 161–196 (2007). https://doi.org/10.3102/0002831206298171
22. Ishizaki, S., Kaufer, D.: Scalable writing pedagogy for strengthening cohesion with interactive visualization. In: 2020 IEEE International Professional Communication Conference (ProComm), pp. 141–145. IEEE (2020). https://doi.org/10.1109/ProComm48883.2020.00029.
23. Knight, S., Shibani, A., Abel, S., Gibson, A., Ryan, P.: AcaWriter: a learning analytics tool for formative feedback on academic writing. J. Writ. Res. https://opus.lib.uts.edu.au/bitstream/10453/141783/2/JoWR_2020_vol12_nr1_Knight_et_al.pdf (2020)
24. Lachner, A., Neuburg, C.: Learning by writing explanations: computer-based feedback about the explanatory cohesion enhances students' transfer. Instr. Sci. **47**(1), 19–37 (2019). https://doi.org/10.1007/s11251-018-9470-4
25. Strobl, C., Ailhaud, E., Benetos, K., Devitt, A., Kruse, O., Proske, A., Rapp, C.: Digital support for academic writing: a review of technologies and pedagogies. Comput. Educ. **131**, 33–48 (2019). https://doi.org/10.1016/j.compedu.2018.12.005
26. Gibson, A., Shibani, A.: In: Lang, C., Siemens, G., Friend Wise, A., Gašević, D., Merceron, A. (eds.) Natural Language Processing-Writing Analytics, 2nd ed, pp. 96–104. SoLAR, Vancouver, Canada. https://solaresearch.org/wp-content/uploads/hla22/HLA22_Chapter_10_Gibson.pdf (2022)
27. Öncel, P., et al.: Automatic student writing evaluation: investigating the impact of individual differences on source-based writing. In: LAK21: 11th International Learning Analytics and Knowledge Conference, pp. 620–625 (2021)
28. Burkhart, C., Lachner, A., Nückles, M.: Assisting students' writing with computer-based concept map feedback: a validation study of the CohViz feedback system. PLoS One. **15**(6), e0235209 (2020). https://doi.org/10.1371/journal.pone.0235209
29. Crossley, S.A.: Linguistic features in writing quality and development: an overview. J. Writ. Res. **11**(3), 415–443 (2020). https://doi.org/10.17239/jowr-2020.11.03.01
30. Latifi, S., Gierl, M.: Automated scoring of junior and senior high essays using Coh-Metrix features: implications for large-scale language testing. Lang. Test. **38**(1), 62–85 (2021). https://doi.org/10.1177/0265532220929918
31. MacArthur, C.A., Jennings, A., Philippakos, Z.A.: Which linguistic features predict quality of argumentative writing for college basic writers, and how do those features change with instruction? Read. Writ. **32**(6), 1553–1574 (2019). https://doi.org/10.1007/s11145-018-9853-6
32. Pezeshki, Z., Mazinani, S.M.: Comparison of artificial neural networks, fuzzy logic and neuro fuzzy for predicting optimization of building thermal consumption: a survey. Artific. Intell. Rev. **52**(1), 495–525 (2019). https://doi.org/10.1007/s10462-018-9630-6
33. Tiruneh, G.G., Fayek, A.R., Sumati, V.: Neuro-fuzzy systems in construction engineering and management research. Auto. Constr. **119**, 103348 (2020). https://doi.org/10.1016/j.autcon.2020.103348
34. Romero, C., Ventura, S.: Educational data science in massive open online courses. Wiley Interdisc. Rev. Data Min. Knowl. Discov. **7**(1), e1187 (2017). https://doi.org/10.1002/widm.1187
35. Piety, P.J., Hickey, D.T., Bishop, M.J.: Educational data sciences: framing emergent practices for analytics of learning, organizations, and systems. In: Proceedings of the fourth international conference on learning analytics and knowledge, pp. 193–202 (2014, March)

36. Baker, R.S., Martin, T., Rossi, L.M.: Educational data mining and learning analytics. In: The Wiley Handbook of Cognition and Assessment: Frameworks, Methodologies, and Applications, pp. 379–396 (2016)
37. Ribeiro, A.C., Sizo, A., Lopes Cardoso, H., Reis, L.P.: Acceptance decision prediction in peer-review through sentiment analysis. In: EPIA Conference on Artificial Intelligence, pp. 766–777. Springer, Cham (2021)
38. Crossley, S.A., Allen, L.K., Snow, E.L., McNamara, D.S.: Incorporating learning characteristics into automatic essay scoring models: what individual differences and linguistic features tell us about writing quality. J. Educ. Data Min. **8**(2), 1–19 (2016)
39. Erbert, J.R.: What is An Abstract? http://employees.oneonta.edu/ebertjr/what_is_an_abstract.htm (2014)
40. Hyland, K.: Disciplinary Discourses: Social Interactions in Academic Writing. Longman, London (2000)
41. Kim, J., Kasabov, N.: HyFIS: adaptive neuro-fuzzy inference systems and their application to nonlinear dynamical systems. Neural Net. **12**(9), 1301–1319 (1999). https://doi.org/10.1016/S0893-6080(99)00067-2
42. Gunning, R.: The Technique of Clear Writing. McGraw-Hill, New York, NY (1952)
43. Couture, B.: Effective ideation in written text: a functional approach to clarity and exigence. In: B. Couture. Functional Approaches to Writing: Research Perspectives. https://digitalcommons.unl.edu/cgi/viewcontent.cgi?article=1066&context=englishfacpubs (1986)
44. Gernsbacher, M.A.: Language Comprehension as Structure Building. Psychology Press (2013)
45. Crossley, S.A., Kyle, K., McNamara, D.S.: The development and use of cohesive devices in L2 writing and their relations to judgments of essay quality. J. Sec. Lang. Writ. **32**, 1–16 (2016). https://doi.org/10.1016/j.jslw.2016.01.003
46. Chanyoo, N.: Cohesive devices and academic writing quality of Thai undergraduate students. J. Lang. Teach. Res. **9**(5), 994–1001 (2018). https://doi.org/10.17507/jltr.0905.13
47. Bridgeman, B., Carlson, S.: Survey of academic writing tasks required of graduate and undergraduate foreign students. ETS Res. Rep. Ser. **1983**(1), i–38 (1983). https://www.ets.org/research/policy_research_reports/rr-83-18_toefl-rr-15
48. Castro, C.D.: Cohesion and the social construction of meaning in the essays of Filipino college students writing in L2 English. Asia Pac. Educ. Rev. **5**(2), 215 (2004). https://link.springer.com/content/pdf/10.1007/BF03024959.pdf
49. Harman, R.: Literary intertextuality in genre-based pedagogies: building lexical cohesion in fifth-grade L2 writing. J. Sec. Lang. Writ. **22**(2), 125–140 (2013). https://doi.org/10.1016/j.jslw.2013.03.006
50. Hinkel, E.: Matters of cohesion in L2 academic texts. Appl. Lang. Learn. **12**(2), 111–132 (2001)
51. Jin, W: A Quantitative Study of Cohesion in Chinese Graduate Students' Writing: Variations across Genres and Proficiency Levels. https://eric.ed.gov/?id=ED452726 (2001)
52. Stotsky, S.: Types of lexical cohesion in expository writing: implications for developing the vocabulary of academic discourse. Coll. Comp. Commun. **34**(4), 430–446 (1983). https://doi.org/10.2307/357899
53. Struthers, L., Lapadat, J.C., MacMillan, P.D.: Assessing cohesion in children's writing: development of a checklist. Assess. Writ. **18**(3), 187–201 (2013). https://doi.org/10.1016/j.asw.2013.05.001
54. Jones-Mensah, I., Tabiri, M.O.: Review of literature on coherence and cohesion in text quality among ESL students. GlobELT & GLOBETS. **2020**, 48 (2020)
55. McNamara, D.S.: Reading both high and low coherence texts: effects of text sequence and prior knowledge. Canadian J. Exper. Psychol. **55**, 51–62 (2001). https://doi.org/10.1037/h0087352
56. Hartley, J.: Academic Writing and Publishing: A Practical Handbook. Routledge, London (2008)
57. Pho, P.D.: Research article abstracts in applied linguistics and educational technology: a study of linguistic realizations of rhetorical structure and authorial stance. Discour. Stud. **10**(2), 231–250 (2008)

58. Dronberger, G.B., Kowitz, G.T.: Abstract readability as a factor in information systems. J. Am. Soc. Inf. Sci. **26**(2), 108–111 (1975)
59. King, R.: A comparison of the readability of abstracts with their source documents. J. Am. Soc. Inf. Sci. **27**(2), 118–121 (1976)
60. Hartley, J., Trueman, M.: Some observations on using journal articles in the teaching of psychology. Psychol. Teach. Rev. **1**(1), 46–51 (1992)
61. Hartley, J.: Three ways to improve the clarity of journal abstracts. Br. J. Educ. Psychol. **64**(2), 331–343 (1994)
62. Plavén-Sigray, P., Matheson, G.J., Schiffler, B.C., Thompson, W.H.: The readability of scientific texts is decreasing over time. Elife. **6**, e27725 (2017) Chicago
63. Dowell, N.M., Graesser, A.C., Cai, Z.: Language and discourse analysis with Coh-Metrix: applications from educational material to learning environments at scale. J. Learn. Anal. **3**(3), 72–95 (2016). https://doi.org/10.18608/jla.2016.33.5
64. McCarthy, P.M., Lightman, E.J., Dufty, D.F., McNamara, D.S.: Using Coh-Metrix to assess cohesion and difficulty in high school textbooks. In: Proceedings of the 28th Annual Conference of the Cognitive Science Society, p. 2556. https://citeseerx.ist.psu.edu/viewdoc/download?doi=10.1.1.530.7721&rep=rep1&type=pdf (2006)
65. Hoehndorf, R., Queralt-Rosinach, N.: Data science and symbolic AI: Synergies, challenges and opportunities. Data. Sci. **1**(1-2), 27–38 (2017)
66. Romero, C., Ventura, S.: Educational data mining and learning analytics: an updated survey. Wiley Interdisc. Rev. Data Min. Knowl. Discov. **10**(3), e1355 (2020). https://doi.org/10.1002/widm.1355
67. Zadeh, L.A.: Fuzzy sets. Inf. Control. **8**(3), 338–353 (1965)
68. Kleene, S.C.: In: Shannon, C., McCarthy, J. (eds.) Automata Studies (1956)
69. Kang, D., Ammar, W., Dalvi, B., van Zuylen, M., Kohlmeier, S., Hovy, E., Schwartz, R.: A dataset of peer reviews (PeerRead): collection, insights and NLP applications. arXiv preprint arXiv:1804.09635 (2018). https://doi.org/10.48550/arXiv.1804.09635
70. Batista, G.E., Prati, R.C., Monard, M.C.: A study of the behavior of several methods for balancing machine learning training data. ACM SIGKDD Explor. Newslett. **6**(1), 20–29 (2004). https://doi.org/10.1145/1007730.1007735
71. Gosain, A., Saha, A., Singh, D.: Measuring harmfulness of class imbalance by data complexity measures in oversampling methods. Int. J. Intell. Eng. Inf. **7**(2–3), 203–230 (2019). https://www.inderscienceonline.com/doi/abs/10.1504/IJIEI.2019.099089
72. Dattagupta, S.J.: A performance comparison of oversampling methods for data generation in imbalanced learning tasks. Unpublished doctoral dissertation, Universidade Nova de Lisboa http://hdl.handle.net/10362/31307 (2018)
73. Arthurs, N.: Structural features of undergraduate writing: a computational approach. J. Writ. Anal. **2**, 138–175 (2018). https://wac.colostate.edu/docs/jwa/vol2/arthurs.pdf
74. Zahavi, J., Meiri, R.: KDnuggets. https://www.kdnuggets.com/2016/08/winner-stepwise-regression.html (2016)
75. Zhang, Z.: Variable selection with stepwise and best subset approaches. Ann. Transl. Medi. **4**(7) (2016). https://doi.org/10.21037/atm.2016.03.35
76. Wang, L.X., Mendel, J.M.: Generating fuzzy rules by learning from examples. IEEE Trans. Syst. Man Cyber. **22**(6), 1414–1427 (1992). https://doi.org/10.1109/21.199466
77. Zarei, M.: Spike discharge prediction based on neuro-fuzzy system. BioRxiv, p. 133967 (2017)
78. Blanchard, G., Bousquet, O., Massart, P.: Statistical performance of support vector machines. Ann. Stat. **36**(2), 489–531 (2008). https://doi.org/10.1214/009053607000000839
79. Wu, Y., Liu, Y.: Robust truncated hinge loss support vector machines. J. Am. Stat. Assoc. **102**(479), 974–983 (2007). https://doi.org/10.1198/016214507000000617
80. Heinze, G., Wallisch, C., Dunkler, D.: Variable selection–a review and recommendations for the practicing statistician. Biometr. J. **60**(3), 431–449 (2018). https://doi.org/10.1002/bimj.201700067

81. Jenei, G.: Referential cohesion in academic writing: a descriptive and exploratory theory- and corpus-based study of the text-organizing role of reference in written academic discourse. Unpublished doctoral dissertation, Eötvös University, Budapest (2014)

82. Liu, M., Braine, G.: Cohesive features in argumentative writing produced by Chinese undergraduates. System. **33**(4), 623–636 (2005). https://doi.org/10.1016/j.system.2005.02.002

83. Bhatia, V.K.: Analysing Genre: Language Use in Professional Settings. Longman, London (1993)

84. Biber, D., Conrad, S., Reppen, R.: Corpus Linguistics: Investigating Language Structure and Use. Cambridge University Press, Cambridge (2005). https://doi.org/10.1017/CBO9780511804489

85. Afful, J.B.A., Nartey, M.: Cohesion in the abstracts of undergraduate dissertations: an intra-disciplinary study in a Ghanaian University. J. ELT Appl. Ling. **2**(1), 93–108 (2014). https://ir.ucc.edu.gh/xmlui/bitstream/handle/123456789/6499/Cohesion%20in%20the%20Abstracts%20of%20Undergraduate%20Dissertations.pdf?sequence=1&isAllowed=y

86. Luthfiyah, L., Alek, A., Fahriany, F.: An investigation of cohesion and rhetorical moves in thesis abstracts. Indonesian J. English Educ. **2**(2), 145–159 (2015)

87. Suwandi, S.: Coherence and cohesion: an analysis of the final project abstracts of the undergraduate students of PGRI Semarang. Indo. J. Appl. Ling. **5**(2):253–261. undergraduate foreign students. ETS Res. Rep. Ser. **1983**(1), i–38 (2016). https://doi.org/10.17509/ijal.v5i2.1349

88. Alluqmani, A., Shamir, L.: Writing styles in different scientific disciplines: a data science approach. Scientometrics, 1–15 (2018). https://doi.org/10.1007/s11192-018-2688-8

89. Crossley, S., Russell, D., Kyle, K., Römer, U.: Applying natural language processing tools to a student academic writing corpus: how large are disciplinary differences across science and engineering fields? J. Writ. Anal. **1** (2017). https://wac.colostate.edu/docs/jwa/vol1/crossley.pdf

Chapter 6
Sequential Pattern Mining in Educational Data: The Application Context, Potential, Strengths, and Limitations

Yingbin Zhang and Luc Paquette

Abstract Increasingly, researchers have suggested the benefits of temporal analyses to improve our understanding of the learning process. Sequential pattern mining (SPM), as a pattern recognition technique, has the potential to reveal the temporal aspects of learning and can be a valuable tool in educational data science. However, its potential is not well understood and exploited. This chapter addresses this gap by reviewing work that utilizes sequential pattern mining in educational contexts. We identify that SPM is suitable for mining learning behaviors, analyzing and enriching educational theories, evaluating the efficacy of instructional interventions, generating features for prediction models, and building educational recommender systems. SPM can contribute to these purposes by discovering similarities and differences in learners' activities and revealing the temporal change in learning behaviors. As a sequential analysis method, SPM can reveal unique insights about learning processes and be powerful for self-regulated learning research. It is more flexible in capturing the relative arrangement of learning events than the other sequential analysis methods. Future research may improve its utility in educational data science by developing tools for counting pattern occurrences as well as identifying and removing unreliable patterns. Future work needs to establish a systematic guideline for data preprocessing, parameter setting, and interpreting sequential patterns.

Keywords Sequential pattern mining · Pattern recognition · Educational data science · Learning process · Temporal analysis · Educational technology

Y. Zhang (✉)
Institute of Artificial Intelligence in Education, South China Normal University, Guangdong, China

L. Paquette
The Department of Curriculum and Instruction, College of Education, University of Illinois at Urbana-Champaign, Champaign, IL, USA
e-mail: lpaq@illinois.edu

© The Author(s), under exclusive license to Springer Nature Singapore Pte Ltd. 2023
A. Peña-Ayala (ed.), *Educational Data Science: Essentials, Approaches, and Tendencies*, Big Data Management,
https://doi.org/10.1007/978-981-99-0026-8_6

Abbreviations

EDS   Educational data science
LSA   Lag-sequential analysis
MOOC   Massive Open Online Course
SPM   Sequential pattern mining
SRL   Self-regulated learning

6.1   Introduction

Learning is the acquisition process of knowledge and skills, which takes time to manifest in behavioral changes [1]. Thus, temporality is innate in learning, and an increasing number of researchers have suggested the temporal analysis of learning [2–5]. In particular, Molenaar and Wise [4] summarized four distinctive values of temporal analysis: detecting transitions between learning events, identifying variation in learning processes, explaining variation in learning outcomes, and boosting the emergence of new questions.

In general, two types of temporal properties of learning have been formed: the passage of time and the order in time [4]. The passage of time concerns when, how often, or how long learning events of interest occur. A limitation of this temporal property is that it omits events before and after the events of interest, i.e., the contextual information [6]. By contrast, the order in time addresses this limitation by focusing on the relative arrangement of learning events; for example, a sequence of events indicating that a learner reads relevant material after viewing the results of a quiz. The relative arrangement of learning events has attracted increasing interest [7–12]. Sequential pattern mining (SPM), as a pattern recognition technique, is powerful for uncovering this temporal property of learning [13, 14]. Particularly, SPM can uncover hidden patterns of ordered events with interesting properties, such as being common in high-performing learners but rare in low-performing learners.

This chapter aims to assist educational researchers in understanding the basics of SPM, its potential in educational data science (EDS), and how to achieve the potential. Section 6.1 introduces SPM and several terminologies. Section 6.2 highlights how SPM can be applied to educational data from various channels for various purposes. Section 6.3 uses example studies conducted on a computer-based learning environment to illustrate general ways in which SPM can contribute to theories of engagement and learning. Section 6.4 discusses the strengths, limitations, and future directions of applying SPM to educational data.

Table 6.1 Examples of transaction sequences

Customer ID	Customer sequence (transaction history)
1	$<\{a,b\}, \{c\}, \{d\}>$
2	$<\{a,c\}>$
3	$<\{b\}, \{c,d,e\}, \{f,b\}, \{c\}>$
4	$<\{b,c\}, \{d\}, \{g\}>$

Note. A pair of curly braces represents one transaction. Items in a pair of curly braces are products purchased in the transaction.

6.1.1   Sequential Pattern Mining (SPM)

SPM was first proposed by Agrawal and Srikant [15] as the problem of finding interesting subsequences in a sequence dataset. The interestingness of a subsequence can be defined in various ways, such as its frequency and length. Mathematically, a sequence can be denoted by $\{i_1, i_2, \ldots, i_n\}$, where i_j is an itemset. An itemset is a non-empty set of items, which can be various things. For example, in educational data, an item may be an action that a student executes in a learning management system, such as downloading a lecture note and submitting an assignment solution. Agrawal and Srikant [15] used SPM in the context of customer transactions, where an item is a product and an itemset is the set of products purchased in one transaction. A customer's ordered transactions form a sequence. Table 6.1 provides an example of sequences for hypothetical customers. In the table, customer 1's sequence has two transactions. The first itemset contains two items, a and b, indicating that products a and b are purchased in transaction 1, while transaction 2 indicates that c is purchased. The order of items in an itemset is not meaningful.

A sequence $\{i_1, i_2, \ldots, i_n\}$ contains a subsequence $\{p_1, p_2, \ldots, p_m\}$ if there are integers $k_1 < k_2 < \ldots < k_n$ such that $\{p_1 \subseteq i_{k_1}, p_2 \subseteq i_{k_2}, \ldots, p_m \subseteq i_{k_n}\}$. For example, sequences 1 and 3 in Table 6.1 contain the subsequence $<\{b\}, \{c\}>$, but sequences 2 and 4 do not. Note that these integers do not need to be consecutive, i.e., $k_{j+1} - k_j$ can be larger than 1. $k_{j+1} - k_j$ is named the gap between i_{k_j} and $i_{k_{j+1}}$. Many SPM algorithms allow users to set the maximum gap to reduce noisy patterns (patterns that are identified as frequent due to random error) and limit the number of patterns returned [16, 17]. For example, if the maximum gap is 1, the algorithm will infer that customer 1's sequence does not contain $<\{b\}, \{d\}>$, because the gap between b and d in this sequence is 2. It is especially important to remove noisy patterns because they cannot provide reliable information about the processes of interest (e.g., learning processes) that generated the sequences.

A customer supports a subsequence if the customer's sequence contains the subsequence [15]. The support value of a subsequence is defined as the proportion of sequences that contain this subsequence. If the support value of a subsequence is no less than a prespecified threshold (typically called the minimum support), this subsequence is a frequent sequential pattern. For example, let us set the support threshold as 0.5. In Table 6.1, the support values of $<\{b\}, \{c\} >$, $< \{b\}, \{d\} >$, and $< \{c\}, \{d\}>$ are 0.5, 0.75, and 0.5, respectively, so these subsequences are frequent sequential patterns. In contrast, $<\{a\}, \{c\}>$ is not a frequent sequential pattern because its support is 0.25 < 0.5. Similarly to using a maximum gap, selecting appropriate minimum support helps reduce noisy patterns because the more sequences that contain a specific pattern, the less likely that this pattern occurs due to random error.

Another metric frequently used in educational research is the instance value [18], which refers to the number of occurrences of a sequential pattern in a sequence. For example, if we do not consider the maximum gap, the instance values of $<\{b\}, \{c\}>$ in sequences 1–4 are 1, 0, 2, and 1, respectively. Sequence 3 has two occurrences of $<\{b\}, \{c\}>$ rather than three because occurrences are usually counted according to the nonoverlapping rule: an item(set) cannot occur at the same position in two instances of a pattern [19, 20]. Some studies have used F-support to refer to the proportion of sequences that contain a pattern and I-support to refer to the occurrences of a pattern (e.g., [9, 21]).

6.1.1.1 SPM Algorithms

Agrawal and Srikant [15] developed three SPM algorithms: AprioriAll, AprioriSome, and DynamicSome. Subsequent studies have proposed better-performing algorithms that allow users to filter patterns using constraints beyond the support value. For example, the Generalized Sequential Patterns (GSP) algorithm allows users to restrict the maximum and minimum gaps between itemsets [16]. Various algorithms have been applied to educational data, such as GSP, constrained Sequential PAttern Discovery using Equivalence classes (cSPADE; [17]), Prefix-projected Sequential pattern mining (PrefixSpan; [22]), Sequential PAttern Mining (SPAM; [23]), and Protein Features EXtractor using SPAM (Pex-SPAM; [24]). For readers interested in the details of SPM algorithms, we recommend the review articles by Mooney and Roddick [25] and Fournier-Viger et al. [26]. It is noteworthy that all of the different algorithms will produce the same set of sequential patterns, given the same constraints [26]. Thus, researchers can use any SPM algorithms that they are familiar with unless the algorithms do not allow constraints in patterns that may have a critical influence on the results.

For readers new to SPM, we recommend cSPADE [17]. It is a fast algorithm, although not the fastest. It allows various constraints on patterns and has been implemented in Python and R,[1] which are tools that EDS researchers may be most familiar with. The next section uses synthetic behavioral sequences to demonstrate the application of cSPADE.

[1]The implementation of cSPADE in Python and R: https://pypi.org/project/pycspade/, https://CRAN.R-project.org/package=arulesSequences

Table 6.2 Four synthetic behavioral sequences

Student	Event ID					
	i1	i2	i3	i4	i5	i6
S1	Read	Hint	Attempt	Note	Attempt	Attempt
S2	Read	Attempt	–	–	–	–
S3	Hint	Read	Attempt	Note	Attempt	–
S4	Hint	Note	Read	Attempt	–	–

Table 6.3 Frequent patterns in the synthetic behavioral sequences

Patterns	Support
Attempt → note	0.50
Note → attempt	0.50
Read → attempt	0.75
Attempt → note → attempt	0.50

6.1.1.2   An Example of Applying SPM to Synthetic Behavioral Sequences

Table 6.2 displays the four synthetic sequences on which cSPADE was applied. We used a maximum gap of 1 and minimum support of 0.5 without other constraints. Note that we decided these constraints arbitrarily for the purpose of this example. Readers should set the constraints according to their research context and purposes. The appendix presents the Python and R code that we used in this example. In both cases, the code first creates the sequences and formats the data. Then, it applies the cSPADE algorithm. Finally, it outputs the identified patterns.

While cSPADE regards individual events as patterns, we discarded these single-event patterns because they do not capture the relative arrangement of events. cSPADE identified four frequent patterns with a support value greater or equal to the minimum support (Table 6.3). *Attempt → note, note → attempt*, and *attempt → note → attempt* had a support value of 0.5, meaning that two students (S1 and S3 in Table 6.2) contained these patterns. *Read → attempt* had a support value of 0.75, meaning that three students showed this pattern. Student S1 did not show this pattern because their read action (i1) was followed by a *hint* action (i2). However, since S1's third action (i3) is an attempt, if we used a maximum gap of two, cSPADE would have considered that S1 executed the *read → attempt* pattern because the second action after *read* (i1) was *attempt* (i3).

This section showed a simple application example of SPM to student behavioral sequences. Subsequent sections will discuss more advanced applications of SPM: how to combine SPM with theoretical considerations and other EDS methods to analyze educational data in various modes for various research purposes.

6.2   What Modes of Educational Data and Research Purposes Is SPM Applicable to?

While SPM may be less popular than other EDS methods, such as predictive modeling and clustering, it has been applied for a variety of educational purposes using data from different modes across a wide range of educational settings. In particular, computer-based educational systems, such as Learning Management Systems (LMS), Massive Open Online Courses (MOOC), Intelligent Tutoring Systems (ITS), computer-supported collaborative learning environments, educational games, and course enrollment systems, are the most common educational settings for the application of SPMs.

This section discusses some of the most common data modes and research purposes for the application of SPM. These two factors were selected as they are the main factors that determine the procedures of SPM. We discuss the two factors separately because there is no consistent one-to-one match between data modes and research purposes. For instance, researchers may apply SPM to students' event logs from an educational game for feature engineering or understanding how students interact with the environment. Similarly, SPM may be used to analyze students' verbal conversations rather than the event logs to reveal how students collaborate in a computer-supported collaborative environment.

6.2.1   Data Modes

In general, any sequence dataset where sequences are composed of ordered itemsets can be the input of the SPM analysis. This allows researchers to apply SPM to educational data from various channels. It is important to note that itemsets for SPM must be composed of categorical variables. Thus, SPM cannot be applied to sequences of numerical variables unless researchers code the variables into discrete categories.

6.2.1.1   Event Logs

Event logs provide detailed traces of students' interaction with digital learning environments [27–29]. Such logs can be collected unobtrusively and without the need for specialized equipment (e.g., microphones, webcam, motion sensor) and are easily accessible. For these reasons, event logs are the most common type of educational data to which SPM has been applied. For example, Jiang et al. [30] applied SPM to students' event logs in Virtual Performance Assessments, a computer-based environment that assessed science inquiry skills. In this context, SPM was used to identify how students without experience with the environment showed more sequential patterns indicative of exploration behaviors than experienced students. Emara et al. [8] compared students' behaviors between working individually and collaboratively in a computer-based science learning environment. Those working individually showed sequential patterns that mainly consisted of reading behaviors, while those working collaboratively showed that patterns consisted of more diverse activities. Kia et al. [31] mined students' interactions with a learning dashboard. They found that low-performing students showed more frequent sequential behavior patterns in the learning dashboard compared to other students.

An itemset in event logs usually has only one event because students rarely initiate two actions simultaneously in a learning environment. Note that raw event logs may contain noises and be at a small grain size [29, 32, 33]. Researchers need to preprocess the raw event logs before using them as input for SPM algorithms. Otherwise, the algorithm may generate many meaningless or uninterpretable sequential patterns. Five main ways of preprocessing event log data for SPM emerge in existing studies: filtering, collapsing, contextualizing, abstraction, and breaking. The choice of which preprocessing methods to use (and how to apply them) relies on the purpose of the SPM analysis, as preprocessing will significantly impact the nature of the patterns that SPM identifies as interesting.

Filtering refers to removing events that may be necessary for students' interactions with a learning environment but irrelevant or meaningless in the context of the studied educational concept [32, 33]. For example, a study may focus on students' behaviors after taking a quiz in a MOOC course. The MOOC platform may automatically display the quiz results and record the action of viewing quiz results. In this case, the action of viewing quiz results may not be meaningful because it is required by the platform, and this action may need to be removed.

Collapsing represents condensing consecutive and qualitatively similar events as one event. Collapsing events reduces redundant patterns. For instance, Kinnebrew et al. [32, 33] condensed a chain of consecutive *reading page* events into a *reading page-MULT* event. In this way, chains of reading events with different lengths will be regarded as the same event (a *reading page-MULT* event) by SPM algorithms, and patterns only differing in the chain length would be considered as the same pattern. For example, both *view quiz results → reading page → reading page* and *view quiz results → reading page → reading page → reading page* would be considered as *view quiz results → reading page-MULT*. Meanwhile, the suffix *MULT* keeps the information, indicating that multiple events occurred and distinguishes a chain of reading events from a single reading event.

Contextualizing refers to labeling events with contextual information, such as the duration of an event. For instance, Emara et al. [8] added *long* or *short* suffixes to *reading page* events based on whether the read events were longer than 3 s. This distinction was meaningful in that context because short reading might indicate that students were skimming pages, while long reading was more likely to indicate that students were trying to understand the page content.

Abstraction refers to translating raw event subsequences into more abstract behaviors. This can be achieved by matching raw event subsequences with prespecified patterns, representing higher order behavior. For instance, in Zhou et al.'s [29] work, a subsequence, *selecting content* → *opening a note window* → *choosing the option "critique note"* → *filling information* → *closing the note window*, was coded as the behavior of *making a critique note*. Abstraction can be especially useful in situations where a single raw event in the logfile rarely represents a meaningful action.

Breaking refers to breaking raw sequences based on research interests. For instance, a group's event sequence in a collaborative learning environment may be split into student-based sequences where each sequence contains ordered events from a group member [34]. Sequential patterns in student-based sequences reflect common behaviors within this group. A student-based sequence may be further split into session-based or day-based sequences, with each containing events from a single learning session or day [29, 35]. Comparing discovered sequential patterns from different sessions or days may reveal how learning behaviors evolve across sessions and days.

6.2.1.2   Discourse Data

SPM has been applied to the analysis of different forms of discourse data, including students' verbal conversations [36], online chat messages [12, 37], and forum posts [38]. Discourse data need to be coded into a set of categories based on coding schemes, where the codes represent abstract behaviors. In this case, an itemset is a code, and a sequence is composed of codes ordered in time. For example, the work by Zheng et al. [37] coded chat messages into eight categories based on the theory of self-regulated learning (SRL) and socially shared regulated learning (SSRL): SRL- and SSRL-task analysis, SRL- and SSRL-planning, SRL- and SSRL-elaborating, as well as SRL- and SSRL-monitoring. Such categorization allowed them to use SPM to explore patterns of socially shared regulation from chat logs. Similarly, Chen et al. [38] studied schemes of knowledge-building discourse using SPM to analyze forum posts coded into six categories: questioning, theorizing, obtaining information, working with information, syntheses and analogies, and supporting discussion.

6.2.1.3   Resource Accessing Traces

For data from this channel, a sequence consists of learning resources ordered based on the timestamp at which a student accessed it. A learning resource may be a course [39, 40], a lecture video [41, 42], a book [43, 44], a web page [45, 46], etc. The application of SPM to these data has mainly been in the context of building recommender systems, providing suggestions about which resources a student should access next. For instance, Chen et al. [46] used collaborative filtering algorithms to generate candidate webpages and applied SPM to filter these candidates according to a learner's historical webpage visiting sequence. The remaining web pages were recommended to this learner. Other studies have applied SPM to resource accessing traces to understand how students learn the material. For example, Wong et al. [47] applied SPM to sequences of video lectures watching events to check whether learners watched videos in the order that instructors planned.

6.2.1.4 Other Data Sources

SPM is applicable to data in other modes as long as the data can be formatted as a sequence of ordered itemsets. For example, Mudrick et al. [21] used SPM to analyze students' eye-tracking data in a multimedia learning environment. In this study, an itemset was a fixation in a screen area, and a sequence was defined as a student's fixation trace. Knight and Martinez-Maldonado et al. [48] explored how SPM could provide insights about rhetoric moves in students' written text. A sequence was defined as a student's written text, but an itemset could be defined at various levels, such as a sentence, a paragraph, and a section. McBroom et al. [49] proposed a method that uses SPM to help teachers identify common mistakes in programming tasks. In this case, an itemset was defined as a submitted answer labeled with a code state (e.g., having syntax errors versus no syntax errors), and a sequence was composed of a student's submissions on a problem. Overall, few studies have explored how to apply SPM to educational data from these uncommon channels, but existing studies have shown the potential of SPM.

6.2.2 Research Purposes

With educational data in the above modes, SPM has been used in descriptive, relational, and predictive research for various purposes, including (1) mining learning behaviors, (2) analyzing and enriching educational theories, (3) evaluating the efficacy of interventions, (4) generating features for prediction and classification models, and (5) filtering learning resources for building recommender systems.

6.2.2.1 Mining Learning Behaviors

Learning process data, such as event logs and discourse data, record detailed information about students' interactions with the learning environments, peers, and instructors. Frequent sequential patterns in such data indicate common behavior patterns across students [29]. These behavior patterns may reveal how students navigate their activities within a learning environment and inform how to update better the design of the learning experience [50]. For example, Kang et al. [51] applied the cSPADE algorithm to gameplay logs in Alien Rescue, a serious game for teaching middle school students scientific problem-solving skills. The study aimed to analyze how sequential patterns of action may differ across multiple days of using the educational game. They observed how sequential patterns in the first few days represented exploration behaviors, while sequential patterns in the remaining days represented scientific problem-solving behaviors, including background search, generating and testing hypotheses, and constructing solutions. Emara et al. [8] used SPM to compare the behaviors of a group of sixth-grade students working individually to another group working in pairs within Betty's Brain, a hypermedia learning environment teaching scientific phenomena. Differences in sequential patterns between the two groups suggested that students were better at fixing errors in solutions when they worked collaboratively than working individually.

6.2.2.2   Enriching Educational Theories

Some studies have linked the interpretation of sequential patterns to educational theories. For instance, Taub and Azevedo [52] applied SPM to investigate how hypothesis-testing behaviors and emotion levels together influenced learning and gameplay within Crystal Island, an educational game that teaches scientific inquiry skills and microbiology. In this game, students needed to collect and test various food items to find the cause of an illness. The hypothesis-testing behaviors were conceptualized as metacognitive monitoring strategies based on the information processing theory of SRL [53], while the emotion levels were conceptualized as the appraisals of events based on the component process model of emotions [54]. The results indicated that students who were less emotional and solved the task with one attempt showed more patterns of hypothesis-testing behaviors indicative of monitoring strategies, suggesting that they might be better at monitoring cognitive activities than the others. Thus, Taub and Azevedo's study enriched the understanding of the link between the information processing theory of SRL and the component process model of emotions. Kinnebrew et al. [55] argue that SPM may discover sequential patterns that do not match a theoretical problem-solving model, and these patterns represent new learning strategies used by students and can inform us about ways to refine the theoretical model.

6.2.2.3   Evaluating the Efficacy of Interventions

Studies have explored whether sequential learning behavior patterns could capture the effect of interventions. For example, Wong et al. [47] designed weekly prompt videos to facilitate students to think about their plans, monitor, and reflect on learning in a Coursera course. They used SPM to compare students who watched at least one prompt video (prompt viewers) and those not watching any prompt video (nonviewers). The group of prompt viewers shared more sequential behavior patterns than nonviewers. In particular, prompt viewers tended to watch videos in the order that instructors planned.

6.2.2.4 Building Predictive Models

Sequential patterns from learning process data have been used to build predictive models. For example, Fatahi et al. [56] used sequential patterns from students' event logs in Moodle to predict students' personality types, which were assessed via the Myers-Briggs Type Indicator (MBTI) questionnaire. The classification accuracies were 8–22% higher than chance. Jaber et al. [57] argue that the position of a sequential pattern in students' event logs is important for classification tasks in education. Based on this idea, they proposed a SPM-based classification framework that segmented each sequence into n bins with equal sizes. SPM was conducted within each bin, so a sequential pattern corresponded to n binary features, i.e., whether it appears in the first bin, second bin, and so on. The researchers applied the method to detect each student's role (executives, managers, or members) in collaborative projects. The SPM-based method had better performance than other methods in precision, recall, and F-1 score.

6.2.2.5 Developing Educational Recommender Systems

Many studies have used SPM, together with other algorithms, to build learning resource recommender systems. A general procedure of using SPM for this purpose is the following: for a learner's resource accessing sequence $S = \{i_1, i_2, \ldots, i_n\}$, where i_j is a learning resource, (1) in experts' [58] or peers' [46] resource accessing sequences, discover frequent sequential patterns that contain subsequences identical or close to S; (2) resources that appear in these sequential patterns and are visited after S are recommended to the learner [59]; or (3) these resources are combined with recommendations by other methods, such as collaborative filtering algorithms, to generate the final recommendation [60].

Instead of directly building a recommender system, some studies have applied SPM to course enrollment data to advise students on course registration and schools on designing and refining professional programs' course paths. Such studies usually compare different academic performance groups' course enrollment paths and try to discover sequential course patterns frequent in the high-performing group but rare in the low-performing group. For example, Slim et al. [61] applied SPM to electrical engineering undergraduates' course enrollment sequences. The result showed that students graduating with a high GPA tended to follow a course enrollment pattern distinct from low GPA students' course enrollment sequences.

6.3 How Can SPM Contribute to Understanding the Temporal Aspects of Learning?

The previous section discusses the potential of SPM in EDS. The current section discusses how to achieve this potential. Particularly, this section focuses on how SPM can assist us in understanding learning processes, including mining learning behaviors, enriching educational theories, and evaluating the efficacy of interventions. Sequential patterns are ordered itemsets and capture the temporal relationship between these itemsets. In education, this temporal relationship may reflect how learners arrange their activities or interact with peers, teachers, and learning environments. Knowledge and cognition are highly contextually dependent [62]. Thus, a sequential pattern or temporal relationship in one setting may not generalize to another. Even if it does, the meaning is likely to be different. Nevertheless, studies have shown general ways in which SPM can contribute to the conceptual understanding of learning by discovering individuals' similarities, differences, and changes in learning behaviors.

This section uses studies on Betty's Brain, an open-ended learning environment, as examples to illustrate how SPM can be used to achieve these contributions. We focused on this environment because the Betty's Brain team regularly used SPM as a key method for their studies. As such, previous research on Betty's Brain has used SPM to achieve multiple research purposes related to understanding learning processes (i.e., mining learning behaviors, enriching educational theories, and evaluating the efficacy of interventions) and have used SPM in different ways to achieve these purposes (i.e., discovering similarities, differences, changes in learning behaviors). Additionally, two of the Betty's Brain studies were the first to combine SPM with statistical analyses for the purpose of discovering differences and changes in learning processes [32, 63]. This section finally summarizes the results of studies applying SPM in other learning environments.

6.3.1 An Example: Combining SPM with Statistical Analyses to Understand Learning in an Open-Ended Learning Environment

Biswas and colleagues have conducted a series of studies that applied SPM to middle schoolers' event logs in Betty's Brain [8, 32, 33, 55, 63–65]. They have typically used the Pex-SPAM algorithm with a minimum support of 0.5 and a maximum gap of 2 to obtain frequent sequential patterns and applied statistical tests to these patterns to investigate differences and changes in students' learning processes. Before diving into these studies, it is helpful to introduce Betty's Brain.

6.3.1.1 Betty's Brain

Betty's Brain is an open-ended computer-based learning environment [66]. Students learn about scientific phenomena, such as climate change, by teaching a virtual pedagogical agent, Betty. Students teach Betty by building a causal map of the scientific phenomenon, in which a causal (cause-and-effect) relationship is represented by a pair of concepts connected by a directed causal link (see the bottom right of Fig. 6.1). To build this map, students can access hypermedia resource pages (Science Book in Fig. 6.1) on relevant scientific concepts. Students can evaluate their causal modeling progress by asking Betty to take quizzes graded by a mentor agent, Mr. Davis. The top right of Fig. 6.1 shows the questions, answers, and grades of a quiz. A gray grade means Betty could not answer the question because the question involved concepts or links that had not been added to the map. The student selected the second question, which was answered incorrectly, and the concepts and links that Betty used to answer this question (i.e., Betty's explanations) were highlighted in the causal map. Betty's quiz grades (correct and incorrect answers), along with her explanations, can help the student keep track of Betty's progress and, in turn, their own progress because Betty's correct and incorrect answers inform problems in the causal map. Students can then improve their understanding of the topic by re-reading the science book and tracking Betty's explanations to correct their perceived problems with their causal map.

Students' activities in Betty's Brain can be grouped into three large categories: (1) information seeking, including reading (read a page in the resources), search (search pages containing entered keywords), and note taking (create, view, or edit a

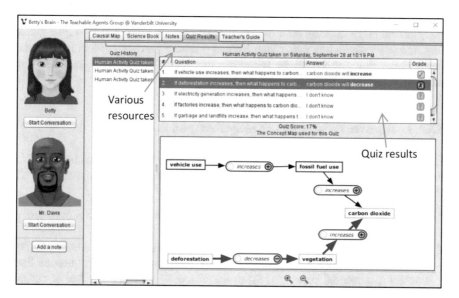

Fig. 6.1 Screenshot of viewing quiz results and checking the chain of links Betty used to answer a quiz question

note); (2) solution construction, including edits (adding or deleting a concept, adding, deleting, or modifying a causal link) and markings (marking a causal link as correct or incorrect); and (3) solution assessment, including queries (ask Betty a cause-and-effect question based on the concepts and links on the map so far) and quizzes (assess the state of the map by having Betty take a quiz, viewing the quiz results, and Betty's explanation on the results).

6.3.1.2 Discovering Similarities in Engagement and Learning

Discovering similarities in learning processes is the most straightforward application of SPM in educational data because popular SPM algorithms, such as GSP, SPAM, and cSPADE, aim to find common sequential patterns across sequences. The discovered similarity may be among a group of learners or a single student's different learning sessions [29]. The similarity may reflect strategies common in students' learning [13], regardless of whether they are effective or ineffective.

In Betty's Brain, researchers developed a task model that specifies strategies students may use [67]. SPM has been used to discover candidate strategies that may increase the task model coverage. For instance, Kinnebrew et al. [55] found three meaningful sequential patterns that were not covered by the initial task model: *search* → *search*, *note* → *read* → *note*, and *quiz* → *adding link* → *quiz*. *Search* → *search* meant that students searched resource pages consecutively, suggesting that students might have difficulty finding a page on the first try. This pattern had a practical implication that the search module could be redesigned to help students quickly find relevant resource pages. *Note* → *read* → *note* meant that students switched between reading pages and taking notes frequently, indicating that students were tracking their understanding. *Quiz* → *adding link* → *quiz* meant that students added a link rather than reading resource pages after taking a quiz and checked the link correctness by taking another quiz. This pattern suggested that students might use a guess-and-check strategy.

6.3.1.3 Discover Differences in Engagement and Learning

In addition to behaviors common among learners, researchers are also interested in identifying sequential patterns that differ across learners. This can be achieved via three steps:

1. Use a SPM algorithm to discover frequent candidate patterns within each group of learners.
2. Calculate frequency metrics of all candidate patterns for each group. This may include metrics such as the support value (the proportion of sequences that contain a sequential pattern; [15]) or the instance value (the number of occurrences of a sequential pattern in a sequence; [18]).

3. Apply statistical tests to identify candidate patterns that occur with statistically significant different frequencies between groups. For support values of a sequential pattern, which are at the group level, researchers have used contingency table-based tests, such as Pearson chi-square tests, to examine differences in the support value [68]. t-tests and the analysis of variances (ANOVA) have been used to examine instance differences [32, 33], which are at the individual or sequence level. Studies in Betty's Brain mainly examined sequential patterns with different instance values between groups.

For example, Kinnebrew et al. [32, 33] used the Pex-SPAM algorithm in step 1 to obtain candidate patterns from students with high and low performance in Betty's Brain. Then, they used t-tests (step 3) to examine the two groups' differences in the instance values (step 2) of the candidate patterns. High-performing students had greater instance values in *adding an incoherent link → quiz → removing the incoherent link*. An incoherent link meant that students added this link without reading any relevant resource pages (possibly based on prior knowledge or guesses). This pattern indicates that high-performing students were more likely to take quizzes to check an incoherent link after adding it, and when the quiz result showed that the link was incorrect, they would remove it. In addition, the high-performing group was more likely to read pages that contained information about incorrect links in the map indicated by quiz results, while low-performing students tended to read irrelevant pages after viewing the quiz results. These results suggested that high-performing students more effectively used quizzes to track their understanding and navigate the subsequent activities.

Emara et al. [64] let students work in pairs in Betty's Brain. Then, they grouped pairs into bothStudents or oneStudent groups based on the quality of their collaboration. Students in the bothStudents group showed many collaborative behaviors, such as elaborating on their partner's initiatives with an alternative, while students in the oneStudent group mainly worked individually. SPM revealed that the bothStudents group was more likely to read resource pages after quizzes than the oneStudent group. The bothStudents group tended to switch between editing, reading, and quizzes, while the oneStudent group tended to switch between editing and quizzes, suggesting that students in this group might solve the task with a trial-and-error strategy. Emara et al. [8] further investigated how learning behavior differed when students worked individually and collaboratively in Betty's Brain. They let a group of students work individually while other students work in pairs. SPM found that, in comparison to the individual group, the collaboration group made the following patterns more often: (1) *adding a relevant and correct link → quiz*, (2) *quiz → removing incorrect links*, (3) *quiz → read → adding a relevant and correct link*, and (4) *read → adding a relevant and correct link*. In contrast, the individual group tended to do multiple readings consecutively. These differences suggested that the collaboration group might be better at using quizzes to monitor their understanding and guide the next activity, and the individual group had difficulty in translating what they read into map construction.

Kinnebrew et al. [63] investigated the efficacy of scaffolding in Betty's Brain. Two groups of students used two versions of Betty's Brain with different scaffolding, while a control group used a version without scaffolding. In terms of the test scores, no treatment effect was found. However, SPM revealed sequential patterns that were used differently across groups. These differences were in line with the function of scaffolding that different groups received.

6.3.1.4 Discovering Changes in Engagement and Learning Over Time

Investigating how sequential learning behavior patterns change over time can be achieved in five steps:

1. Use a SPM algorithm to discover frequent candidate patterns.
2. Break raw sequences into bins with equal sizes or break sequences based on a natural cutoff, such as learning sessions and days.
3. For each candidate pattern, calculate the metric of interest per learner per bin. As usual, the metric may be the support value or instance value.
4. For each candidate pattern, calculate a metric that characterizes its variation across bins. The metric may be information gain [32, 33] or effect size [69].
5. Rank sequential patterns based on the metric computed in step 4. The top-ranked patterns have the largest variation over time.
6. Alternatively, researchers may omit steps 4 and 5 and apply repeated ANOVA to identify sequential patterns whose support or instance values statistically significantly differ across bins.

Kinnebrew et al. [63] segmented each student's event logs into five bins with an equal number of events and examined how the usage of frequent sequential patterns changed across bins. Results showed three clusters among these patterns in terms of the change of their usage: in cluster 1, the usage of sequential patterns decreased quickly over time; in cluster 2, the usage of sequential patterns decreased slowly; in cluster 3, the usage of sequential patterns increased quickly. Further analysis showed that the change of sequential patterns was in line with the scaffolding students received. For example, patterns representing knowledge construction behaviors were used steadily over time by students who received knowledge construction support, while the usage of such patterns decreased over time among students who received support for evaluating solutions or did not receive any support. Kinnebrew et al. [65] found that the overall frequencies over time of some patterns were similar across groups that received different support, but these patterns showed significant differences across groups in the frequency change over time.

6.3.1.5 Summary

The above studies in Betty's Brain illustrate the application of SPM for various research purposes, such as extending the understanding of how students approach a learning task, evaluating the effect of scaffolding, and revealing how learners change behaviors over time. The ways that SPM achieves these purposes are summarized into three types: discovering individuals' similarities, differences, and changes in learning. The following section shows examples of studies where SPM achieves similar purposes in a wide range of learning environments beyond Betty's Brain.

6.3.2 Findings from Applications of SPM in Other Learning Environments

Table 6.4 summarizes 20 example studies applying SPM to understand learning in environments other than Betty's Brain. A comprehensive review is out of the scope of this chapter, so we selected the following studies from prior work. We searched prior work in ERIC and Engineering Village using terms related to (a) sequential pattern and (b) computer-supported learning. We used the AND operator and restricted that the terms must appear in the main text. The search was done in April 2022 and returned 209 articles. We excluded 24 articles that did not use sequential pattern mining in computer-supported learning. We selected the example studies from the remaining 189 articles. The selection is based on three criteria. First, the study used SPM to understand the temporal aspects of learning. Thus, we did not include studies that have used SPM only for predictive modeling or building recommender systems. Second, the study was conducted under a clear theoretical framework or topic. Third, we wanted to cover various topics and learning environments. Thus, we kept only one of the studies that shared the learning environment, topics, and the contributions of SPM (i.e., discovering individuals' similarities, differences, and changes in learning processes). Some of the selected studies have used analyses beyond SPM; however, we report only the results from the SPM analyses.

The context in these studies varied from computer-supported individual learning to collaborative learning and from simple learning environments, such as multimedia slides and online forums, to advanced technologies, such as educational games and ITS. The learning content mainly belonged to science, engineering, technology, and mathematics (STEM). Note this is not because SPM is only suitable for analyzing the STEM learning processes. The reason is perhaps that most educational technologies are designed for teaching STEM.

These studies have used SPM to explore various research topics, such as SRL, collaborative learning, scientific inquiry, at-risk students, knowledge building, and critical thinking. Among these topics, the dominant one is SRL. Some studies have used SRL as an external variable and investigated its relations with sequential patterns. For example, Sabourin et al. [70] classified students as high, medium, and low SRL based on students' self-reported status during an educational game. They compared SRL groups' differences in behavioral patterns during the game. Some studies have coded raw events into SRL behaviors and analyzed sequential

Table 6.4 Summary of selected studies where SPM contributes to the understanding of learning

References	Learning context	Framework/topic	Contributions of SPM	Main results based on SPM
1. Sabourin et al. [70]	Crystal Island, an educational game teaching microbiology	SRL	Detecting differences	• High SRL students showed patterns indicative of recording information after receiving it more frequently than low SRL students • Low SRL students showed patterns indicative of random hypothesis-testing behavior more often than high SRL students • Low SRL students' patterns suggested that they had difficulty in connecting different concepts
2. Taub and Azevedo [52]		SRL; component process model of emotion	Detecting differences	• Students at low-efficiency levels and high levels of facial expressions of emotions showed more sequential patterns representing less strategic hypothesis-testing behavior • Students at high-efficiency levels and emotion levels had more sequential patterns indicative of strategic testing behavior
3. Taub et al. [71]		SRL	Detecting similarities and differences	• Students showed common hypothesis-testing behavior patterns • Students solving the task more efficiently had fewer testing behavior patterns • Students with less efficiency had higher instance values in an ineffective behavior pattern
4. Taub and Azevedo [72]	MetaTutor, ITS teaching the human circulatory system	SRL	Detecting differences	• High prior-knowledge students engaged in sequential patterns containing accurate metacognitive judgement, while low prior-knowledge students engaged in patterns with inaccurate metacognitive judgments • In both groups, few sequential patterns contained both cognitive and metacognitive events • High prior-knowledge students' sequential patterns contained more cognitive events than metacognitive events, while low prior-knowledge students' sequential patterns did not have any cognitive events

5. Martinez-Maldonado et al. [73]	Cmate, a tabletop application for drawing concept maps	Collaborative learning; concept mapping	Detecting differences	• The more collaborative groups showed more sequential patterns that contained verbal discussions in conjunction with physical actions. In contrast, the less collaborative groups showed more sequential patterns that only had physical actions without speech • The less collaborative groups had more patterns where a student's brief speech did not get a response from the other members • The more collaborative groups showed more patterns that represented accessing solutions to trigger discussion
6. Jiang et al. [30]	Virtual performance assessments for assessing science inquiry skills	Scientific inquiry; expert and novice	Detecting differences	• Novices in the learning environment showed more sequential patterns indicative of exploration behaviors than experienced students
7. Chen et al. [38]	Knowledge forum in primary school science courses	Knowledge building	Detecting differences	• In productive inquiry threads, students showed more sequential patterns indicative of (1) explaining after proposing a question, (2) questioning an explanation, (3) sustaining an explanation for questions, and (4) obtaining and analyzing information to propose, support, or improve an explanation • In improvable inquiry threads, students showed more sequential patterns of responding to explanations or new information with an opinion
8. Mudrick et al. [21]	Twelve multimedia science content slides	Cognitive theory of multimedia learning	Detecting similarities and differences	• Learners showed some common sequential eye fixation patterns • Learners in different information discrepancy conditions showed differences in eye fixation patterns
9. Kang et al. [51]	Alien rescue, a serious game for teaching middle schoolers scientific problem-solving skills	Scientific inquiry; expert and novice	Detecting similarities and differences	• The sequential patterns representing exploration behaviors were more frequent in the first few days • Sequential patterns in the remaining days were indicative of scientific problem-solving behaviors • Low-performing students showed a larger number of different frequent sequential patterns than high-performing students

(continued)

Table 6.4 (continued)

References	Learning context	Framework/topic	Contributions of SPM	Main results based on SPM
10. Kang and Liu [35]		Scientific inquiry; at-risk students	Detecting differences and changes	• On the first day, both at-risk and non-at-risk students showed many frequent sequential behavior patterns • From the second to the fourth days, the number of different frequent sequential patterns decreased dramatically in both groups and then increased • In the fifth and sixth days, the non-at-risk group had much more sequential patterns than the at-risk group
11. Zhu et al. [12]	Teaching teamwork, a computer-supported collaborative learning environment	Collaborative inquiry learning	Detecting similarities and differences	• In the successful condition, groups of learners had more regulation activity patterns than in the unsuccessful condition • In the successful condition, groups showed a sequential pattern indicative of maintaining a shared understanding of the task • In both conditions, groups engaged in a sequential pattern indicative of trial-and-error behaviors
12. Zheng et al. [37]	Teaching teamwork	SRL; SSRL	Detecting differences	• In the condition of successfully solving a task, students had more sequential patterns of regulation activities than in the unsuccessful condition • In the successful condition, groups tended to monitor progress after task analysis
13. Zheng et al. [11]	BioWorld, a computer-based environment for practicing clinical reasoning skills	SRL; scientific inquiry	Detecting similarities and differences	• Efficient students focused on fewer clinical reasoning patterns than less efficient students • Less efficient students showed disorganized clinical reasoning patterns
14. Wong et al. [47]	A course in Coursea teaching serious game	SRL	Detecting differences	• Learners who watched at least one SRL-prompt video showed more sequential patterns than those not watching any SRL-prompt video • The SRL-prompt viewers had a sequential pattern indicative of following the order of the videos presented in the MOOC

15. Kia et al. [31]	MyLA, a learning dashboard embedded in canvas	SRL	Detecting differences	• Low-performing students showed more sequential patterns than high-performing students • Sequential patterns indicative of monitoring and planning strategies were more frequent in high-SRL students than in low-SRL students
16. Malekian et al. [74]	A MOOC course teaching discrete optimization	Assessment readiness	Detecting differences	• Students that failed an assessment showed sequential patterns of consecutive attempts on assessments and forum viewing • Students that passed an assessment showed sequential patterns of viewing and reviewing lectures
17. Chen and Wang [45]	A web-based inquiry science environment	Inquiry-based learning	Detecting differences	• Low-performing students showed more sequential patterns of accessed course nodes in line with the designed learning sequence than high-performing students • High-performing students showed more sequential behavior patterns indicative of logical operation in a simulation experiment
18. Mishra et al. [10]	ENaCT, a computer-based environment for learning critical thinking	Critical thinking	Detecting similarities	• Frequent patterns that contained different types of actions did not mark good critical thinking • Investigating the semantic coherence between actions in a pattern may generate insights into critical thinking
19. Liu and Israel [9]	Zoombinis, a puzzle-based game for learning mathematics concepts and cognitive skills	Problem-solving in games	Detecting similarities and differences	• Students showed three common problem-solving strategies • Applying SPM within each problem-solving phase generated insights about how the usage of strategies differed across phases and which strategies might contribute to the transitions from lower to advanced phases
20. Sun et al. [75]	A massive private online course teaching computer science	Online learning	Detecting differences and changes	• Action patterns differed across course phases • High- and low-performing groups showed distinct action patterns

patterns of the SRL behaviors. For instance, Zheng et al. [37] coded chat messages into eight categories of SRL behaviors. The others have mapped sequential patterns to SRL strategies. For example, Kia et al. [31] linked sequential patterns to monitoring and planning strategies. We discuss why SPM is a popular method for the study of SRL in Sect. 6.4.1.3.

Most of the studies have used SPM to understand differences in learning. This is not surprising given that two of the distinctive values of temporal analyses are identifying differences in learning processes and explaining variations in learning outcomes [4]. The studies have grouped students by various factors, including students' backgrounds (e.g., prior domain knowledge and SRL proficiencies; [70, 72]), experimental conditions (e.g., the information discrepancy type; [21]), during task behaviors (e.g., the use of SRL scaffolding and the collaboration levels; [47, 73]), and final performance (e.g., the problem-solving correctness and efficacy; [12, 52]).

Many studies have also used SPM to uncover similarities in learning behaviors. For example, Taub et al. [71] found that learners showed common patterns of hypothesis-testing behaviors in learning scientific inquiry. Liu and Israel [9] identified three problem-solving strategies from students' frequent sequential patterns during learning mathematics concepts and cognitive skills.

Two of the 20 studies have used SPM to investigate the changes in learning behaviors. Kang and Liu [35] investigated how students' behavior patterns changed from day to day. They found no differences between the first and second days. However, from the second to the fourth days, the number of frequent sequential patterns first decreased dramatically and then increased. Sun et al. [75] compared high- and low-performing students' changes in behavior patterns over course phases. High-performing students' behavioral patterns became more complex from the initial stage to the stage of group learning and indicated group construction. By contrast, low-performing students' behavioral patterns in the initial stage were not distinct from the stage of group learning.

Note that when investigating the differences and changes in sequential patterns, some studies have solely relied on descriptive statistics and qualitative observations. We do not recommend this practice because it makes evaluating the reliability of the results challenging. We suggest that researchers should use statistical tests to provide robustness information about the results whenever the statistical tests are applicable (see Sects. 6.3.1.3 and 6.3.1.4).

Overall, the 20 selected studies and research in Betty's Brain depict a broad picture of how to use SPM for EDS research and the potential of SPM in understanding learning processes. It is noteworthy that the use of SPM should be based on research context and purposes. Specifically, researchers need to consider the theoretical underpinnings of their studies when preprocessing the data, setting the parameters of a SPM algorithm, grouping learners for investigating differences in sequential patterns, choosing the temporal unit for investigating the change in patterns over time, and interpreting the sequential patterns.

6.4 The Strengths, Limitations, and Future Directions of SPM in EDS

Previous sections provided an overview of how to apply SPM for various research purposes. This section discusses the strengths and weaknesses of the application of SPM in EDS. In some cases, the strengths and weaknesses are built on the comparison with other sequential analysis methods. We highlight factors that lead to the weaknesses and limit the application of SPM and suggest research directions.

6.4.1 Strengths

Although educational researchers may be unfamiliar with SPM, it is easy to learn. As a sequential analysis method, it can reveal meaningful information about learning processes and is a powerful tool for investigating SRL. Compared with the other sequential analysis approaches, SPM is more flexible.

6.4.1.1 Easy to Learn and Accessible

The rationale of SPM is straightforward: finding sequential patterns that frequently occur in a sequence dataset. Using SPM correctly does not require a detailed understanding of how SPM algorithms efficiently find the number of sequences containing a pattern and count pattern occurrences. Moreover, given the same parameter configuration, different algorithms discover the same sequential patterns [26]. Educational researchers simply need to know the meaning of the different parameters that can be used to adjust what types of patterns found by the algorithm. SPM tools are also accessible to educational researchers. SPM libraries are available in C++, Python, data mining tools such as WEKA and RapidMiner, and data analysis tools such as R. There is also a specialized tool, SPMF, which has a simple graphical user interface and is fast and lightweight [76]. It offers over 50 SPM algorithms.

6.4.1.2 Revealing Meaningful Information That May Be Ignored Otherwise

SPM focuses on the temporal relationship between events. It can bring insights about learning that are different from those obtained using counts, proportions, and durations of individual events. Combining SPM with statistical inference methods may discover more differences between groups of learners than the analyses of individual events [52, 63, 73]. For instance, a beforementioned study on Betty's Brain did not find a treatment effect on test scores but discovered meaningful differences in sequential patterns between groups receiving different treatments [63]. Similarly, Taub and Azevedo [52] did not find statistically significant differences in the frequency of individual hypothesis-testing behaviors between groups at different performance and emotion levels, but they found interesting differences in sequential behavior patterns (see study 2 in Table 6.4).

6.4.1.3   A Powerful Tool for SRL Research

SRL refers to the learners' process of actively managing and regulating their behavior, cognition, emotion, and motivation toward their learning goals [77]. SPM has been used as a powerful tool for investigating SRL because some operationalizations of SRL theories explicitly involve sequences of events unfolding over time.

For example, SRL events may be operationalized at three levels [6]: the occurrence level, the contingency level, and the patterned contingency level. In this operationalization, the occurrence level considers the features of individual events, e.g., the frequency of taking a quiz. The contingency level considers the conditional probability of a subsequent event given a prior event, e.g., the probability of reading a resource page after taking a quiz. The patterned contingency level considers the arrangement of events that repeatedly occurs, e.g., the frequency of reading a book page after taking a quiz. Sequential patterns are at the patterned contingency level because they are temporally ordered events. Thus, SPM serves well the purpose of investigating SRL at the patterned contingency level. This is why nearly half (8) of the 20 studies in Table 6.4 are under the framework of SRL.

6.4.1.4   More Flexible than Other Sequential Analysis Methods

Process mining and epistemic network analyses are also powerful EDS tools for understanding the temporal property of learning [78, 79]. However, these approaches aggregate individual sequences at group levels and depict the holistic learning process. In contrast, SPM can capture local learning patterns.

Another sequential analysis method, the lag-sequential analysis (LSA), also focuses on local patterns [80]. However, LSA restricts the gap between events of a sequential pattern to be fixed (in LSA, the gap is named lag). For example, when counting the occurrences of *quiz* → *read*, if the gap is fixed to 1, reading a page immediately after taking a quiz is counted as an instance, but reading a page after making a note after taking a quiz is not. When the gap is fixed to 2, reading a page after making a note after taking a quiz is counted as an instance of *quiz* → *read*, but reading a page immediately after taking a quiz is not. By contrast, SPM allows the gap to vary. Both the sequence of reading a page immediately after taking a quiz and the sequence of reading a page after making a note after taking a quiz can be counted as instances of *quiz* → *read*. Similarly, the Markov chain model and transition metrics [81, 82] suffer the same issue as with LSA.

T-pattern analysis can also find frequent sequential patterns and does not suffer the issues of the above methods [83]. However, T-pattern analysis requires that the dataset contains the timestamp of each event, and the interval between events must be meaningful. Many educational datasets, particularly event logs, do not meet this requirement. For example, a student may leave for a while after taking a quiz before reading a page. The interval between taking the quiz and reading is not the time that the student spends on the quiz.

6.4.2 Limitations and Future Directions

SPM is accessible and easy to learn, but it does not mean educational researchers can master this technique easily. Educational researchers may find various challenges when applying SPM to understand learning. Some challenges are shared by other sequence analysis methods, while others are unique to SPM.

6.4.2.1 No Available SPM Libraries for Computing Instance Values

Although various tools are available for conducting SPM, they usually compute the support values and do not return instance values. However, when applying SPM to learning process data, the support value may not be sufficient because it only counts whether a learner's event log contains a sequential pattern without considering how often a pattern is repeated. When a learner's event log contains hundreds or thousands of events, only using support value losses much information. For example, using only support value, a learner who does the *quiz → read* pattern 10 times while using a learning environment would be regarded as the same as a learner who does *quiz → read* only one time.

Researchers have proposed efficient algorithms to compute instance values and share their programs [20, 84]. However, these programs were developed using C++ and Java code. Users need to know at least the basics of these programming languages in order to use these programs. Many educational researchers may not possess the necessary computer programming expertise. SPMF implements one algorithm that counts the instance value, but it only gives the sum of pattern instances across sequences. Users do not have access to pattern instances in each sequence, preventing further analysis, such as conducting statistical tests to examine whether a pattern occurs differentially between groups. This may be one factor that limits the application of SPM in education. Thus, a SPM application that is friendly to users without programming backgrounds and meets the unique needs of EDS researchers is worth developing.

6.4.2.2 Lack a Guideline for Preprocessing and Parameter Setting

Preprocessing is necessary for feeding event logs to SPM algorithms. As mentioned in Sect. 6.1.1, educational researchers have used five ways to preprocess the data: filtering, collapsing, contextualizing, abstracting, and breaking. However, it is unclear how different preprocessing decisions would impact the results. For instance, should we collapse multiple *read* events? If so, should we collapse these events to *read* or *read-MULT* to distinguish a single read event from multiple read events? How would the results differ between decisions? These questions await future work.

Similarly, there is no clear guideline for parameter setting. What should the minimum support be to consider a pattern interesting? In other words, how many learners are required to show a sequential pattern for it to be safe to say a sequential pattern is frequent? What should the maximum gap between itemsets of a sequential pattern be? How would the results differ under different parameter configurations? And so on.

Unlike hyperparameters in machine learning models, which can be finely tuned via validation, researchers decide SPM parameters based on theories, prior studies, and experiences [14]. Researchers can compare the results under different SPM parameters and choose values that produce the best result. However, without validation, the result may be overfitting the data. Moreover, the meaning of *best* is up to researchers. Researchers may choose parameters that produce the results they agree with most or that are in line with their hypotheses. Consequently, the results may be biased. A systematic guideline about preprocessing data, setting SPM parameters, and being transparent about these procedures will facilitate the application of SPM and benefit the field of EDS.

6.4.2.3 Excessive Sequential Patterns

SPM algorithms may generate excessive sequential patterns, most of which are uninteresting or irrelevant to the research purpose [29]. Educational researchers have applied interestingness metrics, such as lift and Jaccard coefficients, to rank sequential patterns and select the top-ranked for further analyses [30, 85]. Several studies have investigated the match between these metrics and expert judgment of interestingness on association rules in the field of education [86, 87]. To what extent these metrics can serve as evidence for the interestingness of sequential patterns is unclear. Besides, most interestingness metrics are based on support values and ignore the multiple occurrences of a sequential pattern within individual sequences. Only using support values may cause a lot of information loss. Interestingness metrics based on instance values may be more informative and produce sequential pattern rankings different from those based on support values.

Setting constraints, such as the maximum gap, the maximum pattern length, and the maximum window (which is the maximum gap between the first and last item of a pattern), also reduce the number of patterns. For example, a smaller maximum gap and maximum window lead to fewer patterns. However, as the interestingness metrics are based on support values, reducing the number of identified patterns by adjusting these parameters risks missing meaningful patterns and information loss. We recommend adjusting the parameters mainly based on theoretical considerations. For example, Liu and Israel [9] set the minimum pattern length as three because a single activity typically came along with three events in the learning environment of their study.

Another option is to keep only generator patterns or closed patterns. A pattern is a generator if its support is not equal to any of its super-pattern, while a pattern is closed if its support is not equal to any sub-pattern [26]. Generator patterns and closed patterns are regarded as a concise and representative subset of all patterns, and thus, some algorithms only return either generator or closed patterns [88, 89]. However, we argue that it is not a wise option to keep only one kind of pattern in education. A pattern may have the same support but different instance values with its super-patterns or sub-patterns. Moreover, although some super-patterns and sub-patterns have the same meaning (e.g., *watching a lecture video → watching a lecture video → taking a quiz* vs. *watching a lecture video → taking a quiz*), others do not (e.g., *watching a lecture video → watching a lecture video → taking a quiz* vs. *watching a lecture video → watching a lecture video*). In addition, the sub-pattern of an understandable pattern may be difficult to interpret because of missing some critical events [83]. In summary, there are a few solutions for addressing the issue of excessive patterns, but these solutions have various disadvantages in educational research. New approaches to reduce patterns in educational data are necessary.

6.4.2.4 Interpreting Sequential Pattern Differences Is Challenging

Researchers must be cautious when interpreting differences in sequential patterns. First, when comparing two groups, if group A has more sequential patterns than group B, it does not necessarily mean that group A employs more strategies. For instance, in Zhu et al. [12] study, the number of sequential patterns was greater when students successfully solved a task for those failing a task. The authors concluded that students used more strategies in the success condition than in the failure condition. However, students might be more engaged with the task and execute more events in the success condition than in the failure condition. The longer a student's event log is, the more sequential patterns the log might contain. Consequently, there were more sequential patterns in the success condition than in the failure condition.

Second, interpreting differences in the support and instance values of sequential patterns also entails the same caution. Student A having a higher instance value in *quiz → read* than student B only means that student A uses this pattern more frequently than student B. It does not necessarily imply that student A is more likely to do reading after a quiz than student B. For example, assume that student A did a sequence of 200 actions in which *quiz* → *read* appears four times, while student B did 100 actions in which *quiz* → *read* appears twice. The instance value of *quiz* → *read* for A is 4, larger than that for B, i.e., 2. However, it would be wrong to claim that A is more likely to use *quiz* → *read* than B because if B also did 200 actions, the instance value might become 4.

It seems that the issue might be solved by controlling for the sequence length. However, this would still not guarantee the claim that student A used this pattern more frequently than student B. The base rates of *quiz* and *read* also need to be controlled. Assume that students A and B both did a total of 100 actions in which *quiz* → *read* appears four times for each student. However, student A did the *quiz* action 10 times and *read* 10 times, while Student B did *quiz* five times and *read* five times. In this case, the instance value of *quiz* → *read* and the sequence length are the same for A and B. However, it would be wrong to claim that A and B used *quiz* → *read* at the same frequency because the base rate of *quiz* and *read* are different for A and B.

Researchers may use other sequence analyses, such as LSA, to address these limitations. LSA characterizes a sequential pattern by the conditional probability that the second pattern event occurs given the first pattern event, with the simple probabilities of the second pattern event being controlled. However, as discussed in Sect. 6.4.1.4, these metrics are only applicable when sequential patterns contain two events and the gap between events is fixed. Given the relative pros and cons of different sequence analyses, we follow the recommendation of prior studies: combining SPM and other sequence analyses may generate better insights about learning [9].

6.4.2.5 Sequential Patterns Do Not Imply Causality

Sequential patterns are a set of temporally ordered events that co-occur regularly. Significant co-occurrences of events may suggest causality, but this kind of understanding of causality is not quite helpful [90]. When the context or the condition changes, the co-occurrence may disappear. Researchers should further investigate the mechanism behind the co-occurrence that explains why the previous events trigger the subsequent events [90]. This way, SPM can better contribute to understanding engagement and learning. Thus, educational researchers need to interpret the results of SPM based on proper learning theories if the purpose is to understand engagement and learning.

6.5 Conclusion

Sequential pattern mining (SPM) is a useful educational data science (EDS) method for uncovering the relative arrangement of learning events. It can be applied to common educational sequence datasets, such as event logs, discourse data, and resource-accessing traces. Through discovering similarities and differences in learners' activities and revealing the temporal change of learning behaviors, SPM can be used for mining learning behaviors, analyzing and enriching educational theories, evaluating the efficacy of instructional interventions, generating features for prediction models, and building educational recommender systems. Many SPM algorithms are publicly available and easy to use. As a sequential analysis method, SPM can reveal unique insights about learning processes and be powerful for SRL research. It is more flexible in capturing the relative arrangement of learning events than the other sequential analysis methods. Nevertheless, it is noteworthy that the use of SPM should be based on research context and purposes. Future research may improve the potential of SPM in EDS by developing tools for counting pattern occurrences as well as identifying and removing unreliable patterns. Another direction is to establish guidelines for data preprocessing, parameter setting, and interpretation of sequential patterns.

Acknowledgments This research was partially funded by the China Scholarship Council (grant number 201806040180).

Appendix

R Code for the Synthetic Example

install and load the R package that implements cSPADE

```
# install.packages('arulesSequences')
library(arulesSequences)
```

Create the sequences. Each row of data contains an event, its event ID, and sequence ID.

```
data <- data.frame(
sequenceID = c(rep(1, 6), rep(2,2), rep(3, 5), rep(4, 4)),
events = c(
"read", "hint", "attempt", "note", "attempt", "attempt",
"read", "attempt",
"hint", "read", "attempt", "note", "attempt",
"hint", "note", "read", "attempt")
```

```
)
data$eventID <- 1:nrow(data)
```

Convert the data to the basket format that cspade can handle
Note that the first two columns should represent sequence ID and event ID

```
write.table(data[,c("sequenceID", "eventID", "events")],
file = "formated_event_data.txt",
sep=";",
row.names = FALSE, col.names = FALSE,
quote = FALSE)
data_baskets <- read_baskets("formated_event_data.txt",
sep = ";",
info=c("sequenceID","eventID"))
```

Apply cspade to data_baskets

```
freq_patterns <- cspade(data_baskets,
parameter = list(support = 0.5,
maxgap = 1))
```

Inspect the frequent patterns. The minimum length of patterns in cspade is 1, so some patterns in freq_patterns are actually individual events.

```
inspect(freq_patterns)
```

Convert the freq_patterns to a data.frame

```
freq_patterns_df <- as(freq_patterns, "data.frame")
```

Python Code for the Synthetic Example

install and load the Python package that implements cSPADE

```
# !pip install Cython pycspade
from pycspade.helpers import spade, print_result
```

Create a list to represent the sequences
The first, second, and third columns are the sequence ID, event ID, and events, respectively.
At the time we wrote this example code, pycspade cannot handle events in string types.
So we converted the events to integers: 1—read, 2—hint, 3—attempt, 4—note.

```
data = [
  [1, 1, [1]],
  [1, 2, [2]],
  [1, 3, [3]],
  [1, 4, [4]],
  [1, 5, [3]],
  [1, 6, [3]],
  [2, 7, [1]],
  [2, 8, [3]],
  [3, 9, [2]],
  [3, 10, [1]],
  [3, 11, [3]],
  [3, 12, [4]],
  [3, 13, [3]],
  [4, 14, [2]],
  [4, 15, [4]],
  [4, 16, [1]],
  [4, 17, [3]]
]
```

Apply cSPADE to the data

```
result = spade(data=data, support=0.5, maxgap = 1)
```

Print the frequent patterns and interestingness measures

```
print_result(result)
```

References

1. Soderstrom, N.C., Bjork, R.A.: Learning versus performance. Perspect. Psychol. Sci. **10**(2), 176–199 (2015)
2. Hadwin, A.F.: Commentary and future directions: what can multi-modal data reveal about temporal and adaptive processes in self-regulated learning? Learn. Instr. **72**(101), 287 (2021)
3. Knight, S., Wise, A.F., Chen, B.: Time for change: why learning analytics needs temporal analysis. J. Learn. Anal. **4**(3), 7–17 (2017)
4. Molenaar, I., Wise, A.F.: Temporal aspects of learning analytics - grounding analyses in concepts of time. In: Lang, C., Siemens, G., Wise, A.F., Gašević, D., Merceron, A. (eds.) The Handbook of Learning Analytics, 2nd edn, pp. 66–76. SoLAR (2022)
5. Reimann, P.: Time is precious: variable- and event-centred approaches to process analysis in CSCL research. Int. J. Comput. Supp. Collab. Learn. **4**(3), 239–257 (2009)
6. Winne, P.H.: Improving measurements of self-regulated learning. Educ. Psych. **45**(4), 267–276 (2010)
7. Caglar Ozhan, S., Altun, A., Ekmekcioglu, E.: Emotional patterns in a simulated virtual classroom supported with an affective recommendation system. Br. J. Educ. Technol. **53**(6), 1724–1749 (2022)
8. Emara, M., Rajendran, R., Biswas, G., Okasha, M., Elbanna, A.A.: Do students' learning behaviors differ when they collaborate in open-ended learning environments? Proc. ACM Hum. Comput. Interact. **2**(CSCW), 49 (2018)

9. Liu, T., Israel, M.: Uncovering students' problem-solving processes in game-based learning environments. Comput. Educ. **182**(104), 462 (2022)
10. Mishra, S., Majumdar, R., Kothiyal, A., Pande, P., Warriem, J.M.: Tracing embodied narratives of critical thinking. In: Roll, I., McNamara, D., Sosnovsky, S., Luckin, R., Dimitrova, V. (eds.) Artificial Intelligence in Education. AIED 2021 Lecture Notes in Computer Science, vol. 12749, pp. 267–272. Springer (2021)
11. Zheng, J., Li, S., Lajoie, S.P.: Diagnosing virtual patients in a technology-rich learning environment: a sequential mining of students' efficiency and behavioral patterns. Educ. Inf. Technol. **27**(3), 4259–4275 (2022)
12. Zhu, G., Xing, W., Popov, V.: Uncovering the sequential patterns in transformative and non-transformative discourse during collaborative inquiry learning. Internet High. Educ. **41**, 51–61 (2019)
13. Moon, J., Liu, Z.: Rich representations for analyzing learning trajectories: systematic review on sequential data analytics in game-based learning research. In: Tlili, A., Chang, M. (eds.) Data Analytics Approaches in Educational Games and Gamification Systems, pp. 27–53. Springer, Cham (2019)
14. Van Laer, S., Elen, J.: Towards a methodological framework for sequence analysis in the field of self-regulated learning. Front. Learn. Res. **6**(3), 228–249 (2018)
15. Agrawal, R., Srikant, R.: Mining sequential patterns. In: Proceedings of the 11th International Conference on Data Engineering. IEEE (1995)
16. Srikant, R., Agrawal, R.: Mining sequential patterns: generalizations and performance improvements. In: Apers, P., Bouzeghoub, M., Gardarin, G. (eds.) Advances in Database Technology. EDBT 1996 Lecture Notes in Computer Science, vol. 1057, pp. 1–17. Springer, Cham (1996)
17. Zaki, M.J.: Sequence mining in categorical domains: incorporating constraints. In: Proceedings of the Ninth International Conference on Information and Knowledge Management, pp. 422–429. ACM (2000)
18. Lo, D., Khoo, S., Liu, C.: Efficient mining of recurrent rules from a sequence database. In: Haritsa, J.R., Kotagiri, R., Pudi, V. (eds.) Database Systems for Advanced Applications. DASFAA 2008 Lecture Notes in Computer Science, vol. 4947, pp. 67–83. Springer (2008)
19. Ding, B., Lo, D., Han, J., Khoo, S.: Efficient mining of closed repetitive gapped subsequences from a sequence database. In: 2009 IEEE 25th International Conference on Data Engineering, pp. 1024–1035. IEEE (2009)
20. Wu, Y., Zhu, C., Li, Y., Guo, L., Wu, X.: NetNCSP: nonoverlapping closed sequential pattern mining. Knowl. Based Syst. **196**(105), 812 (2020)
21. Mudrick, N.V., Azevedo, R., Taub, M.: Integrating metacognitive judgments and eye movements using sequential pattern mining to understand processes underlying multimedia learning. Comput. Hum. Behav. **96**, 223–234 (2019)
22. Jian, P., Jiawei, H., Behzad, M.-A., Jianyong, W., Helen, P., Qiming, C., Umeshwar, D., Mei-Chun, H.: Mining sequential patterns by pattern-growth: the PrefixSpan approach. IEEE Trans. Knowl. Data Eng. **16**(11), 1424–1440 (2004)
23. Ayres, J., Flannick, J., Gehrke, J., Yiu, T.: Sequential pattern mining using a bitmap representation. In: Proceedings of the Eighth ACM SIGKDD International Conference on Knowledge Discovery and Data Mining, pp. 429–435. ACM (2002)
24. Ho, J., Lukov, L., Chawla, S.: Sequential pattern mining with constraints on large protein databases. In: Proceedings of the 12th International Conference on Management of Data, pp. 89–100. Computer Society of India (2005)
25. Mooney, C.H., Roddick, J.F.: Sequential pattern mining—approaches and algorithms. ACM Comput. Surv. **45**(2), 19 (2013)
26. Fournier-Viger, P., Lin, J.C., Kiran, R.U., Koh, Y.S., Thomas, R.: A survey of sequential pattern mining. Data Sci. Patt. Recogn. **1**(1), 54–77 (2017)

27. Paquette, L., Bosch, N.: The invisible breadcrumbs of digital learning: how learner actions inform us of their experience. In: Matthew, M. (ed.) Handbook of Research on Digital Learning, pp. 302–316. IGI Global, Pennsylvania (2020)
28. Winne, P.H.: Construct and consequential validity for learning analytics based on trace data. Comput. Hum. Behav. **112**(106), 457 (2020)
29. Zhou, M., Xu, Y., Nesbit, J.C., Winne, P.H.: Sequential pattern analysis of learning logs: methodology and applications. In: Romero, C., Ventura, S., Pechenizkiy, M., Baker, R.S.J.D. (eds.) Handbook of Educational Data Mining, pp. 107–121. CRC Press, Boca Raton (2010)
30. Jiang, Y., Paquette, L., Baker, R.S., Clarke-Midura, J.: Comparing novice and experienced students within virtual performance assessments. In: Proceedings of the Eighth International Conference on Educational Data Mining, pp. 136–143. International Educational Data Mining Society (2015)
31. Kia, F.S., Teasley, S.D., Hatala, M., Karabenick, S.A., Kay, M.: How patterns of students dashboard use are related to their achievement and self-regulatory engagement. In: Proceedings of the Tenth International Conference on Learning Analytics and Knowledge, pp. 340–349. SoLAR (2020)
32. Kinnebrew, J.S., Loretz, K.M., Biswas, G.: A contextualized, differential sequence mining method to derive students' learning behavior patterns. J. Educ. Data Min. **5**(1), 190–219 (2013)
33. Kinnebrew, J., Mack, D., Biswas, G.: Mining temporally-interesting learning behavior patterns. In: Proceedings of the Sixth International Conference on Educational Data Mining, pp. 252–255. International Educational Data Mining Society (2013)
34. Perera, D., Kay, J., Koprinska, I., Yacef, K., Zaiane, O.R.: Clustering and sequential pattern mining of online collaborative learning data. IEEE Trans. Knowl. Data Eng. **21**(6), 759–772 (2009)
35. Kang, J., Liu, M.: Investigating navigational behavior patterns of students across at-risk categories within an open-ended serious game. Technol. Knowl. Learn. (2020)
36. Swiecki, Z., Lian, Z., Ruis, A., Shaffer, D.W.: Does order matter? Investigating sequential and cotemporal models of collaboration. In: Lund, K., Niccolai, G.P., Lavoué, E., Hmelo-Silver, C., Gweon, G., Baker, M., Bailey, J. (eds.) The 13th International Conference on Computer Supported Collaborative Learning (CSCL), pp. 112–119. International Society of the Learning Sciences (2019)
37. Zheng, J., Xing, W., Zhu, G.: Examining sequential patterns of self- and socially shared regulation of STEM learning in a CSCL environment. Comput. Educ. **136**, 34–48 (2019)
38. Chen, B., Resendes, M., Chai, C.S., Hong, H.: Two tales of time: uncovering the significance of sequential patterns among contribution types in knowledge-building discourse. Interact. Learn. Environ. **25**(2), 162–175 (2017)
39. Elbadrawy, A., Karypis, G.: UPM: discovering course enrollment sequences associated with success. In: Proceedings of the Ninth International Conference on Learning Analytics and Knowledge, pp. 373–382. SoLAR (2019)
40. Jin, S.Y., Yei-Sol, W., Sang, J.P.: Mining course trajectories of successful and failure students: a case study. In: 2017 IEEE International Conference on Big Knowledge (ICBK), pp. 270–275. IEEE (2017)
41. Bhatt, C., Cooper, M., Jian, Z.: SeqSense: video recommendation using topic sequence mining. In: MultiMedia Modeling. MMM 2018 Lecture Notes in Computer Science, vol. 10705, pp. 252–263. Springer, Berlin (2018)
42. Wang, Y., Li, T., Geng, C., Wang, Y.: Recognizing patterns of student's modeling behaviour patterns via process mining. Smart Learn. Environ. **6**(1), 26 (2019)
43. Anwar, T., Uma, V., Shahjad: Book recommendation for eLearning using collaborative filtering and sequential pattern mining. In: 2020 International Conference on Data Analytics for Business and Industry: Way Towards a Sustainable Economy (ICDABI), pp. 1–6. IEEE (2020)

44. Sitanggang, I.S., Husin, N.A., Agustina, A., Mahmoodian, N.: Sequential pattern mining on library transaction data. In: Proceedings of the 2010 International Symposium on Information Technology (ITSim 2010), pp. 1–4. IEEE (2010)
45. Chen, C., Wang, W.: Mining effective learning behaviors in a web-based inquiry science environment. J. Sci. Educ. Technol. 29(4), 519–535 (2020)
46. Chen, W., Niu, Z., Zhao, X., Li, Y.: A hybrid recommendation algorithm adapted in e-learning environments. World Wide Web. 17(2), 271–284 (2014)
47. Wong, J., Khalil, M., Baars, M., de Koning, B.B., Paas, F.: Exploring sequences of learner activities in relation to self-regulated learning in a massive open online course. Comput. Educ. 140, 43–56 (2019)
48. Knight, S., Martinez-Maldonado, R., Gibson, A., Shum, S.B.: Towards mining sequences and dispersion of rhetorical moves in student written texts. In: Proceedings of the Seventh International Learning Analytics and Knowledge Conference, pp. 228–232. SoLAR (2017)
49. McBroom, J., Yacef, K., Koprinska, I., Curran, J.R.: A data-driven method for helping teachers improve feedback in computer programming automated tutors. In: Artificial Intelligence in Education. 19th International Conference, AIED 2018. Proceedings: LNAI 10947, pp. 324–337. Springer (2018)
50. Mirzaei, M., Sahebi, S.: Modeling students' behavior using sequential patterns to predict their performance. In: Artificial Intelligence in Education, AIED 2019 Lecture Notes in Artificial Intelligence, vol. 11626, pp. 350–353. Springer (2019)
51. Kang, J., Liu, M., Qu, W.: Using gameplay data to examine learning behavior patterns in a serious game. Comput. Hum. Behav. 72, 757–770 (2017)
52. Taub, M., Azevedo, R.: Using sequence mining to analyze metacognitive monitoring and scientific inquiry based on levels of efficiency and emotions during game-based learning. J. Educ. Data Min. 10(3), 1–26 (2018)
53. Winne, P.H., Hadwin, A.F.: The weave of motivation and self-regulated learning. In: Schunk, D.H., Zimmerman, B.J. (eds.) Motivation and self-regulated learning: theory, research, and application, pp. 297–314. Lawrence Erlbaum Associates Publishers (2008)
54. Scherer, K.: Emotions are emergent processes: they require a dynamic computational architecture. Philos. Trans. R. Soc. 364, 3459–3474 (2009)
55. Kinnebrew, J.S., Segedy, J.R., Biswas, G.: Integrating model-driven and data-driven techniques for analyzing learning behaviors in open-ended learning environments. IEEE Trans. Learn. Technol. 10(2), 140–153 (2017)
56. Fatahi, S., Shabanali-Fami, F., Moradi, H.: An empirical study of using sequential behavior pattern mining approach to predict learning styles. Educ. Inf. Technol. 23(4), 1427–1445 (2018)
57. Jaber, M., Wood, P.T., Papapetrou, P., Papapetrou, P.: A multi-granularity pattern-based sequence classification framework for educational data. In: 2016 IEEE International Conference on Data Science and Advanced Analytics (DSAA), pp. 370–378. IEEE (2016)
58. Fournier-Viger, P., Faghihi, U., Nkambou, R., Nguifo, E.M.: Exploiting sequential patterns found in users' solutions and virtual tutor behavior to improve assistance in ITS. Educ. Technol. Soc. 13(1), 13–24 (2010)
59. El-Ramly, M., Stroulia, E.: Analysis of web-usage behavior for focused web sites: a case study. J. Softw. Main. Evol. 16(1–2), 129–150 (2004)
60. Tarus, J.K., Niu, Z., Yousif, A.: A hybrid knowledge-based recommender system for e-learning based on ontology and sequential pattern mining. Fut. Gen. Comput. Syst. 72, 37–48 (2017)
61. Slim, A., Heileman, G.L., Al-Doroubi, W., Abdallah, C.T.: The impact of course enrollment sequences on student success. In: Proceedings of 2016 IEEE 30th International Conference on Advanced Information Networking and Applications (AINA), pp. 59–65. IEEE (2016)
62. Brown, J.S., Collins, A., Duguid, P.: Situated cognition and the culture of learning. Educ. Res. 18(1), 32–42 (1989)
63. Kinnebrew, J.S., Segedy, J.R., Biswas, G.: Analyzing the temporal evolution of students' behaviors in open-ended learning environments. Metacogn. Learn. 9(2), 187–215 (2014)

64. Emara, M., Tscholl, M., Dong, Y., Biswas, G.: Analyzing students' collaborative regulation behaviors in a classroom-integrated open ended learning environment. In: Smith, B.K., Borge, M., Mercier, E., Lim, K.Y. (eds.) Making a Difference: Prioritizing Equity and Access in CSCL, 12th International Conference on Computer Supported Collaborative Learning (CSCL), pp. 319–326. International Society of the Learning Sciences (2017)
65. Kinnebrew, J.S., Mack, D.L.C., Biswas, G., Chang, C.: A differential approach for identifying important student learning behavior patterns with evolving usage over time. In: Trends and Applications in Knowledge Discovery and Data Mining, pp. 281–292. Springer (2014)
66. Biswas, G., Segedy, J.R., Bunchongchit, K.: From design to implementation to practice a learning by teaching system: Betty's Brain. Int. J. Artif. Intell. Educ. 26(1), 350–364 (2016)
67. Segedy, J.R., Biswas, G., Sulcer, B.: A model-based behavior analysis approach for open-ended environments. J. Educ. Technol. Soc. 17(1), 272–282 (2014)
68. He, Z., Zhang, S., Wu, J.: Significance-based discriminative sequential pattern mining. Exp. Syst. Appl. 122, 54–64 (2019)
69. Zhang, Y., Paquette, L.: An effect-size-based temporal interestingness metric for sequential pattern mining. In: Rafferty, A.N., Whitehill, J., Cavalli-Sforza, V., Romero, C. (eds.) Proceedings of the 13th International Conference on Educational Data Mining (EDM 2020), pp. 720–724. International Educational Data Mining Society (2020)
70. Sabourin, J., Mott, B., Lester, J.: Discovering behavior patterns of self-regulated learners in an inquiry-based learning environment. In: Lane, H.C., Yacef, K., Mostow, J., Pavlik, P. (eds.) Artificial Intelligence in Education. AIED 2013 Lecture Notes in Computer Science, vol. 7926, pp. 209–218. Springer, Cham (2013)
71. Taub, M., Azevedo, R., Bradbury, A.E., Millar, G.C., Lester, J.: Using sequence mining to reveal the efficiency in scientific reasoning during STEM learning with a game-based learning environment. Learn. Instr. 54, 93–103 (2018)
72. Taub, M., Azevedo, R.: How does prior knowledge influence eye fixations and sequences of cognitive and metacognitive SRL processes during learning with an Intelligent Tutoring System? Int. J. Artif. Intell. Educ. 29(1), 1–28 (2019)
73. Martinez-Maldonado, R., Dimitriadis, Y., Martinez-Monés, A., Kay, J., Yacef, K.: Capturing and analyzing verbal and physical collaborative learning interactions at an enriched interactive tabletop. Int. J. Comput. Supp. Collab. Learn. 8(4), 455–485 (2013)
74. Malekian, D., Bailey, J., Kennedy, G.: Prediction of students' assessment readiness in online learning environments: the sequence matters. In: Proceedings of the Tenth International Conference on Learning Analytics and Knowledge, pp. 382–391. SoLAR (2020)
75. Sun, D., Cheng, G., Luo, H.: Analysing the evolution of student interaction patterns in a Massive Private Online Course. In: Interactive Learning Environments, pp. 1–14 (2022)
76. Fournier-Viger, P., Lin, J.C., Gomariz, A., Gueniche, T., Soltani, A., Deng, Z., Lam, H.T.: The SPMF open-source data mining library version 2. In: Berendt, B., Bringmann, B., Fromont, É., Garriga, G., Miettinen, P., Tatti, N., Tresp, V. (eds.) Machine Learning and Knowledge Discovery in Databases. ECML PKDD 2016 Lecture Notes in Computer Science, vol. 9853, pp. 36–40. Springer, Cham (2016)
77. Zimmerman, B.J.: Self-regulated learning and academic achievement: an overview. Educ. Psych. 25(1), 3–17 (1990)
78. Bannert, M., Reimann, P., Sonnenberg, C.: Process mining techniques for analysing patterns and strategies in students' self-regulated learning. Metacog. Learn. 9(2), 161–185 (2014)
79. Paquette, L., Grant, T., Zhang, Y., Biswas, G., Baker, R.: Using epistemic networks to analyze self-regulated learning in an open-ended problem-solving environment. In: Advances in Quantitative Ethnography, pp. 185–201. Springer, Cham (2021)
80. Bakeman, R., Quera, V.: Sequential Analysis and Observational Methods for the Behavioral Sciences. Cambridge University Press, Cambridge (2011)
81. Bosch, N., Paquette, L.: What's next? Sequence length and impossible loops in state transition measurement. J. Educ. Data Min. 13(1), 1–23 (2021)

82. Matayoshi, J., Karumbaiah, S.: Adjusting the L statistic when self-transitions are excluded in affect dynamics. J. Educ. Data Min. **12**(4), 1–23 (2020)
83. Magnusson, M.S.: Discovering hidden time patterns in behavior: T-patterns and their detection. Behav. Res. Methods Instr. Comput. **32**(1), 93–110 (2000)
84. Wu, Y., Tong, Y., Zhu, X., Wu, X.: NOSEP: non-overlapping sequence pattern mining with gap constraints. IEEE Trans. Cyber. **48**(10), 2809–2822 (2018)
85. Kinnebrew, J.S., Killingsworth, S.S., Clark, D.B., Biswas, G., Sengupta, P., Minstrell, J., Martinez-Garza, M., Krinks, K.: Contextual markup and mining in digital games for science learning: connecting player behaviors to learning goals. IEEE Trans. Learn. Technol. **10**(1), 93–103 (2017)
86. Bazaldua, D.L., Baker, R., Pedro, M.O.: Comparing expert and metric-based assessments of association rule interestingness. In: Stamper, J., Pardos, Z., Mavrikis, M., McLaren, B.M. (eds.) Proceedings of the Seventh International Conference on Educational Data Mining (EDM 2014), pp. 44–51. International Educational Data Mining Society (2014)
87. Merceron, A., Yacef, K.: Interestingness measures for association rules in educational data. In: Baker, R.S.J.D., Barnes, T., Beck, J.E. (eds.) Proceedings of the First International Conference on Educational Data Mining, pp. 57–66. International Educational Data Mining Society (2008)
88. Gao, C., Wang, J., He, Y., Zhou, L.: Efficient mining of frequent sequence generators. In: Proceedings of the 17th International Conference on World Wide Web, pp. 1051–1052. ACM (2008)
89. Wang, J., Han, J.: BIDE: efficient mining of frequent closed sequences. In: Proceedings of the 20th International Conference on Data Engineering, pp. 79–90. IEEE (2004)
90. Reimann, P., Markauskaite, L., Bannert, M.: e-Research and learning theory: what do sequence and process mining methods contribute? Br. J. Educ. Technol. **45**(3), 528–540 (2014)

Chapter 7
Sync Ratio and Cluster Heat Map for Visualizing Student Engagement

Konomu Dobashi

Abstract In the current learning management system, it is difficult for even experienced teachers to grasp the learning situation and to engage in a timely manner for each individual, and the response to this problem remains inadequate. In this study, in order to improve the learner's engagement and the teacher's help with the lesson, a cluster heat map of student engagement, teaching material browsing sync ratio, and experimental results of outlier detection were examined. Sync ratios for browsing teaching materials were generated on-site in real time, and teachers could refer to them when teaching lessons. From the analysis of the descriptive statistics in the learning log, the material clickstreams, the quiz scores, and Mahalanobis' generalized distance were obtained and the engagement cluster heat map was generated based on the weekly learning pattern. As a result, it became possible to clearly discuss the relationship between the appearance frequency of learning patterns and the appearance frequency of abnormal values in teaching material clickstreams and quiz scores. It was clarified that some of the frequency of the appearance of the learning pattern correlated with the frequency of the occurrence of abnormal values of the teaching material clickstream and the quiz score. The results of this study help to find learners who repeat inappropriate learning patterns early and to support appropriate teacher interventions.

Keywords Learning analytics · Cluster heat map · Sync ratio · Prerequisite knowledge · Outlier detection · Correlation analysis · Academic involvement

Abbreviations

EDM   Educational data mining
LA   Learning analytics

K. Dobashi (✉)
Aichi University, Nagoya, Aichi, Japan
e-mail: dobashi@vega.aichi-u.ac.jp

LMS   Learning management system
MGD   Mahalanobis' generalized distance
MOOCs   Massive open online courses
RTTSCS   Real-time time-series cross-section
VBA   Visual Basic for Applications

7.1 Introduction

In recent years, with the increased spread of online courses utilizing learning management system (LMS) and web conferencing tools, application research on learning logs has been ongoing for the purpose of improving education based on the concept of educational data mining (EDM) [1] and learning analytics (LA) [2, 3]. Additionally, due to the COVID-19 global pandemic in 2020, many educational institutions have used LMS and web conferencing tools to teach online [4]. Therefore, the accumulation of learning logs that record the state of learners has become larger than ever before, and the concept of big data has become necessary for EDM and LA [5].

In this reality, the current LMS learning log analysis function is gradually being improved. However, the analysis results only roughly show the learner's tendencies, and it is difficult to understand the relationship between the teacher's instructions and the learner's behavior. Such analytical functions are inadequate, even though LMS need analytical functions such as statistical software. In order to improve the quality of education, teachers must measure the learner's individual efforts in real time, not only face-to-face but also during online lessons.

Currently, the utilization of LMS and the analysis of learning logs are becoming more and more important, and the development of useful utilization methods for both teachers and learners is an urgent issue. The accumulated learning log records the learning behavior of each learner and is valuable data for EDM and LA. Depending on the analysis method, it is possible to obtain knowledge that is useful for individual instruction for learners, such as abnormal values, and analysis results that lead to support for class progress, such as the sync ratio of reading materials. Such analysis is effective when time-based observation is possible, and because LMS records events that occur in chronological order, it becomes easier to understand changes in learning behavior. By chronologically analyzing the clickstream for viewing teaching materials, one can investigate changes in learners' reactions to the teacher's instructions and changes in their engagement with the class.

In this situation, in simultaneous classes where dozens of students gather in a classroom, it is necessary to have a function to perform learning analytics in real time in addition to the functions of the current LMS. For example, regardless of whether it is a face-to-face blended class or an online distance class and whether or not the student is opening the teaching materials as instructed by the teacher, it is difficult to immediately determine which student is opening which page. Thus, current methods and functions for immediately measuring the student's reaction are extremely insufficient.

However, in a large class, it is difficult for even an experienced teacher to properly measure the individual state of the learner [6]. In addition, there are several methods for measuring a learner's condition, such as video recording in a classroom, recording of conversations, eye movement measurement, and responses that differ in various educational environments. The issue of how to measure the efforts of individual learners in this way remains unsolved, even in the current LMS, and future research is needed [7, 8]. Therefore, this research seeks to answer the following research question:

RQ: How should analysis and visualization be carried out in order to quantitatively grasp the learner's engagement with the lesson by using the material clickstream on the LMS and the learning log of the quiz?

Recent e-book systems are used not only to record the page movement of teaching materials but also to record writing such as markers and memos, and these latest technologies will likely be used in many classes in the future. However, there are many nonspecialists in computers and software among teachers, and in order to master the use of LMS and e-book system, the teachers themselves need to study how to use such technologies. The reality is that the hurdles to go further into learning analysis are high. In the future, it is also likely that teachers and learners will share the results of LA, so it is necessary to develop LA algorithms that are easy to understand both for teachers and learners and to develop visualization methods as well that are easy to see and understand.

The author is continuing research to accumulate learning logs using the Moodle LMS and to utilize the analysis results in classes. VBA (Visual Basic for Applications) and Excel have been used to analyze learning logs, and technology has been researched to visualize the analysis results of learning logs in tables and graphs based on time-series cross-section analysis (TSCS). Recently, the analysis of learning logs has been sped up, making it possible to refer to the sync ratio of browsing materials during classes in an actual classroom. Educational institutions generally have new learners when the semester or academic year changes, so the results of education effects and results of LA also change. Therefore, it is necessary to improve teaching materials that can be applied to learners in various situations and to constantly collect and analyze learning logs. In this research, the analysis process was improved to generate a cluster heat map from the data of material browsing clickstream and quiz score. Then, the learning logs collected in the new experimental class were analyzed and the analysis results obtained using the improved method were discussed.

7.2 Review Baseline

Moodle is now a widely used LMS worldwide [9] and has learning log collection functions. Moodle's page view log displays a record of when and how learners accessed course materials, as well as access start times, IP addresses, users' names, actions, and items accessed. This event log is only a chronological list, and its usage is limited because it displays only a broad view of the learner. On the other hand, in order to overcome the lack of data analysis functions in LMS such as Moodle, data mining and dashboard development research in EDM and LA have been actively conducted, and new learning log analysis methods have been proposed [10]. In addition, the user interface to visualize the analysis results has been developed [11].

In this field, Dierenfeld and Merceron [12] showed that various analyses were possible by utilizing pivot tables in Excel for analysis of Moodle learning logs. They also showed that Moodle events, content access counts, and test scores for each learner could be displayed in graphs. Furthermore, Konstantinidis et al. [13] developed an offline process to analyze Excel learning logs downloaded from Moodle by utilizing Excel macros and VBA. Their system allowed analysis of logs for specific time periods and student groups. It was possible to combine metrics such as total page views, unique users, unique actions, IP addresses, unique pages, and average session length for relevance analysis.

Subsequently, in the preprocessing of the downloaded Moodle learning log, an Excel macro was developed to additionally generate time-series data and visualize the learning material page views of learners during the lesson in time series with tables and graphs [14, 15]. In the research, a pivot table was automatically generated, and by combining the elements of the Moodle log, they were able to analyze the entire class and individual learners from various perspectives. In addition, research was conducted to generate a heat map by performing data mining using Moodle logs and Excel [16]. The heat map was based on the classification of learning patterns, and it was possible to find underperforming students by dividing them into four learning patterns according to the deviation of teaching material page views and quiz scores.

Furthermore, in order to speed up the analysis of learning logs, automation by web scraping was added, enabling teachers to conduct lessons while checking learners' teaching material page views in the classroom. When the analysis system first loads, it downloads the Moodle log, performs the preprocessing necessary for time-series processing, generates pivot tables and graphs, and visualizes the learner's teaching material page views. It is now possible for the teacher to proceed with the class while also observing data [17].

In recent years, technology development has become increasingly important for analyzing learning logs, such as EDM and LA, and various methods other than the development examples mentioned above are being actively researched [18]. This chapter continues the analysis of learning logs with Moodle and focuses on the analysis of new experimental lessons. In today's educational institutions, there are

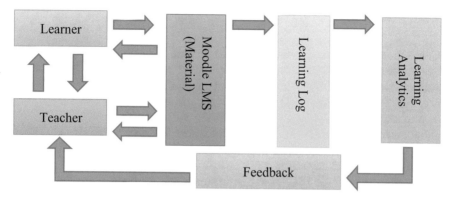

Fig. 7.1 Overview of data processing for learning log analytics

numerous types of classes and educational purposes, so it is necessary to clarify the type of classes that this research analyzes.

In the experimental class focused by this research, before digitized online teaching materials were used, students learned about computers using paper textbooks. However, today, physical textbooks have morphed into PDF files, which can be downloaded from the Moodle LMS to digital devices such as personal computers. The materials used in this chapter were intended for reading purposes [19], and therefore they did not include video or audio commentary. However, diagrams and charts for explanation were provided in the same way as conventional paper materials. The quiz was also based on a five-point multiple-choice question printed on paper, and the quiz was answered by clicking the radio button displayed on Moodle. Instead of opening conventional paper textbooks, learners opened PDF files in the Moodle LMS and participated in class.

According to experiences so far, even in lessons using an LMS, some learners opened the teaching materials later than the teacher's instructions, and others did not open the teaching materials at all. In the classes targeted by this research, however, it was essential for students to open and read the teaching materials. In addition, if it was possible for the teacher to immediately notice any students' inappropriate viewing of teaching materials during class, the teacher could call attention to students, devise explanation methods, improve teaching materials, and devise ways to deal with students (Fig. 7.1).

Therefore, this chapter aimed to develop a method that could collect learners' reactions to a teacher's instructions to browse teaching materials, visualize them in real time using numerical values and graphs, and list them in a class in which many learners participated [20]. In blended learning that utilizes LMS in the classroom and online remote lectures, research is being conducted to utilize timestamp information accumulated in learning logs to perform time-series cross-section analysis and visualization of the learning process [21].

7.2.1 Method

Conventional time-series cross-sections express data that include time elements in a two-dimensional tabular format and are frequently used in research fields such as comparative politics and the empirical analysis of international relations [22, 23]. The data handled in the time-series cross-section analysis are time-series data collected by continuously observing the same subject over time [24]. These data allow both time-series and cross-section analyses. However, time-series data are often fixed timelines representing years, months, or other timeframes, and it is difficult to flexibly change them.

The basic property of TSCS is to handle numerical quantitative data, but it is also possible to quantify and handle qualitative data. The data are quantified by treating them as a multiple-choice question, in a similar manner to answering a qualitative data questionnaire using a multiple-choice format. For example, if the item is selected once, it is quantified as 1, and if it is not selected, it is quantified as 0. This concept can be used as a basic method for aggregating clickstreams in the Moodle log [25, 26]. A general TSCS is created in the form of a cross table, in which observation targets are arranged in rows and the time-series data are arranged in columns. Alternatively, a table in which time-series data are arranged in rows and observation objects are arranged in columns is also used.

Excel has many functions for processing qualitative data. Among them, the pivot table function is specialized for counting qualitative data such as character strings, quantifying them, and creating a cross-table (Fig. 7.2). Therefore, in this research, these Excel functions and VBA were used to perform preprocessing to generate

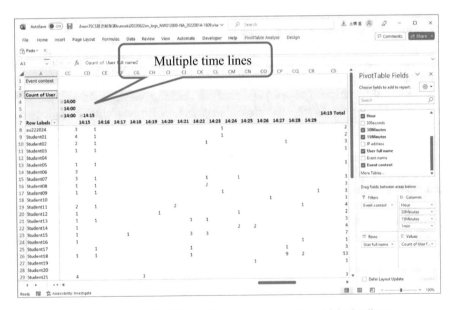

Fig. 7.2 Pivot table and example of time-series cross-section with multiple timelines

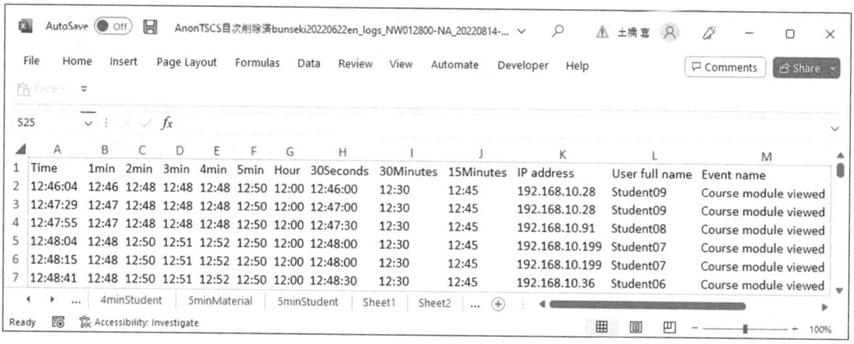

	A	B	C	D	E	F	G	H	I	J	K	L	M
1	Time	1min	2min	3min	4min	5min	Hour	30Seconds	30Minutes	15Minutes	IP address	User full name	Event name
2	12:46:04	12:46	12:48	12:48	12:48	12:50	12:00	12:46:00	12:30	12:45	192.168.10.28	Student09	Course module viewed
3	12:47:29	12:47	12:48	12:48	12:48	12:50	12:00	12:47:00	12:30	12:45	192.168.10.28	Student09	Course module viewed
4	12:47:55	12:47	12:48	12:48	12:48	12:50	12:00	12:47:30	12:30	12:45	192.168.10.91	Student08	Course module viewed
5	12:48:04	12:48	12:50	12:51	12:52	12:50	12:00	12:48:00	12:30	12:45	192.168.10.199	Student07	Course module viewed
6	12:48:15	12:48	12:50	12:51	12:52	12:50	12:00	12:48:00	12:30	12:45	192.168.10.199	Student07	Course module viewed
7	12:48:41	12:48	12:50	12:51	12:52	12:50	12:00	12:48:30	12:30	12:45	192.168.10.36	Student06	Course module viewed

Fig. 7.3 Example of preprocessing for splitting time data for process mining

time-series data (Fig. 7.3). Furthermore, the TSCS table was created by quantifying the qualitative data using Excel's worksheet and pivot table functions. The major difference between this research and the conventional TSCS is that the author included a timeline that can flexibly change the concept of time. This is for use in classrooms and online distance learning and for the real-time analysis of continuous data generation over time.

In a face-to-face class attended by a large number of students, time-series data corresponding to the number of teaching materials and students are generated when students browse teaching materials in Moodle and take quizzes. In order to efficiently process the time-series data generated in this way, it is necessary to be able to analyze the data from various viewpoints while switching between the overall viewpoint and the partial viewpoint. In addition, it is necessary to employ statistical software with an easy-to-use user interface in order to efficiently process the data according to the purpose of analysis.

Excel pivot tables were used in this research. The advantage of this software is its easy-to-use and easy-to-understand user interface that can be mastered in a short period of time. Teachers today almost always use computers, but many teachers are not experts in data science or computer science, so complex analysis of learning logs may be difficult for many such professionals. However, if teachers have information processing skills, Excel and pivot tables are easy to learn how to operate, and it is expected that they can be used to analyze learning logs.

The pivot table function in Excel is mainly used to aggregate the frequency of the appearance of multiple discrete data and to create a frequency distribution cross table. From the cross table, various graphs can be created according to the purpose by using the graph creation function of Excel. Furthermore, by using the macro function of Excel, it is possible to build a system that automates a series of processes, such as creating cross tables and graphs that previously were performed manually. In this research, VBA and pivot tables were used to operate the real-time time-series cross-section (RTTSCS) table from various viewpoints.

The pivot table is primarily suitable for summarizing discrete data, and Moodle's event log consists entirely of character string and discrete data. Therefore, although time data are recorded as time-series data, character string processing can be performed using VBA or Excel functions. As described below, by dividing the time data into appropriate time units and intervals, it is possible to generate TSCS in pivot tables and perform various analyses. The more types of data items that are handled through cross tabulation, the more complex and detailed the tabulation tends to become, increasing the possibility for multifaceted analysis of the Moodle logs.

Looking at the format of Moodle's event log, time data are accumulated in a format that combines the date and time when the material was opened, such as "6/5/2022, 13:45:35." These data are also recorded in the same format when posting to a forum or performing other tasks in the LMS. Although it is possible to use the pivot table for analysis in this form, it is extremely limited for generating a time-series cross-section table, which is the purpose of this chapter.

Therefore, in order to visualize the analysis results in class, a method was devised to process the original time data from Moodle and divide them at regular time intervals. In this research, character string processing was applied to the time data, and units representing time—such as year, month, date, day of the week, hour, minute, and time—were generated. This enabled time-series cross-tabulation with pivot tables (Fig. 7.2). For example, for individual students or individual teaching materials, categories such as date and time could be used, combined, and cross-tabulated (Fig. 7.3). In addition, various narrowing aggregations were possible, such as the aggregation for a specific time period on a specific day [27].

7.2.2 Background

This section describes the background of the research and the issues involved in learning, using the current LMS and related research. Even before the advent of LMS, research on learning patterns was underway [28]. Assessing a learner's psychological state via a questionnaire—and discovering various characteristics of the learning pattern via factor analysis—has been widely practiced [29]. In recent years, LMS learning logs of various functions attached to e-book systems have become widespread, including learning logs that can be analyzed [30, 31]. E-book systems can accumulate learning logs by tracking underlining, movement among pages, commenting, and highlighting. Various methods of these logs and learner behaviors are then analyzed and have been shown to be useful for improving lessons [32, 33].

Li et al. [34] conducted an experiment in which graduate students in an educational technology course were asked to read academic papers using a web-based digital textbook system to analyze learning behavior patterns from the log of their page operations. The analysis was based on the lag-sequential analysis method. This method can investigate and summarize the interdependencies that occur in complex interactive sequences of actions. The frequency of behavioral patterns such as "page back-and-forth movement," "make underline," and "make highlight" were analyzed, revealing that certain tools and actions could improve the use of digital textbooks and future teaching materials.

Research on learner performance prediction and detection of outliers is also well underway [35–37]. Based on the traffic volume of the campus network, Pytlarz et al. [38] explored student GPA, attendance, class engagement, and study time outside class and proposed a model to predict student success.

In massive open online courses (MOOCs), researchers are aiming to identify struggling students at earlier stages using machine learning and hybrid algorithms, along with analyzing the dropout behavior of students [39–41]. Dropout predictions by Gitinabard et al. [42] analyzed student access to materials and logs from forum posts, revealing potential applications for early learner intervention and learning guidance by finding learners who were at risk of potentially poor grades. Many learners' logs have been accumulated, such as those via MOOCs [43–45]. In addition, many methods have been proposed for identifying the factors associated with at-risk learners, along with various data mining methods for finding important knowledge and data [46–48]. Estacio and Raga [49] developed a data mining algorithm that applied a vector space model to the analysis of Moodle logs, showing a correlation between action logs and final grades. They created a dashboard-style interface to quantify and visualize the learner's behavioral level, demonstrating that learners with inadequate lesson engagement could be identified immediately.

Additionally, the research and development of systems equipped with an easy-to-operate user interface called a dashboard have been actively carried out [50–53]. These researchers have sought useful knowledge for class management and teaching materials. Moreover, statistical analysis plugin software such as GISMO for analyzing learning logs has been released for Moodle [54, 55]. By installing this software on Moodle, a user can analyze results such as learner activities, participation, course completion time, and quizzes [56].

Another useful technology is that of heat mapping. This is a clustering technique that uses shades of multiple colors for the purpose of visually expressing the magnitude of relationships of numerical values in order to visualize the meaning and relationships among multivariate data [57]. Conventionally, it has been used to visualize air temperature distribution, sea surface temperature distribution, human body temperature distribution and so on. There are various forms of data visualization for heat maps, such as MapBox for choropleth mapping, Qlucore for gene expression analysis, HeatmapGenerator, Wi-Fi Network Heatmap generator, and other commercial heat map tools, some of which are available for free. In this study, the main items related to the cluster heat map, which is closely related to this research, are taken up.

Wilkinson and Friendly [58] conducted a historical literature survey from the nineteenth century on cluster heat maps. They revealed that the cluster heat map was a composite of several different graphical representations developed by statisticians over a century. Currently, the cluster heat map is used in various fields including bioinformatics. The cluster heat map is an excellent way to simultaneously graph the hierarchical cluster structure of rows and columns within a data matrix. Each row and column is arranged so that rows and columns with similar values are closer together and colored with a color scale corresponding to the cell value. Many of the graphs have a clustering tree that represents the classification hierarchy displayed around them.

In addition, the statistical software R and the programming language Python have functions for creating heat maps, and cluster heat maps can be created by preparing appropriate data. However, using R or Python to create cluster heat maps requires programming knowledge, which is a high hurdle for many teachers who do not have computer or software skills. Thankfully, Excel has a function to examine cell values and color cells that match specified conditions. This function allows even nonexperts to create simple heat maps, but clustering must be done separately. Against this background, Heatmapper [59] was developed with the goal of being an easy-to-use system for both experts and nonexperts. This system is a general-purpose heat map generation tool and has been published as a web server. An easy-to-use graphical interface allows users to interactively graph data in the form of heat maps. Heatmapper is a versatile tool that works with many different data types and applications, making it easy to create a wide variety of heat maps.

7.2.3 Resources

The subject of this research was experimental classes, which are regular courses offered during every spring and fall semester at Aichi University in Japan. Classes are held every week in a computer classroom, and digital teaching materials in PDF format uploaded to Moodle are the primary class resources. The learning logs examined in this research were the learning logs collected by Moodle for 104 days from April 11, 2022 to July 20, 2022. Of the 51 students, who initially registered for the class, 42 took the final quiz, and the average weekly attendance was 43.7. The gender distribution of registrants was 29.4% female and 70.6% male, and most of the participants were 18–22 years old. Many of the participants had previously taken classes aimed at acquiring the basic skills of typing, Word, and Excel.

In this study, the online teaching materials used in class were reading materials in PDF format. These teaching materials corresponded to half a year of university classes and were prepared for 15 weeks. The contents were divided into sections, and PDF files were created for each section. The materials were then uploaded using Moodle's topic mode and used in weekly classes. Exercises and quizzes were also provided for each chapter. A quiz was given at the beginning of class regarding the contents of the chapters studied the previous week. All teaching materials used in the class were prepared on Moodle, and the teacher explained and demonstrated the materials to the students. The students then used the teaching materials for preparation and review.

In this research, the gateway page of the course was prepared using the topic mode of Moodle, and the table of contents of the course was created on it. Teachers and students clicked on links on the gateway page to view online materials. In the Moodle log, when a student joined a course for the first time, he or she registered for the class, and the recording of the learning log started from that time. Then, when the students in the class opened these online teaching materials, the data of the following items were recorded: resource start time (time), student ID and name (user full name), information viewed by the student (event context), operations performed in Moodle (event name), and IP address of the computer accessed (IP address).

Course administrators could download these learning logs not only during class but also after class and use them freely in Excel or other programs. Additionally, at the beginning of the class, the teacher logged into Moodle in the same way as the students and conducted the class while giving instructions for students to browse the online teaching materials. In the case of this research, both the learning log of the student and the operation log of the teacher in charge were recorded. From these operation log records, it was possible to ascertain at what point during the lesson the teacher instructed the learners to open the teaching materials. Furthermore, it was also possible to count how many students responded to the teacher's instructions. One of the purposes of this study was to examine the synchronization between the teacher's instructions and the learner's reactions.

7.2.4   Web Scraping for On-Site Learning Analytics

Because the teacher was often busy during class, it was better to operate the system as little as possible during the lecture. In this research, web scraping technology was adopted to automate the process of downloading Moodle logs, performing preprocessing, and displaying TSCS tables and graphs (Fig. 7.4). VBA was used for programming, Internet Explorer was used for web browsing, and Moodle log downloads could be automatically operated. The first major step in this process was the automation of the operations, from login to Moodle to downloading the course logs. Step 1-1 was to start Excel. Step 1-2 was to start the macro. Step 1-3 was to start Moodle from VBA. Step 1-4 was to log into Moodle as a course administrator. Step 1-5 was to open the log page of the course used in class. Step 1-6 was to select the date to download. Finally, Step 1-7 was to select the type of log to download.

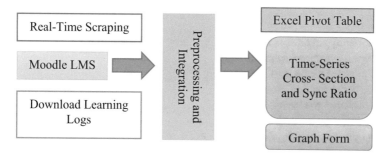

Fig. 7.4 Integration by web scraping and data mining

The second major process was the generation of time-series cross-section tables and graphs from the downloaded course logs and visualization of the sync ratio of the course materials. Step 2-1 was to download the log in an Excel format file. Step 2-2 was to start Excel and open the log file. Step 2-3 was to perform preprocessing and divide the time stamps. Step 2-4 was to start a pivot table and generate the TSCS table. Step 2-5 was to calculate the sync ratio. Finally, Step 2-6 was to generate a graph.

7.3 Sync Ratio of Opening Teaching Materials

In the 2022 spring semester, in order to conduct an experiment for this research, online teaching materials on Moodle were used, and face-to-face classes were held in the classroom. The class was entitled "Introduction to Social Data Analysis," in which students learned the basics of data science using statistical data published on the Internet. The contents covered in the class were use of Excel worksheets, creating graphs, use of functions, representative values, probability and probability distribution, random numbers, basics of simulation, frequency distribution and histograms, deviation, variance, standard deviation, normal distribution, cross tabulation, attribute correlation, covariance, correlation coefficient, and regression analysis. These topics were covered in 15 sessions.

The teaching materials consisted of 13 chapters and 95 sections, totaling 177 pages. The average number of pages per chapter was 12.6, the average number of sections was 6.8, and the average number of pages per section was 1.86. There were 15 quizzes, 14 file submissions for practice assignments, and 14 external links. Each quiz was a fill-in-the-blank question created from the teaching materials and was administered at the beginning of the next week's class after each chapter. The quiz had a time limit of 5 min and consisted of five multiple-choice questions. In the latter half of the class, the learners worked on their own exercises and were asked to submit a file summarizing their exercises by the following week.

7.3.1 Student Sync Ratio

Conventionally, in classes using computers, teachers prepare teaching materials and deliver demonstrations from the teacher's desk. In such cases, it is desirable that the teacher's instructions and the learner's reactions are synchronized as much as possible. If the teacher proceeds with a lesson unilaterally without considering the reaction of the students, there is a high risk that more learners will struggle. However, due to several factors related to the equipment of the classroom, it is difficult to fully grasp the state of the learners in the current computer classroom. For example, it is difficult to check the monitors used by the learners unless the teacher physically walks around the classroom.

Therefore, online teaching materials in this study were distributed from Moodle so that all students could view the teaching materials on a personal computer. Then, the teaching material clickstream was analyzed in real time, and the sync ratio was calculated. At that time, the teacher instructed the learners to open the teaching materials as necessary and encouraged the learners to engage. Teaching materials uploaded to Moodle made it possible to collect material clickstreams, increasing the possibility of a more detailed analysis of learner reactions. In addition, in this research, the clickstream was collected only when the material on Moodle was clicked, rather than the clickstream of the entire computer operation.

7.3.2 Visualization Sync Ratio by Table and Graph

In this section, an analysis of the teaching material clickstream, which was collected in the class when about 50–60 learners were gathered in the computer classroom, was examined. It was expected that the reactions of learners during class would change from moment to moment. Therefore, the RTTSCS table was used to aggregate the time-series data for each learner, and this was collected in a table. By reediting the data items of the Moodle log into TSCS, it was possible to determine in real time who opened which material and when and from which IP address it was used. In this research, it was possible to download the teaching materials, and the clickstream viewed after downloading from Moodle could not be counted.

Figure 7.5 shows a sample of the teaching material clickstream of the class that started at 13:00 on June 22, 2022. The preprocessing program automatically reedited the Moodle log for each learner every minute and aggregated it in the TSCS table. The bar graph at the bottom of Fig. 7.5 is a sync ratio graph created from the numerical data at the top, and the time series matches. The table and graph in Fig. 7.5 show the clickstream of the 5-min quiz, and the two teaching materials were taken immediately after class began. Cell A3 shows the teacher's ID, and we can see the teacher's activity in Row 3 as the teacher opened teaching materials and Excel and explained and demonstrated the lesson.

At 13:00, 13:10, and 13:19, the time data of the clickstream are displayed when the teacher opened the teaching material. At 13:00, instructions were given to take the quiz, and the teacher opened the teaching materials on the teacher's computer and instructed the learners to open the teaching materials for the day. The topic of the lesson that day was discrete variables and crosstabs, and the first teaching material "10.01 statistical independence" was opened at 13:10. Next, at 13:19, the second teaching material "10.02 Phi coefficient" was opened, and the lesson proceeded.

Approximate sync ratios can be read from the table and graph in Fig. 7.5. Regarding the quiz, the sync ratio was between 80% and 90% for up to 3 min, but the sync ratio was about 70% approximately 1 min before the end. We can see that about 30% had finished answering by then. At 13:05, the sync ratio rose again to 87% because the learners opened the page to check their quiz scores.

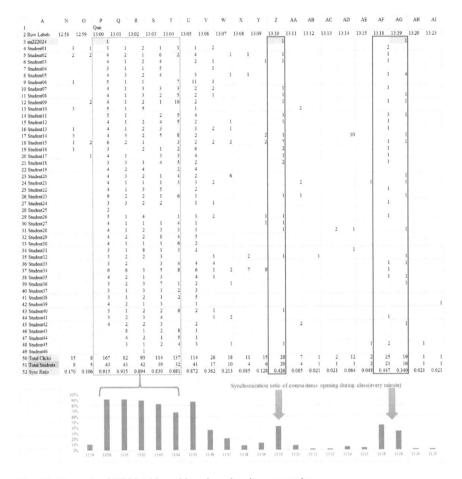

Fig. 7.5 Example of TSCS table and bar chart showing sync ratio

A closer look at the changes in the numerical data in Row 52 in the table and the graph shows that the material sync ratio was considerably lower than the quiz sync ratio. The sync ratio at 13:10 was 42.6%, and it was 55.4% if the sync ratio immediately before was added. In addition, the 13:19 teaching material sync ratio was 34.0%. The low sync ratio at 13:10 was probably because the learners had already opened the material after the end of the quiz because they knew in advance the materials to be used for the day. In this way, the table and graph in Fig. 7.5 show the sync ratio of multiple teaching materials, and it is thought that the state of synchronization for the entire class is shown. Incidentally, the time in Fig. 7.5 is 13:00 to 13:23, but the average sync ratio during this period was 29.0%.

7.3.3   Each Material Sync Ratio and Learner's Browsing Process

In addition, since learners could open multiple teaching materials at the same time, it was also important to calculate the sync ratio for each teaching material. If the sync ratio for each teaching material could be found in this way, it would be possible to know what percentage of learners were synchronizing by the time the teacher gave an instruction to open the teaching material. The data in Fig. 7.5 can be narrowed down by teaching material, and the sync ratio by teaching material can be obtained. The materials opened at 13:10 and 13:19 were narrowed down, and examples of the regenerated tables and graphs were then examined.

Figure 7.6 shows the clickstream of the teaching material "10.01 statistical independence" that was opened first on the day of the class. The sync ratio at 13:10 was 40.4%, but the table and graph show that many learners had already opened the teaching material before this time. It was found that the early learners had opened the material before the class even started, and by 13:10, there were 36 people with

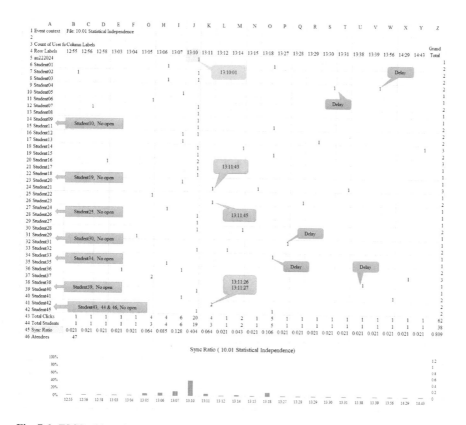

Fig. 7.6 TSCS table and sync ratio by one material

the material open (76.6%). Furthermore, from the total column on the right end of the table, it can be seen that the same learner opened the same teaching material multiple times. Additionally, learners with no data prior to 13:10 and data afterwards more than 2 min apart (Students 4, 31, 33, and 38) were marked with the label "delay." They were learners who opened the material late.

The quiz ended at 13:05, and the teacher opened the score confirmation page immediately after that and presented it to the learners. Because there were no data at 13:08 and 13:09, no materials were opened by the students. Additionally, 19 students opened the teaching materials at 13:10, almost at the same time as the teacher. At the next time point, i.e., 13:11, three students opened the material, and 22 (47%) learners opened the material within 2 min.

In Fig. 7.6, only the learners who opened the material, including the teacher, were counted, and the learners who did not open the material were not shown. There were 47 students (Cell B46) in attendance, but only 38 students (Cell Z44) opened the material "10.01." The IDs of the nine learners whose ID numbers do not appear in column A (Students 10, 19, 25, 30, 34, 39, 43, 44, and 46) are not displayed because they did not open the corresponding teaching materials. The course participants item in the Moodle log records the number of times each learner opened each material, and according to that data, 11 learners did not open the material "10.01" at all during the semester. Therefore, it was confirmed that the aforementioned nine learners never opened the teaching material "10.01" during the semester.

It is not clear why these learners did not open the teaching materials, but we suggest the following possibilities. First, one teaching material monitor is installed for every two students, and learners can always see the teaching materials and demonstrations displayed by the teacher. Therefore, it is possible that they did not feel the need to open the materials by themselves since they were looking at the teaching materials and demonstrations presented by the teacher. Second, it is possible that they were able to understand the class because they already had the prerequisite knowledge related to the content of the class. Third, it is conceivable that they did not open the teaching materials because they were not interested in the contents of the class for some reason.

As mentioned here, it is possible to find the sync ratio for the teaching material "10.02 Phi coefficient" in the same way and to investigate when the teaching material was opened late or not opened at all.

7.4   Student Engagement and Cluster Heat Map

7.4.1   Cluster Heat Maps and Outlier Detection

A student engagement heat map involves the application of heat map research that has been conducted in other fields to the field of student engagement [60–62]. The authors conducted research that utilized heat maps to visualize class efforts in relation to LA [16]. This chapter then examined both the method of improving the heat map in our LA so far and the latest analysis results.

The assumption for using engagement heat maps in learning analysis is that the use of LMS in classes has made it easier to collect various learning logs [63]. From the analysis of the learning logs accumulated in the LMS, there are many cases where the characteristic learning behavior of the student is found, and these characteristics can be grasped as the learning pattern [64]. For example, using TSCS analysis, it is possible to extract learning patterns for each learner, such as which teaching materials were viewed and when, as described above.

Conventional research on learning patterns has been conducted by preparing question items, requesting answers through questionnaires, and analyzing those answers using techniques such as factor analysis [65, 66]. In the field of management, questionnaire items have been prepared to investigate employee engagement, and the responses to questionnaires were statistically analyzed and used for staffing plans. Conventional learning pattern analysis results were often displayed in tables and graphs as numerical data, but there are also examples of using heat maps in other fields such as employee engagement.

In contrast to the conventional methods described above, it has become possible to extract some behavioral patterns of learners from data accumulated in LMS and e-book systems from learning logs. In other words, the digital transformation of learning pattern analysis has begun [67, 68]. In the learning log, various learning behaviors that occur when the learner utilizes the LMS are automatically accumulated, including time-series data. By analyzing these learning logs, it is becoming possible to make analyses from a new perspective and use methods different from the conventional learning pattern analysis. The digital transformation of learning pattern analysis is expected to continue into the future.

When extracting learning patterns from LMS learning logs, it is possible to easily obtain log data that are easy to quantify, so these data can be used as bases for grouping learners' learning patterns. Furthermore, it is possible to consider the generation of a heat map based on learning patterns and visualizing outliers and abnormal values. If the learning patterns are reflected in the heat map in this way and lead to the detection of outliers and abnormal values, it will be possible to support the teacher's class management and make appropriate interventions that reflect the learning patterns.

The main purpose of extracting learning patterns and detecting outliers in class is to detect as early as possible learners who repeatedly engage in inefficient and inappropriate behaviors and prevent them from dropping out of class. Taking the example of the class in this chapter that used reading materials, the occurrence of abnormal values was assumed in the following states: (a) did not open the teaching materials instructed by the teacher; (b) opened the teaching materials after the teacher instructed them; (c) did not read the teaching materials even if they opened them; and (d) did not understand the content of the teaching materials. If these conditions continued, the test scores tended to be low. Thus, such situations are abnormal behavior that cause learning impediments, which are likely to lead to abnormal values.

Of the above, States (a) and (b) can be detected from the clickstream of viewing teaching materials accumulated in Moodle, and the TSCS analysis taken up in this study can be utilized. States (c) and (d) can be handled by combining the quiz scores and clickstreams. In addition, the learning logs accumulated in the LMS were observed variables, and the learning logs themselves could be treated as feature values for statistical analysis. Therefore, clustering using descriptive statistics was examined as a basic data analysis method.

Many data mining methods have been proposed to discover abnormal values and important findings and symptoms that can cause class failures when learning logs are accumulated on a large scale, such as in MOOCs [69]. In this study, groupings of learning patterns that indicated the level of the learner's engagement were examined using the standard normal distribution of clickstreams and quiz scores. In addition, the created heat map could be used to visualize learning patterns to identify learners who were having trouble learning, and the results of the analysis are presented later [70]. Furthermore, this research compared the extraction of outliers using the standard normal distribution with outlier detection via the square of the MGD (Mahalanobis' generalized distance) [71, 72].

7.4.2 Taxonomy of Learning Patterns via Normal Distribution

Standard tests are usually created so that scores are normally distributed. In this study, reading materials were used in class, and quizzes were created from the contents of the teaching materials. Because scores are expected to be normally distributed in weekly quizzes and the final quiz, it was assumed that learners' clickstreams for teaching materials were also normally distributed, and the standards for data analysis were unified. Then, in order to investigate the effects of teaching materials and quizzes on learners, the classification of learning patterns using the standard normal distribution was considered. The average and standard deviation from each of the teaching material clickstreams and quiz scores were obtained. The three-sigma method was then used in normal distribution, and these numerical values were divided as follows and applied to the classification of learning patterns.

When the random variable Z follows the standard normal distribution, the probability distribution of Z is as follows, where μ is the mean and σ is the standard deviation. That is, the probability that $\mu - \sigma \leq Z \leq \mu + \sigma$ is 68%; the probability that $\mu - 2\sigma \leq Z \leq \mu + 2\sigma$ is 95%; and the probability that $\mu - 3\sigma \leq Z \leq \mu + 3\sigma$ is 99.7%. By applying these distributions to the material clickstreams and quiz scores, the following six learning patterns were identified.

The occurrence probability of the part corresponding to $\mu - 3\sigma \leq Z$ and the part corresponding to $Z \geq \mu + 3\sigma$ is extremely low, at 0.3%. Therefore, they were included in $\mu - 2\sigma < Z$ or $\mu + 2\sigma \> Z$, respectively. As a result, data with a probability distribution of less than 5% were determined to be abnormal values. In addition, the square of the MGD was similarly judged to be an abnormal value of less than 5%, and the two were compared.

Additionally, as described below, the outlier values and abnormal values were treated separately, based on the probability distribution:

1. $\mu < Z \leq \mu + \sigma$ is a high-normal value ("+normal").
2. $\mu + \sigma < Z \leq \mu + 2\sigma$ is a high-outlier value ("+outlier").
3. The range corresponding to $\mu + 2\sigma > Z$ is a high-abnormal value ("+abnormal").
4. $\mu > Z \geq \mu - \sigma$ is a low-normal value ("−normal").
5. $\mu - \sigma > Z \geq \mu - 2\sigma$ is a low-outlier value ("−outlier").
6. The range corresponding to $\mu - 2\sigma < Z$ is a low-abnormal value ("−abnormal").

In actual calculations, material clickstreams and quiz scores were used for the above classification, but since the units of these data were different, it was difficult to compare them as real data. Therefore, by calculating the *z-score*, the data group with different units was standardized to have an average of 0 and a standard deviation of 1.

7.4.3 Integration of Learning Patterns

In addition, using the teaching material clickstream and the *z-score* of the quiz score, the learning patterns from G1 to G4 were classified as follows, and the characteristics of each learner's pattern were expressed. For the classification of the learning patterns described below, when a scatter diagram of teaching material clickstreams and quiz scores was created, the data were distributed from the first quadrant to the fourth quadrant:

G1: A learning pattern with a higher-than-average material clickstream and higher-than-average quiz scores.

G2: A learning pattern with below-average material clickstreams and higher-than-average quiz scores.

G3: A learning pattern with below-average material clickstreams and below-average quiz scores.

G4: A learning pattern with above-average material clickstreams but below-average quiz scores.

In order to investigate the effectiveness of teaching using reading materials, only material clickstreams and quiz scores were used here, and other data such as forums were not used. Therefore, a two-variable analysis was considered, with the teaching material clickstream as the independent variable and the quiz score as the dependent variable. The classification of learning patterns was generally performed according to the following procedure:

1. Moodle, online teaching materials, and quizzes were used to conduct classes and collect learning logs.

2. TSCS was used to aggregate the weekly quiz range material clickstreams.
3. In order to apply the probability distribution in the normal distribution, the *z-score* was obtained and normalized for the teaching material clickstream and the quiz score.
4. The teaching material clickstream and the *z-score* of the quiz score were classified by applying the ranges of the six learning patterns described above.
5. The square of the MGD was calculated, and the outliers were confirmed and compared.
6. Focusing on the positive/negative sign of the *z-score*, the pattern of G1 to G4 was determined.

7.4.4 Detection of Abnormal Values

In Table 7.1, based on the above procedure, the results of the material clickstream and quiz scores for the class held on April 22, 2022 are shown. Column A shows student ID; Column B is material clickstream; Column C is quiz score; Column D is the *z-score* of the clickstream; Column E is the *z-score* of the quiz; Column F is the taxonomy of the clickstream; Column G is the taxonomy of the quiz; Column H is MGD squared (D^2); Column I is judgment $(x > 0.95)$; and Column J is the learning pattern (G1–G4).

Looking at Column F, three instances of "+abnormal" appeared, indicating that there were an extremely large number of teaching material clickstreams. Student 21 clicked 138 times, Student 26 clicked 137 times, and Student 33 clicked 155 times. In addition, an outlier in the quiz score appeared in Column G where Student 16 scored zero points.

Moreover, looking at Columns H and I, there were three data points that corresponded to the abnormal values via MGD. Interestingly, they overlapped with two abnormal data points in the clickstream. Column J shows the classification of the learning patterns based on the material clickstream and quiz scores and indicates to which pattern the abnormal value corresponds. Student 33 had a quiz score of 10 points, which was classified as a G1 pattern, and the other abnormal values were G4 patterns.

As stated above, G1 is a pattern with an above-average clickstream and quiz score, while G4 is a pattern with a higher-than-average clickstream but a below-average quiz score. Figure 7.7 shows a scatterplot of the clickstreams and *z-scores* of the quiz scores in Table 7.1, and Fig. 7.8 shows a scatterplot of the MGD squared. MGD squared clearly shows the above four abnormal values, so it is easy to identify the abnormal data points; however, in the calculation of Table 7.1, there were three abnormal data points in MGD.

After 13 weekly quizzes and the final quiz, those learning patterns were concatenated to create a semester-wide heat map. As shown in Figs. 7.9 to 7.11, three heat maps were created: material clickstreams, quiz scores, and learning patterns determined from these two variables. The data of all registered students

Table 7.1 Weekly log data aggregation (Chapter 1, clickstream, quiz score, taxonomy, MGD squared, judgment, and learning pattern)

A	B	C	D	E	F	G	H	I	J
	Chapter 1		Z-score		Taxonomy		Mahalanobis		Pattern
Student	Click	Quiz	Click	Quiz	Click	Quiz	D^2	x>0.95	
Student01	79	2	-0.342	-1.814	-normal	-outlier	1.826	0.8234	G3
Student02	65	6	-1.024	-0.235	-outlier	-normal	0.587	0.5563	G3
Student03	71	2	-0.732	-1.814	-normal	-outlier	1.979	0.8405	G3
Student04	88	6	0.096	-0.235	+normal	-normal	0.038	0.1555	G4
Student05	74	6	-0.585	-0.235	-normal	-normal	0.206	0.3500	G3
Student06	79	10	-0.342	1.344	-normal	+outlier	1.121	0.7104	G2
Student07	105	8	0.924	0.554	+normal	+normal	0.588	0.5570	G1
Student08	75	10	-0.537	1.344	-normal	+outlier	1.246	0.7357	G2
Student09	66	6	-0.975	-0.235	-normal	-normal	0.534	0.5350	G3
Student10	84	8	-0.098	0.554	-normal	+normal	0.182	0.3307	G2
Student11	77	8	-0.439	0.554	-normal	+normal	0.306	0.4197	G2
Student12	107	6	1.022	-0.235	+outlier	-normal	0.639	0.5758	G4
Student13	99	8	0.632	0.554	+normal	+normal	0.354	0.4479	G1
Student14	95	8	0.437	0.554	+normal	+normal	0.250	0.3829	G1
Student15	72	6	-0.683	-0.235	-normal	-normal	0.272	0.3981	G3
Student16	93	0	0.340	-2.603	+normal	-abnormal	3.935	0.9527	G4
Student17	86	6	-0.001	-0.235	-normal	-normal	0.031	0.1392	G3
Student18	63	6	-1.121	-0.235	-outlier	-normal	0.700	0.5973	G3
Student19	108	8	1.070	0.554	+outlier	+normal	0.741	0.6108	G1
Student20	72	6	-0.683	-0.235	-normal	-normal	0.272	0.3981	G3
Student21	138	6	2.531	-0.235	+abnormal	-normal	3.663	0.9444	G4
Student22	72	4	-0.683	-1.024	-normal	-outlier	0.764	0.6181	G3
Student23	92	6	0.291	-0.235	+normal	-normal	0.086	0.2302	G4
Student24	103	4	0.827	-1.024	+normal	-outlier	1.060	0.6968	G4
Student25	76	10	-0.488	1.344	-normal	+outlier	1.211	0.7289	G2
Student26	137	4	2.483	-1.024	+abnormal	-outlier	4.301	0.9619	G4
Student27	66	8	-0.975	0.554	-normal	+normal	0.761	0.6170	G2
Student28	61	6	-1.218	-0.235	-outlier	-normal	0.825	0.6362	G3
Student29	70	8	-0.780	0.554	-normal	+normal	0.558	0.5451	G2
Student30	94	10	0.389	1.344	+normal	+outlier	1.030	0.6897	G1
Student31	75	4	-0.537	-1.024	-normal	-outlier	0.682	0.5912	G3
Student32	99	4	0.632	-1.024	+normal	-outlier	0.879	0.6516	G4
Student33	155	10	3.359	1.344	+abnormal	+outlier	6.774	0.9907	G1
Student34	80	6	-0.293	-0.235	-normal	-normal	0.071	0.2099	G3
Student35	65	8	-1.024	0.554	-outlier	+normal	0.818	0.6343	G2
Student36	92	10	0.291	1.344	+normal	+outlier	1.007	0.6845	G1
Student37	68	6	-0.878	-0.235	-normal	-normal	0.436	0.4910	G3
Student38	93	6	0.340	-0.235	+normal	-normal	0.104	0.2530	G4
Student39	81	8	-0.245	0.554	-normal	+normal	0.219	0.3606	G2
Student40	122	6	1.752	-0.235	+outlier	-normal	1.786	0.8185	G4
Student41	75	2	-0.537	-1.814	-normal	-outlier	1.881	0.8298	G3
Student42	91	10	0.242	1.344	+normal	+outlier	1.000	0.6828	G1
Student43	73	8	-0.634	0.554	-normal	+normal	0.434	0.4901	G2
Student44	65	8	-1.024	0.554	-outlier	+normal	0.818	0.6343	G2
Student45	80	2	-0.293	-1.814	-normal	-outlier	1.818	0.8225	G3
Student46	80	10	-0.293	1.344	-normal	+outlier	1.097	0.7050	G2
Student47	82	10	-0.196	1.344	-normal	+outlier	1.056	0.6958	G2
AVERAGE	86.0	6.6	0.000	0.000					
STDEV	20.5	2.5	1.000	1.000					
Spearman	0.101								

Fig. 7.7 Scatterplot of material clickstream and quiz scores (according to *z-score*)

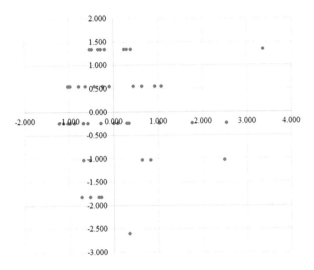

Fig. 7.8 Scatterplot according to distance from origin of MGD (according to *z-score*)

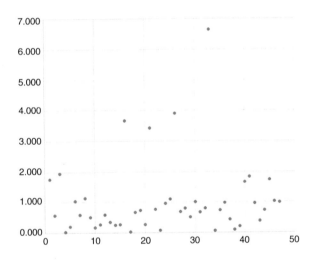

for the class were entered, but if students were absent, a blank was displayed because there were no data. In addition, the teaching material clickstream was aggregated using TSCS every week, corresponding to the time of the quiz. The characteristics of each heat map are summarized below.

Figure 7.9 shows an example of a heat map of the teaching material clickstream. In Fig. 7.9, there are five red cells that correspond to "-abnormal" values with extremely low abnormal clickstreams, and four students (7.8%) correspond. Additionally, some of the learners participated in the class more than once, but there were cases where there were no clickstream data during the class after the middle of the semester. These nine (17.6%) learners seemed to have abandoned the class in the middle of the semester, and three of them corresponded to low abnormal values.

Student	Chap1	Chap2	Chap3	Chap4	Chap5	Chap6	Chap7	Chap8	Chap9	Chap10	Chap11	Chap12	Chap13	Total
Student01	-normal	-normal	-normal	+normal	+normal			+normal	+outlier	+normal				
Student02	-outlier		-normal	-normal	-normal	-outlier	-normal	-normal	-normal			-outlier	-outlier	-outlier
Student03		-abnormal	-outlier		-abnormal									
Student04	-normal		-normal	-outlier	-normal	-outlier	-normal		-outlier	-outlier	+normal	-outlier	-normal	-outlier
Student05	+normal	-normal	+outlier	+normal	+normal	+normal	+outlier	+normal	+normal	+normal	+normal	+normal	-normal	+normal
Student06		-outlier	-normal	-outlier	-normal	-normal	-normal	-normal	-normal	-outlier	-normal	-normal	-outlier
Student07	-normal	-normal	-normal	-normal	+normal	+normal	-normal	-normal	-normal	-normal	+normal	-normal	-normal	-normal
Student08	-normal	+normal	-outlier		-outlier	-normal	-normal	-outlier	-normal	-outlier	-outlier	-outlier	-outlier	-outlier
Student09	+normal	+outlier	+abnormal	-normal	+outlier	+normal	+abnormal	+normal	+outlier	-normal	+normal	-normal	-outlier	+normal
Student10	-normal	-normal	-normal	+outlier	-normal	-normal	+normal	-outlier	-normal	-normal	+normal	-normal	+normal	-normal
Student11	-normal	-outlier	+normal	-normal	-normal	-normal	-normal	+normal	-normal			-outlier	+normal	-normal
Student12	-normal	-normal	-normal	-outlier	+normal	-normal	-normal	-normal	-normal	-normal	-normal	+normal	+normal	-normal
Student13	-normal		-normal		-abnormal	-normal	-outlier	-normal	-normal	-normal	-normal	-outlier	-normal	-normal
Student14	+outlier	+normal	+outlier	+normal	+outlier	+normal	+abnormal	+outlier	+abnormal	+outlier	+abnormal	+abnormal	+abnormal	+abnormal
Student15	+normal	+normal	+normal	-normal		-outlier	+normal	-normal	-normal	-normal	-normal	-normal	-normal	-normal
Student16	+normal	-normal	-outlier	-outlier	-normal	-outlier	-normal	-normal	-normal	-outlier	+normal	+normal	-normal
Student17	-normal	+normal		-normal	-normal	+normal	+outlier	+normal	+normal		+normal	+normal	-normal	-normal
Student18	+normal	+normal	-normal	+normal	+normal	+outlier	+normal	+outlier		-outlier	-normal	-outlier	-outlier	-normal
Student19	-normal	-normal	+normal	+outlier	+outlier	-normal	+normal	-normal	-normal	-normal	-normal	+normal	-normal	-normal
Student20	-outlier	-normal	-normal	-normal	+normal	+normal	+normal	+normal		-outlier	+normal	-normal	+normal	-normal
Student21	+outlier	-normal	+normal	-normal	-normal	-normal	-normal	-outlier	-normal	-normal	+normal	+normal	+normal
Student22	-normal	+normal	-outlier	-normal	-normal	-normal	-normal	+normal	-normal	+normal	-normal	-normal	-normal
Student23	+abnormal		-outlier	-normal	-outlier	-outlier	-normal	-normal	+normal		-outlier			
Student24	-normal	-normal	+normal	+normal	+normal	+normal	+normal	-normal	+normal	+outlier	+normal	+normal	+normal	+normal
Student25	+normal	+outlier	+normal	+normal	+normal	+normal	+normal	+normal	+normal	+normal	+normal	+normal	+outlier	+outlier
Student26	+normal	+normal			-normal	+outlier	+outlier	+normal	+outlier	+normal	+normal	+normal	+normal
Student27	-normal	+normal	+outlier	+normal	+normal	+normal	-normal	+normal	-normal	-normal	-normal	-normal	-normal	-normal
Student28	+abnormal	+outlier	+abnormal	+abnormal	+abnormal	+abnormal	+outlier	+abnormal	+abnormal	+outlier	+abnormal	+outlier	+outlier	+abnormal
Student29	-normal	+normal	+normal	-normal	-normal	+normal	-normal	+normal	-normal	-normal	-normal	+normal	+normal	+normal
Student30	-outlier	-outlier	+normal	-normal	-outlier	-normal	+normal		+normal	-normal	-normal		+normal	-normal
Student31	-normal	-normal	+normal	-normal	-normal	+normal	+normal		-normal	-outlier	-normal	+normal	-normal	-normal
Student32	+normal	-normal	-normal	-normal	+normal	-normal	-normal		-normal		-outlier	-normal	+normal	-normal
Student33	-normal	+outlier	-normal	+normal										
Student34														
Student35	+normal	+normal	-normal	-normal	-outlier	-normal	-outlier	-normal	-normal	-normal	-normal	-normal	-normal	-normal
Student36	+abnormal	+abnormal	+normal	+outlier	+normal	+normal	-normal	+outlier	+normal	-normal	+normal	+normal	+outlier	+outlier
Student37	-normal	-normal	+normal	-normal	-normal	-normal	-normal	+normal	+normal	-normal	+normal	-normal	-normal	-normal
Student38	-outlier	-outlier	-outlier	-normal	-outlier	-outlier		-normal	-normal	-outlier		-outlier	-outlier	-outlier
Student39	+normal	+outlier	+outlier	+outlier	+normal	+outlier		+outlier	+abnormal	+outlier	+outlier	+outlier	+normal	+outlier
Student40	-normal	-normal	-normal	-normal	-normal	-normal	-normal		+normal	+normal	+abnormal	-normal	-normal
Student41	+normal	-normal	+normal	+outlier	+outlier	+outlier	+outlier	+normal	+outlier	+outlier	+normal	+outlier	+outlier	+outlier
Student42	-normal	-normal	-outlier		-normal	+normal	-outlier		-outlier					
Student43	+outlier	+outlier	+normal	+abnormal	+outlier		-outlier	-normal	-normal	+normal	-outlier	-normal	+normal	+normal
Student44	-normal	+outlier		-outlier	+normal		-abnormal	-outlier	-outlier					
Student45	+normal	-normal		-outlier	+normal		+normal	+normal	+normal					
Student46	-normal	+normal		-outlier	-normal	-normal	-normal	-outlier	-normal	+normal	-normal	-normal	-normal	-normal
Student47	-outlier	+normal	+outlier	-normal	-normal	+normal	+outlier	-normal	+normal	-normal	-normal	-normal	+normal	+normal
Student48	-normal	+normal	-normal	+normal	+outlier	+abnormal		-outlier	+normal	+abnormal	+normal	+outlier	+normal	+normal
Student49	-normal	-outlier	-normal	-outlier	-outlier	-outlier	-normal	-normal	-normal	-normal	-normal	-normal	-normal	-outlier
Student50		-abnormal												
Student51	-normal	+normal	+normal	+normal	+normal	+normal	-normal	+normal	-normal	+normal	+normal	+normal	+normal	+normal

Fig. 7.9 A heat map of the teaching material clickstream (blanks are absent)

There was also one student who did not participate in classes at all from the beginning of the semester. "+abnormal" with extremely high clickstreams are indicated by green cells, and there are 25 of them, and nine students (17.6%) correspond. Among them, the most corresponded student was nine times, one student corresponded six times, and the others corresponded one to two times.

Student	Chap1	Chap2	Chap3	Chap4	Chap5	Chap6	Chap7	Chap8	Chap9	Chap10	Chap11	Chap12	Chap13	Final
Student01	-outlier		+normal	-normal	-outlier	-normal		+normal	-normal	-outlier		-normal		
Student02	-normal		-normal	-normal	+outlier	+normal	-normal	-outlier	-outlier			-outlier	-outlier	-normal
Student03		-normal	-outlier		-abnormal	-outlier						+outlier		
Student04	-outlier	-outlier	-abnormal	-outlier	-normal	-normal	-normal		-normal	-normal	-normal	-outlier	+normal	-abnormal
Student05	-normal	-outlier	+normal	+normal	+outlier	-normal	-normal	+normal	+normal	+normal	+outlier	-abnormal	-normal	+normal
Student06		+outlier	-abnormal	+normal	+normal	-normal	-normal	-outlier	+normal	+normal	-normal	+normal	+normal	-normal
Student07	-normal	+outlier	-normal	-outlier	-normal	+outlier	-abnormal	-abnormal	-outlier	-abnormal	-abnormal	+outlier	-outlier	-abnormal
Student08	+outlier	+normal	+normal		-normal	-normal	+outlier	+normal	-normal	+normal	+normal	-normal	+outlier	+normal
Student09	+normal	-normal	+normal	+outlier	+outlier	+normal	+outlier	+normal	+outlier	+normal	+outlier	+outlier	-normal	+outlier
Student10	+outlier	+normal	+outlier	-normal	-normal	-outlier	-normal	-normal	+normal	+normal	-normal	+normal	+outlier	+normal
Student11	-normal		+normal	+outlier	+normal	+normal	+normal	-normal	-normal	+normal		-normal	+outlier	+normal
Student12	+normal	-normal	+outlier	-normal	+normal	-normal	-normal	-normal	+normal	-outlier	-outlier	+normal	+normal	+outlier
Student13	+normal	+outlier	+normal		-normal	-normal	+outlier	+normal	+normal	+outlier	+outlier	+normal	+normal	+normal
Student14	-normal	+outlier	+normal	+outlier	+normal	+normal	+outlier	+outlier	+outlier	-normal	+outlier	+outlier	-normal	+outlier
Student15	+normal	+outlier	-normal	+normal		+normal	+normal	+outlier	+normal	+normal	+normal	+outlier	+outlier	+normal
Student16	+normal	-normal	+normal	-outlier	+outlier	-outlier	-normal	-normal	+normal	-normal	-normal	-normal	+normal	-normal
Student17	-normal	-outlier		+normal	+outlier	-outlier	+outlier	+outlier	+normal		+outlier	-normal	-outlier	-normal
Student18	-abnormal	-normal	+normal	-normal	-outlier	-outlier	-normal	-normal		+normal	-normal	-normal	+normal	-normal
Student19	-normal	+normal	-normal	+outlier	-outlier	-outlier	-normal	+normal	+normal	-normal	-normal	-normal	-normal	-outlier
Student20	-normal	-outlier	-normal	-outlier	-normal	-outlier	-normal	-normal		-normal	-normal	-normal	-abnormal	-outlier
Student21	+normal	-outlier	-outlier	+outlier	+normal	+outlier	-outlier	-normal	+outlier	-outlier	-normal	+normal	-abnormal	-normal
Student22	-normal	+outlier	-normal	-normal	+outlier	-outlier	-abnormal	-outlier	-normal	-normal	-outlier	-normal	+outlier	-normal
Student23	-normal	+normal	-normal	+normal	-normal	+normal	-outlier	-normal	-outlier		-normal	-normal		
Student24	-outlier	-outlier	-outlier	+normal	+outlier	+normal	-normal	-normal	-normal	+normal	+normal	-normal	-normal	-normal
Student25	-normal	-normal	+normal	+normal	+normal	+outlier	+normal	+outlier	+outlier	+normal	+outlier	-normal	+normal	+normal
Student26	-outlier	-outlier	+normal	-normal		-outlier	+normal	-normal	+normal	+normal	-normal	+normal	+normal	+normal
Student27	+outlier	-normal	-normal	+normal	-normal	-outlier	-normal	-outlier	-normal	-normal		-outlier	+normal	
Student28	-outlier	-outlier	-outlier	+normal	+normal	-outlier	+outlier	+outlier	+outlier	+normal	+outlier	+outlier	+outlier	+outlier
Student29	+normal	+normal	+normal	+normal	-normal	-normal	-normal	-normal	+normal	-outlier	-outlier	+normal	+outlier	+normal
Student30	-normal	-normal	+normal	+normal	+outlier	+normal	+normal		+outlier	+outlier	-normal	+normal	+normal	+normal
Student31	+normal	+normal	-normal	+normal	+normal	-outlier	+normal		-outlier	-outlier	+outlier	+outlier	-normal	+normal
Student32	+outlier	+outlier	-normal	+normal	+normal	-normal	+normal		-outlier		-normal	+normal	+normal	+normal
Student33	-outlier	+normal	-normal	+normal		-normal						-outlier		
Student34		+normal				+outlier						+outlier		
Student35	-outlier	-outlier	+outlier	-normal	+normal	+normal	+normal	-normal	-outlier	-outlier	-normal	-normal	-outlier	+normal
Student36	+outlier	-outlier	+normal	-normal	+outlier	+outlier	-normal	+outlier	+outlier	+normal	+normal	+normal	+normal	+normal
Student37	-normal	-normal	+outlier	+normal	+normal		+outlier	+outlier	+normal	+normal	-normal	-normal	+normal	+normal
Student38	+normal	+outlier	-outlier	+normal	+normal	-normal		-normal	-outlier	-normal		+outlier	+normal	-outlier
Student39	+outlier	+normal	+outlier	+normal	-normal	+outlier		+normal	-normal	-normal	+normal	-normal	-outlier	-normal
Student40	-normal	+normal	-outlier	+outlier	+outlier	-outlier	+outlier			-normal	+outlier	-normal	-normal	-normal
Student41	-normal	+outlier	+normal	+normal	+outlier	+outlier	+outlier	+outlier	+normal	+outlier	+outlier	-normal	+outlier	+outlier
Student42	+normal	+normal	-normal		-abnormal	+normal	+normal		-outlier			+normal		
Student43	-normal		+normal	-normal	+normal		-normal	-normal	+normal	+normal	-outlier		+outlier	+normal
Student44	-outlier		-normal	+normal		-normal	-normal	-normal						
Student45	+outlier			-normal	+normal		+outlier	-normal	-normal					
Student46	+normal			+normal	+normal		+outlier	-normal	+outlier	+normal	-normal		-normal	+normal
Student47	+normal		-normal	+normal	+normal		+outlier	+outlier	+normal	+normal	+outlier		+normal	+normal
Student48	-outlier		-normal	+normal	+normal			-outlier	-outlier	+normal	-normal		-outlier	+normal
Student49	+outlier		+normal	+normal	+outlier		+outlier	+outlier	-normal	+normal	+normal		+outlier	+normal
Student50														
Student51	+outlier		-normal	-normal	+normal		-normal	-normal	+normal	+normal	+normal		+normal	-normal

Fig. 7.10 Heat map example of quiz scores (blanks are absent)

Next, Fig. 7.10 shows an example of a heat map of the quiz scores. Abnormal with extremely low quiz scores, "−abnormal," are indicated by the red cells, 15 of which corresponded, and 5 of which were from the same learner. The score of this learner was zero points when it corresponded to the abnormal value. Since the quiz score in the class had a constant upper limit of 10 points, abnormal values with high scores are less likely to occur and did not appear in Fig. 7.10. Compared to the clickstream heat map, clickstreams do not have an upper limit, so they tended to have high abnormal values, but low abnormal values also occurred.

Student	Chap1	Chap2	Chap3	Chap4	Chap5	Chap6	Chap7	Chap8	Chap9	Chap10	Chap11	Chap12	Chap13	Final	G1	G2	G3	G4	Abse	Abno	Click	Final
Student14	G4	G4	G1	G1	G1	G4	G1Ab	G1	G1Ab	G4	G1Ab	G1Ab	G4Ab	G1Ab	9	0	0	5	0	6	1084	28
Student25	G4	G1	G1	G1	G1	G4	G1	G1	G1	G1	G1	G4	G1	G1	11	0	0	3	0	0	770	22
Student26	G4	G1	G1	G4		G2	G1	G4	G1	G1	G4	G4	G1	G1	7	1	0	5	1	0	746	25
Student28	G4Ab	G4	G4Ab	G1Ab	G1Ab	G1Ab	G1	G1Ab	G1Ab	G1	G1Ab	G4	G1	G1Ab	10	0	0	4	0	9	1088	27
Student34															0	0	0	0	14	0	96	-
Student39	G1	G1	G1	G1	G4	G1		G1	G4Ab	G1	G1	G1	G4	G4	9	0	0	4	1	1	811	20
Student09	G1	G1	G1Ab	G2	G1	G1	G1Ab	G1	G1	G2	G1	G2	G3	G1	10	3	1	0	0	2	693	27
Student15	G1	G1	G4	G2		G3	G1	G2	G2	G2	G2	G2	G2	G2	3	8	1	1	1	0	550	26
Student36	G1Ab	G1Ab	G1	G4	G1	G4	G3	G1	G1	G2	G1	G1	G1	G1	10	1	1	2	0	2	876	23
Student41	G4		G1	G1	G1	G1	G1	G1	G1	G1	G1	G1	G1	G1	12	0	1	1	0	0	838	28
Student45	G1	G2		G3	G1		G1	G4	G4						3	1	1	2	7	0	294	-
Student47	G2	G1	G4	G2	G2	G1	G1	G2	G1	G2	G2	G3	G1	G1	6	6	1	1	0	0	609	24
Student50		G3Ab													0	0	1	0	13	1	73	-
Student51	G2	G1	G4	G4	G1	G1	G3	G4	G2	G1	G1	G1	G1	G4	7	2	1	4	0	0	655	19
Student01	G3	G3	G2	G4	G4		G1	G4	G4						1	1	2	4	6	0	397	-
Student05	G4	G3	G1	G1	G1	G4	G4	G1	G1	G1	G1	G1	G3	G1	9	0	2	3	0	0	686	23
Student17	G3	G1		G2	G2	G1	G1	G1	G1		G1	G1	G3	G2	7	3	2	0	2	0	577	25
Student33	G3	G1	G3	G1											2	0	2	0	10	0	220	-
Student37	G3	G2	G1	G2	G2	G3	G2	G1	G1	G2	G4	G2	G2	G2	3	8	2	1	0	0	587	23
Student49	G2	G2	G2	G2	G2	G2	G2	G2	G3	G2	G2	G3	G2	G2	0	12	2	0	0	0	416	24
Student03		G3Ab	G3		G3Ab										0	0	3	0	11	2	117	-
Student11	G2	G3	G1	G2	G2	G2	G3	G4	G2		G2	G1	G2		2	7	3	1	1	0	527	26
Student13	G2		G1	G2	G3Ab	G3	G2	G2	G2	G2	G1	G3	G2	G2	1	8	3	0	2	1	449	24
Student18	G4Ab	G4	G2	G4	G4	G4	G4	G4		G2	G3	G3	G2	G3	0	3	3	7	1	1	564	17
Student24	G3	G3	G4	G1	G1	G1	G4	G3	G4	G1	G1	G1	G4	G4	6	0	3	5	0	0	649	20
Student48	G3	G1	G3	G1	G4Ab		G3	G4	G1Ab	G4	G4	G4	G1		5	0	3	5	1	2	727	23
Student08	G2	G1	G2		G3	G3	G2	G2	G3	G2	G2	G3	G2	G2	1	8	4	0	1	0	429	26
Student10	G2	G2	G2	G4	G3	G3	G4	G3	G2	G2	G4	G3	G1	G2	1	6	4	3	0	0	562	22
Student21	G1	G2	G4	G2	G1	G3	G4	G3	G2	G3	G3	G4	G4Ab	G4	2	3	4	5	0	1	631	19
Student29	G2	G4	G1	G2	G4	G3	G4	G2	G2	G3	G1	G1	G1		4	3	4	3	0	0	617	22
Student30	G3	G3	G1	G2	G2	G3	G1		G1	G2	G3		G1	G2	4	4	4	0	2	0	515	26
Student32	G1	G2	G3	G2	G1	G3	G1		G3		G3	G2	G1	G2	4	4	4	0	2	0	535	22
Student42	G2	G3	G3		G3Ab	G1	G2		G3						1	2	4	0	7	1	297	-
Student43	G4	G4	G1	G4Ab	G1		G3	G3	G2	G1	G3		G1	G1	5	1	4	3	1	1	701	24
Student46	G2	G1		G2	G2	G3	G2	G3	G2	G1	G3	G2	G3	G2	2	7	4	0	1	0	490	25
Student12	G2	G2	G2	G3	G1	G4	G3	G3	G2	G3	G1	G1	G2		3	5	5	1	0	0	587	27
Student23	G4Ab		G3	G2	G2	G2	G3	G3	G4		G3				0	2	5	2	5	1	385	-
Student40	G3	G3	G3	G1	G2	G2	G2		G4	G1	G1Ab	G3	G3		3	3	5	1	2	1	527	18
Student44	G3	G4		G3	G1		G3Ab	G3	G3						1	0	5	1	7	1	242	-
Student16	G1	G2	G2	G3	G2	G2	G3	G3	G2	G3	G3	G1	G1	G3	3	5	6	0	0	0	559	21
Student27	G2	G2	G4	G1	G4	G1	G1	G3	G1	G4	G3	G3	G3	G3	3	1	6	4	0	0	567	19
Student31	G2	G3	G4	G2	G2	G4	G1		G3	G3	G1	G3	G3	G3	2	3	6	2	1	0	535	21
Student35	G4	G4	G2	G3	G2	G2	G2	G3	G3	G3	G2	G3	G2		0	6	6	2	0	0	493	23
Student20	G3	G3	G3	G3	G4	G4	G4	G4		G3	G4	G3	G4Ab	G3	0	0	7	6	1	1	559	14
Student38	G2	G2	G2	G2	G4	G2	G3		G3	G3		G3	G2	G3	0	5	7	0	2	0	350	15
Student19	G3	G3	G4	G1	G1	G4		G4	G2	G3	G3	G4	G3	G3	2	1	8	3	0	0	571	15
Student04	G3		G3Ab	G3	G3	G2	G3		G3	G3	G4		G2	G3	0	2	9	1	2	1	415	12
Student06		G3	G3Ab	G2	G2	G3	G3	G3	G2	G3	G3	G2	G3		0	4	9	0	1	1	429	17
Student22	G3	G4	G3	G3	G2	G3	G3Ab	G4	G3		G4	G3	G3	G3	0	2	9	3	0	1	528	17
Student02	G3		G3	G3	G2	G3	G3	G3		G3		G3		G3	0	1	10	0	3	0	382	18
Student07	G3	G3	G3	G3	G4	G4	G3Ab	G3Ab	G3	G3Ab	G4Ab	G3Ab	G3	G3Ab	0	0	11	3	0	6	494	8

Fig. 7.11 Example of a sorted heat map of learning patterns (right side shows pattern appearance frequency, material clickstream, and final quiz score data)

Figure 7.11 shows an example of a cluster heat map of learning patterns created from the teaching material clickstream and the *z-score* of the quiz score. This cluster heat map determined learning patterns from combinations of positive and negative signs of *z-scores*. In addition, the letters "Ab" were manually added to data with a probability distribution of less than 5% for either the teaching material clickstream or the quiz score. On the right side of Fig. 7.11, the numbers of occurrences of patterns

G1–G4—absence (Abse), abnormal value (Abnor), total material clickstreams (Click), and final quiz scores (Final)—are tabulated.

Figure 7.11 is the result of sorting the heat map in ascending order by the value of Pattern G3 after creating it. This sorting makes it easier for learners who often fall under G3 to gather at the bottom of the heat map, and learners who fall under G1 more often to gather at the top of the heat map. And many learners who correspond to G2 or G4 are gathered in the central part of the heat map. From Figs. 7.9 to 7.11, since the Student ID is made to match, the learning pattern can be confirmed in each figure.

In Fig. 7.11, Student07 at the bottom corresponds only to G3 or G4, and data indicated that this student rarely opened the teaching materials and repeatedly scored low on the quizzes. In addition to this learner, there were many learners who repeatedly corresponded to G3 at the bottom of the heat map. G3 indicates a lower-than-average clickstream for material and a lower-than-average quiz score. In Fig. 7.11, Student 7 corresponded to G3Ab five times and G4Ab once. These results corresponded to the abnormal values of patterns G3 and G4, respectively, but both were cases where the quiz score was zero points.

In this way, learning patterns G2 and G4 in Fig. 7.11 are also sorted, and by comparing the relationship between the material clickstream and the quiz score heat map, it is possible to grasp the characteristics of each learning pattern.

7.5   Discussion

7.5.1   Effect of Real-Time Sync Ratio On-Site

In recent versions of Moodle, a function called Course Participant has been added, and it can be used as a function to aggregate a learner's teaching material clickstreams. The date and time to be aggregated are specified as a day, week, or month, such as 1 day ago or 1 week ago, and the number of times the learner opened the material during that time can be displayed numerically. However, this function cannot be used in real time for lessons currently in progress. RTTSCS, which is the basis of this research, compiles on-site teaching material clickstreams in real time. The benefit to this system is that the actions of all learners are displayed in a graph, and the clickstreams of individual learners are displayed in a tabular format. In addition, this system provides an advantage by displaying not only the result of the number of times the teaching material was opened but also the process in progress.

If the process data can be obtained as shown in this research, it would be easier to grasp the reactions of individual learners and it would be possible to quantify them as the sync ratio. The analysis results of RTTSCS make it easy for a user to understand how many learners are responding to a teacher's instructions at a given time. While confirming learners' responses in real time using tables and graphs, it is possible to flexibly consider the speed of progress by adding ingenuity to the way a teacher explains the material.

A system that shows the sync ratio of opening teaching materials is currently in the testing stage and was actually used in a class that the author led. From the start of the program to the generation of tables and graphs, it takes about 40–45 s to check the sync ratio in real time. However, when the class was busy, there were many times when the sync ratio could not be confirmed. When checking the sync ratio for each individual learner, it is necessary to operate the pivot table with a mouse or keyboard, which may take time. In addition, in order to reduce the burden on teachers in the classroom, web scraping technology is used to automate the operation of Moodle, and the burden of operating the system is being reduced. However, some teachers may not be familiar with LA or computers, so it is necessary to consider a method that is less burdensome, in addition to devising an easy-to-use and easy-to-see user interface.

7.5.2 Sync Ratio and Learner's Reaction

The learners often opened the materials earlier than the teacher instructed them to do so, but other teachers were of the opinion that this occurrence was not a problem. However, when the sync ratio is low, a teacher may have to slow down the progress of the lesson and instruct the learners who do not open the teaching materials to do so. In addition, in a physical classroom and depending on the situation, a teacher may consider measures such as patrolling among the learners' desks and disclosing the sync ratio to the learners.

Moreover, the timelines in Figs. 7.5 and 7.6 show that the sync ratios were every minute. Therefore, even if students opened the teaching material a few seconds earlier or a few seconds later than the teacher's instruction, it would be counted within the adjacent time zone. Therefore, the teacher should wait at least 1 min after giving instructions before moving on. For this reason, it is desirable to adopt not only 1-min but also 2-min, 3-min, and other intervals that are most suitable for the class style.

For both classroom and online remote lectures, learner IDs and names are also used in the sync ratio analysis. Therefore, in order to protect personal information, it is appropriate to prepare a dedicated personal computer for displaying the sync ratio separately from the personal computer for demonstration purposes.

7.5.3 Statistical Analysis Between Engagement Heat Map and Learning Patterns

From the heat map in Fig. 7.11, it was able to aggregate the frequency of the appearance of learning patterns, absences, and abnormal values for each learner from G1 to G4. If there was a correlation among these data, the characteristics of the learning pattern could be explained in more detail. It is also important to determine whether the frequency of occurrence of learning patterns was related to teaching material clickstreams and final quiz scores for the entire semester. Therefore, Table 7.2 shows learning patterns and material clickstreams, and Table 7.3 shows correlation coefficients between learning patterns and final quiz scores. When calculating the correlation coefficient, Spearman's rank correlation coefficient was used in this chapter, and zeros included in the appearance frequency of the learning pattern were deleted. The results of these two analyses are shown in the Appendix.

Because clickstreams have no upper limit, the numbers vary widely, which is a factor in the stronger correlation. On the other hand, the final quiz score has a fixed upper limit, so the score tends to concentrate in a narrower range than the clickstream. Thus, it seems that the correlation coefficient is low. This difference in correlation coefficient may be due to the presence of some learners with relatively high final quiz scores, even though their clickstreams were close to the average value (604.2) among the learners of Pattern G1. Conversely, even if there were relatively many clickstreams, there were multiple learners whose final quiz score was close to the average value (21.5). Considering that the class used the reading material, it would be desirable to have a correlation between the final quiz score and Pattern G1; however, if many learners with prerequisite knowledge took the course, the correlation coefficient likely would be low.

Pattern G2 had a negative correlation with the clickstream, but a weak positive correlation with the final quiz score. This seems to indicate that learners who corresponded to Pattern G2 may have been able to increase their final quiz score by opening more teaching materials. Pattern G3 had a negative correlation with both clickstream and final quiz score. Therefore, learners who fell into G3 tended to have lower final quiz scores if they had fewer clickstreams. By creating a heat map after the weekly quiz, they were able to find learners who fell into G3 early. At the bottom of Fig. 7.11, the learners who corresponded most to G3 are displayed, and these students were international students from overseas.

Pattern G4 had $p > 0.05$ for both clickstream and final quiz scores, so no correlation was observed. However, some of those who corresponded to G1 may also have corresponded to G4, indicating that quiz scores were not always high, even when clickstreams were high.

Absence was negatively correlated with clickstreams, which was an expected correlation due to nonattendance. However, there was no correlation between absences and the final quiz score, indicating that absenteeism did not necessarily result in low scores (see the Appendix).

For abnormalities, correlations between both clickstreams and final quiz scores were identified. Although the number of abnormalities in this sample data was only 16, it was assumed that the correlation was likely to emerge due to the relatively large variation in the numerical values. The final quiz score tended to be high when the clickstream was extremely high, and the final quiz score tended to be abnormally low when the clickstream was extremely low.

In the examples of learning patterns via classification in this study, both the clickstream and the final quiz score were shown to be negatively correlated with

Pattern G3. The bottom of the heat map in Fig. 7.11 shows that there were many learners who corresponded to Pattern G3, but there were also a few cases in which Patterns G1 and G4 applied. Learners who fell under Pattern G3 tended to open less material, so both their clickstreams and quiz scores were lower than average. Learners who repeatedly corresponded to G3 tended to struggle in class, so appropriate teacher guidance would be particularly important for these students.

In addition, in Fig. 7.11, nine students who seemed to have abandoned class in the middle of the semester did not have final quiz score data, and four of them were identified as Pattern G3 outliers. This indicated that such learners need to be addressed early, and it is important to take measures to analyze learning patterns early after each week's class. In addition, there were many learners whose final quiz score corresponded to G1 or G2, even when they corresponded to the abnormal values of Patterns G3 and G4. It is also necessary to consider visualizing and comparing the clickstreams for successful learners with those of learners with abnormal values and to consider a support system for that purpose.

In the clickstreams in this study, extremely large outliers occurred in weekly classes, but none of the learners had low final quiz scores. There was a weak positive correlation between the total clickstream and the final quiz score ($r = 0.353$, $p = 0.011$, $p < 0.05$). Additionally, among the final test takers, there were no learners with low clickstream outliers throughout the period, and only two with high outliers. Furthermore, in both the weekly quiz scores and the final quiz score, there were no extremely high abnormal scores, only low abnormal scores. This result can be explained by the fact that the upper limit of the quiz score was fixed, so if the number of high scorers increases, then the average score increases. Thus, it is likely that abnormal values do not appear.

One possible reason for this outcome is that there were learners who were waiting for the prerequisite knowledge. In other words, the content of the class and the degree of difficulty of the final quiz may have been easy for some learners, which may be a factor influencing those who fell under Pattern G2. In addition, it is possible that the number of learners who understood the content of the class increased, and this raised their quiz scores.

Additionally, at the top of the heat map in Fig. 7.11, the learners who often corresponded to Patterns G1 and G4 are displayed. It is possible that these learners corresponded to Pattern G1 when they read and understood the material well and to Pattern G4 when they did not fully understand it. Figure 7.11 shows that many learners repeated the movements of Patterns G1 to G4. In addition, depending on the appearance frequency of each learning pattern, the final quiz score of learners who frequently corresponded to Pattern G1 tended to be high, and the final quiz score of learners who frequently corresponded to Pattern G3 tended to be low.

Moreover, since this study examined the sync ratio of opening teaching materials, it may be useful to analyze the sync ratio during class for learners who fall under low abnormal values. If classes are conducted using appropriate reading materials that consider the learner's level, many classes will show the same tendency as Pattern G3.

7.5.4 Limitations of This Research

The study analyzed quiz scores and teaching material clickstreams directly related to the quiz. Quiz scores were collected for all test takers. However, since learners could freely download the teaching materials from Moodle, it is impossible to tally the clickstreams of those who opened the materials after downloading. Furthermore, since the quiz was held 1 week after the class, the quiz score may have been influenced not only by the course materials on Moodle but also by the books or websites that the learners used for independent study.

Some studies have pointed out that previous clickstream studies contained uncertainties that were difficult to measure [46], and this observation also applied to our study. In the classroom used in this research, a monitor was set up to project the teaching materials and the teacher's demonstration, so there were many cases where learners only looked at this monitor and did not open the teaching materials on their own computers. Such cases also cannot be included in the teaching material clickstream.

In order to reproduce the contents of this research, it is necessary to prepare online teaching materials, conduct classes, conduct quizzes, preprocess learning logs, and operate pivot tables. The problem is that this requires a considerable amount of work time. However, when the class style is the same as that in this study, there is a strong possibility that the method of this research can be applied to other subjects.

Regarding the correlation analysis of learning patterns, if the number of target learners is small, the analysis results may be affected, and the class content and the difficulty of the quiz also affect the correlation and the occurrence of abnormal values. In this study, classification via normal distribution and MGD was attempted, but there was a slight difference between the two methods in extracting outliers. Since there are differences in calculation methods for classification, it seems necessary to conduct comparative research with other methods such as *k-means* and hierarchical clustering. *k-means* and hierarchical clustering use the MGD, which is the same as the method in this research. However, even with these clustering methods, optimal classification may not always be performed, and manual reexamination is required especially for classification near boundaries. In addition, *k-means* and hierarchical clustering do not have outlier detection functions. Therefore, when performing outlier detection with *k-means* and hierarchical clustering, some ingenuity would be required.

7.6 Conclusion

Learning analytics (LA) is now required in many classes, and this research shows a collection method and analysis examples of learning logs in actual classes. This study presented an example to the teacher in real time, mainly in the forms of tables and graphs, about the sync ratio of students viewing teaching materials during class. In addition, the study showed an example of creating a heat map from the classification of learning patterns by clickstreams and quiz scores and discovering outliers by MGD.

For the sync ratio of viewing teaching materials, this study used a system that integrated Moodle and Excel pivot tables to visualize the analysis results in real time.

In the sync ratio tables and graphs, the material clickstream was displayed for each learner every minute. Furthermore, the table of sync ratio by material showed learners who opened the material later than the teacher's instruction. By conducting lessons while viewing these analysis results, the teacher could match his or her instruction to the learning situation of each student.

For heat maps and outlier detection based on learning patterns, clickstreams and quiz scores were classified by the three-sigma method of normal distribution. Furthermore, outlier detection was performed using z-score and MGD. This research found that the learning patterns and the appearance of abnormal values were affected by the learner's prior knowledge, the contents of the teaching materials, and the difficulty of the quiz. In particular, learners whose quiz scores fell under Pattern G3 and who repeatedly fell under abnormal values were likely to struggle in class. Teachers should consider early intervention for such learners. When teachers employ such interventions, they can utilize the indicators presented in this research, such as clickstreams, for browsing teaching materials in class, quiz scores, and learning patterns [73].

In addition, there was no correlation between the final quiz score and Pattern G1 in this experiment. In classes using reading materials, it is important to maintain the correlation by preparing content at an appropriate level for learners so that quiz scores follow a normal distribution. The content discussed in this study was eluci-dated by analyzing learning logs using Moodle. If learning analysis like that used in this research is performed early, individual learning behaviors that are difficult to notice in the classroom can be clarified, which can be used to improve lessons in the future.

Acknowledgments This work was supported by JSPS KAKENHI (Grant numbers 18K11588 and 21K12183).

Appendix

Table 7.2 Correlation coefficient between clickstream and learning pattern

	G1	G2	G3	G4	Absence	Abnormal
Correlation coefficient	0.743	−0.402	−0.556	0.278	−0.507	0.620
N	32	32	37	30	20	15
t-value	6.072	2.406	3.962	1.530	2.498	2.849
Degrees of freedom	30	30	35	28	18	13
p-value	0.000	0.011	0.000	0.069	0.011	0.007
	$p < 0.05$	$p < 0.05$	$p < 0.05$	$p > 0.05$	$p < 0.05$	$p < 0.05$

Table 7.3 Correlation coefficient between final quiz score and learning pattern

	G1	G2	G3	G4	Absence	Abnormal
Correlation coefficient	0.116	0.403	−0.614	−0.252	−0.212	0.437
N	32	32	37	29	20	16
t-value	0.640	2.415	4.596	1.352	0.918	1.818
Degrees of freedom	30	30	35	27	18	14
p-value	0.264	0.011	0.000	0.094	0.185	0.045
	$p > 0.05$	$p < 0.05$	$p < 0.05$	$p > 0.05$	$p > 0.05$	$p < 0.05$

References

1. Bachhal, P., Ahuja, S., Gargrish, S.: Educational data mining: a review. J. Phys. Conf. Ser. **1950**(1), 012022 (2021)
2. Hernández-de-Menéndez, M., Morales-Menendez, R., Escobar, C.A., Ramírez Mendoza, R.A.: Learning analytics: state of the art. In: International Journal on Interactive Design and Manufacturing (IJIDeM), pp. 1–22 (2022)
3. Yassine, S., Kadry, S., Sicilia, M.A.: A framework for learning analytics in Moodle for assessing course outcomes. In: 2016 IEEE Global Engineering Education Conference (EDUCON), pp. 261–266. IEEE (2016)
4. Paudel, P.: Online education: benefits, challenges and strategies during and after COVID-19 in higher education. Int. J. Stud. Educ. **3**(2), 70–85 (2021)
5. Vaidya, A., Saini, J.R.: A framework for implementation of learning analytics and educational data mining in traditional learning environment. In: ICT Analysis and Applications, pp. 105–114. Springer (2021)
6. Oguguo, B.C., Nannim, F.A., Agah, J.J., Ugwuanyi, C.S., Ene, C.U., Nzeadibe, A.C.: Effect of learning management system on student's performance in educational measurement and evaluation. Educ. Inf. Technol. **26**(2), 1471–1483 (2021)
7. Kokoç, M., Altun, A.: Effects of learner interaction with learning dashboards on academic performance in an e-learning environment. Behav. Inf. Technol. **40**(2), 161–175 (2021)
8. Marticorena-Sánchez, R., López-Nozal, C., Ji, Y.P., Pardo-Aguilar, C., Arnaiz-González, Á.: UBUMonitor: an open-source desktop application for visual e-learning analysis with Moodle. Electronics. **11**(6), 954 (2022)
9. Dougiamas, M.: Moodle: a virtual learning environment for the rest of us. TESL-EJ. **8**(2), 1–8 (2004)
10. Romero, C., Ventura, S.: Educational data mining: a review of the state of the art. IEEE Trans. Syst. Man Cyber. C. **40**(6), 601–618 (2010)
11. Verbert, K., Duval, E., Klerkx, J., Govaerts, S., Santos, J.L.: Learning analytics dashboard applications. Am. Behav. Sci. **57**(10), 1500–1509 (2013)
12. Dierenfeld, H., Merceron, A.: Learning analytics with Excel pivot tables. In: Proceedings of the 1st Moodle Research Conference (MRC2012), pp. 115–121 (2012)
13. Konstantinidis, A., Grafton, C.: Using Excel macros to analyse Moodle logs. In: 2nd Moodle Research Conference (MRC2013), pp. 33–39 (2013)
14. Dobashi, K.: Time series analysis of the in class page view history of digital teaching materials using cross table. Proc. Comput. Sci. **60**, 1032–1040 (2015)
15. Dobashi, K.: Development and trial of Excel macros for time series cross section monitoring of student engagement: analyzing students' page views of course materials. Proc. Comput. Sci. **96**, 1086–1095 (2016)

16. Dobashi, K., Ho, C.P., Fulford, C.P., Lin, M.F.G.: A heat map generation to visualize engagement in classes using Moodle learning logs. In: 2019 4th International Conference on Information Technology (InCIT), pp. 138–143. IEEE (2019)
17. Dobashi, K., Ho, C.P., Fulford, C.P., Lin, M.F.G., Higa, C.: Synchronization ratio of time-series cross-section and teaching material clickstream for visualization of student engagement. In: International Conference on Artificial Intelligence in Education, pp. 125–131. Springer, Cham (2022)
18. Bradley, V.M.: Learning management system (LMS) use with online instruction. Int. J. Technol. Educ. 4(1), 68–92 (2021)
19. Yogev, E., Gal, K., Karger, D., Facciotti, M.T., Igo, M.: Classifying and visualizing students' cognitive engagement in course readings. In: Proceedings of the Fifth Annual ACM Conference on Learning at Scale, pp. 1–10 (2018)
20. Coffrin, C., Corrin, L., de Barba, P., Kennedy, G.: Visualizing patterns of student engagement and performance in MOOCs. In: Proceedings of the Fourth International Conference on Learning Analytics and Knowledge, pp. 83–92 (2014)
21. Cenka, B.A.N., Santoso, H.B., Junus, K.: Analysing student behaviour in a learning management system using a process mining approach. Knowl. Manage. E-Learn. Int. J. 14(1), 62–80 (2022)
22. Beck, N.: Time-series–cross-section data: what have we learned in the past few years? Ann. Rev. Polit. Sci. 4(1), 271–293 (2001)
23. Beck, N., Katz, J.N.: Modeling dynamics in time-series–cross-section political economy data. Ann. Rev. Polit. Sci. 14, 331–352 (2011)
24. Jo, Y., Maki, K., Tomar, G.: Time series analysis of clickstream logs from online courses. arXiv, 13 pages (2018). https://doi.org/10.48550/arXiv.1809.04177
25. Sadagopan, N., Li, J.: Characterizing typical and atypical user sessions in clickstreams. In: Proceedings of the 17th International Conference on World Wide Web, pp. 885–894 (2008)
26. Wang, G., Konolige, T., Wilson, C., Wang, X., Zheng, H., Zhao, B.Y.: You are how you click: clickstream analysis for sybil detection. In: 22nd USENIX Security Symposium (USENIX Security 13), pp. 241–256 (2013)
27. Dobashi, K.: Interactive mining for learning analytics by automated generation of pivot table. In: International Conference on Applied Human Factors and Ergonomics, pp. 66–77. Springer, Cham (2018)
28. Vermunt, J. D., & Vermetten, Y. J.: Patterns in student learning: Relationships between learning strategies, conceptions of learning, and learning orientations. Educ. Psychol. Rev. 16(4), 359-384 (2004)
29. Vermunt, J.D., Donche, V.: A learning patterns perspective on student learning in higher education: state of the art and moving forward. Educ. Psychol. Rev. 29(2), 269–299 (2017)
30. Hsiao, C.C., Huang, J.C., Huang, A.Y., Lu, O.H., Yin, C.J., Yang, S.J.: Exploring the effects of online learning behaviors on short-term and long-term learning outcomes in flipped classrooms. Interact. Learn. Environ. 27(8), 1160–1177 (2019)
31. Matayoshi, J., Cosyn, E.: Identifying student learning patterns with semi-supervised machine learning models. In: Proceedings of the 26th International Conference on Computers in Education, pp. 11–20 (2018)
32. Mouri, K., Ren, Z., Uosaki, N., Yin, C.: Analyzing learning patterns based on log data from digital textbooks. Int. J Dist. Educ. Technol. 17(1), 1–14 (2019)
33. Yin, C., Yamada, M., Oi, M., Shimada, A., Okubo, F., Kojima, K., Ogata, H.: Exploring the relationships between reading behavior patterns and learning outcomes based on log data from e-books: a human factor approach. Int. J. Hum. Comput. Interact. 35(4–5), 313–322 (2019)
34. Li, L., Uosaki, N., Ogata, H., Mouri, K., Yin, C.: Analysis of behavior sequences of students by using learning logs of digital books. In: Proceedings of 26th International Conference on Computers in Education, Manila, Philippines, pp. 26–30 (2018)
35. Elbadrawy, A., Polyzou, A., Ren, Z., Sweeney, M., Karypis, G., Rangwala, H.: Predicting student performance using personalized analytics. Computer. 49(4), 61–69 (2016)

36. Simon, C., Bugusa, Y.: Survey on data mining approach for analysis and prediction of student performance. Int. J. Eng. Technol. **7**(4.5), 467–470 (2018)
37. Chu, Y. W., Tenorio, E., Cruz, L., Douglas, K., Lan, A.S., Brinton, C.G.: Click-based student performance prediction: a clustering guided meta-learning approach. In: 2021 IEEE International Conference on Big Data (Big Data), pp. 1389–1398. IEEE (2021)
38. Pytlarz I., Pu S., Patel M., & Prabhu R.: What can we learn from college students' network transactions? Constructing useful features for student success prediction. In: Proceedings of the 11th International Conference on Educational Data Mining, pp. 444–448 (2018)
39. Chen, C., Sonnert, G., Sadler, P.M., Sasselov, D.D., Fredericks, C., Malan, D.J.: Going over the cliff: MOOC dropout behavior at chapter transition. Dist. Educ. **41**(1), 6–25 (2020)
40. Chen, J., Feng, J., Sun, X., Wu, N., Yang, Z., Chen, S.: MOOC dropout prediction using a hybrid algorithm based on decision tree and extreme learning machine. Math. Prob. Eng. **2019**, Article ID 8404653, 11 pages (2019). https://doi.org/10.1155/2019/8404653
41. Dalipi, F., Imran, A.S., Kastrati, Z.: MOOC dropout prediction using machine learning techniques: review and research challenges. In: 2018 IEEE Global Engineering Education Conference (EDUCON), pp. 1007–1014. IEEE (2018)
42. Gitinabard, N., Khoshnevisan, F., Lynch, C.F., Wang, E.Y.: Your actions or your associates? Predicting certification and dropout in MOOCs with behavioral and social features. arXiv, 7 pages (2018). https://doi.org/10.48550/arXiv.1809.00052
43. Aldowah, H., Al-Samarraie, H., Fauzy, W.M.: Educational data mining and learning analytics for 21st century higher education: a review and synthesis. Tele. Inf. **37**, 13–49 (2019)
44. Deng, H., et al.: Performances: visual analytics of student performance data from an introductory chemistry course. Vis. Inf. **3**(4), 166–176 (2019)
45. Fischer, C., et al.: Mining big data in education: affordances and challenges. Rev. Res. Educ. **44**(1), 130–160 (2020)
46. Baker, R., et al.: The benefits and caveats of using clickstream data to understand student self-regulatory behaviors: opening the black box of learning processes. Int. J. Educ. Tech. High. Educ. **17**(1), 1–24 (2020)
47. Bogarín, A., Cerezo, R., Romero, C.: A survey on educational process mining. Wiley Interdisc. Rev. Data Min. Knowl. Discov. **8**(1), e1230 (2018)
48. Romero, C., Ventura, S.: Educational data mining and learning analytics: an updated survey. Wiley Interdisc. Rev. Data Min. Know. Discov. **10**(3), e1355 (2020). https://doi.org/10.1002/widm.1355
49. Estacio, R.R., Raga Jr., R.C.: Analyzing students online learning behavior in blended courses using Moodle. Asian Assoc. Open Univ. J. **12**(1), 52–68 (2017)
50. Aguilar, S.J., Karabenick, S.A., Teasley, S.D., Baek, C.: Associations between learning analytics dashboard exposure and motivation and self-regulated learning. Comput. Educ. **162**, 104085 (2021)
51. Bennett, L., Folley, S.: Students' emotional reactions to social comparison via a learner dashboard. In: Sahin, M., Ifenthaler, D. (eds.) Visualizations and Dashboards for Learning Analytics, pp. 233–249. Springer (2021)
52. Duan, X., Wang, C., Rouamba, G.: Designing a learning analytics dashboard to provide students with actionable feedback and evaluating its impacts. CSEDU. **2**, 117–127 (2022)
53. Susnjak, T., Ramaswami, G.S., Mathrani, A.: Learning analytics dashboard: a tool for providing actionable insights to learners. Int. J. Educ. Technol. High. Educ. **19**(1), 1–23 (2022)
54. Mazza, R., & Milani, C.: Gismo: a graphical interactive student monitoring tool for course management systems. In: International Conference on Technology Enhanced Learning, Milan, pp. 1–8 (2004)
55. Mazza, R., & Botturi, L.: Monitoring an online course with the GISMO tool: A case study. Journal of Interactive Learning Research, 18(2), 251–265 (2007)
56. Slater, S., Joksimović, S., Kovanovic, V., Baker, R.S., Gasevic, D.: Tools for educational data mining: a review. J. Educ. Behav. Stat. **42**(1), 85–106 (2017)

57. Metsalu, T., Vilo, J.: ClustVis: a web tool for visualizing clustering of multivariate data using principal component analysis and heatmap. Nucleic Acids Res. **43**(W1), W566–W570 (2015)
58. Wilkinson, L., Friendly, M.: The history of the cluster heat map. Am. Stat. **63**(2), 179–184 (2009)
59. Babicki, S., Arndt, D., Marcu, A., Liang, Y., Grant, J.R., Maciejewski, A., Wishart, D.S.: Heatmapper: web-enabled heat mapping for all. Nucleic Acids Res. **44**(W1), W147–W153 (2016)
60. Halverson, L.R., Graham, C.R.: Learner engagement in blended learning environments: a conceptual framework. Online Learn. **23**(2), 145–178 (2019)
61. Henrie, C.R., Halverson, L.R., Graham, C.R.: Measuring student engagement in technology-mediated learning: a review. Comput. Educ. **90**, 36–53 (2015)
62. Redmond, P., Abawi, L.A., Brown, A., Henderson, R., Heffernan, A.: An online engagement framework for higher education. Online Learn. **22**(1), 183–204 (2018)
63. Hussain, M., Zhu, W., Zhang, W., Abidi, S.M.R.: Student engagement predictions in an e-learning system and their impact on student course assessment scores. Comput. Intell. Neurosci. **2018**, Article ID 6347186 (2018). https://doi.org/10.1155/2018/6347186
64. Sinatra, G.M., Heddy, B.C., Lombardi, D.: The challenges of defining and measuring student engagement in science. Educ. Psychol. **50**(1), 1–13 (2015)
65. Schaufeli, W. B., Salanova, M., González-Romá, V., & Bakker, A. B.: The measurement of engagement and burnout: A two sample confirmatory factor analytic approach. Journal of Happiness Studies, 3, 71-92 (2001)
66. Handelsman, M. M., Briggs, W. L., Sullivan, N., & Towler, A.: A measure of college student course engagement. The Journal of Educational Research, 98(3), 184-192 (2005)
67. Matt, C., Hess, T., Benlian, A.: Digital transformation strategies. Bus. Inf. Syst. Eng. **57**(5), 339–343 (2015)
68. Vial, G.: Understanding digital transformation: a review and a research agenda. In: Managing Digital Transformation, pp. 13–66 (2021)
69. Deng, R., Benckendorff, P., Gannaway, D.: Learner engagement in MOOCs: scale development and validation. Br. J. Educ. Technol. **51**(1), 245–262 (2020)
70. Kloft, M., Stiehler, F., Zheng, Z., Pinkwart, N.: Predicting MOOC dropout over weeks using machine learning methods. In: Proceedings of the EMNLP 2014 Workshop on Analysis of Large Scale Social Interaction in MOOCs, pp. 60–65 (2014)
71. Iida, T., Fukushima, T., Shinozaki, N.: Detection of abnormality using squared Mahalanobis distance when both continuous and dichotomous variables exist. Jpn. J. Appl. Statist. **37**(2), 55–76 (2008) (in Japanese)
72. Qunigoh, M.: An introduction to Mahalanobis' generalized distance for MTS methods. Q. Eng. **9**(1), 13–21 (2001) (in Japanese)
73. Morrison, C., Doherty, G.: Analyzing engagement in a web-based intervention platform through visualizing log-data. J. Med. Internet Res. **16**(11), e3575 (2014)

Corrections to: Educational Data Science: Essentials, Approaches, and Tendencies

Alejandro Peña-Ayala 🄳

Correction to:
A. Peña-Ayala (ed.), *Educational Data Science:*
Essentials, Approaches, and Tendencies,
https://doi.org/10.1007/978-981-99-0026-8

Owing to an unfortunate oversight during the production process, the affiliation of the author Dr. Alejandro Peña-Ayala was incorrectly presented in the book's front matter and Chapter 3 in the initially published version. The book has been updated with the corrections.

The updated version of this book can be found at
https://doi.org/10.1007/978-981-99-0026-8
https://doi.org/10.1007/978-981-99-0026-8_3

Index

Printed in the United States
by Baker & Taylor Publisher Services